Sinusoidal
Analysis and Modeling of
Weakly Nonlinear Circuits

Van Nostrand Reinhold
Electrical/Computer Science and Engineering Series

Edited by Sanjit Mitra

HANDBOOK OF ELECTRONIC DESIGN AND ANALYSIS PROCEDURES USING PROGRAMMABLE CALCULATORS, by Bruce K. Murdock

COMPILER DESIGN AND CONSTRUCTION, by Arthur B. Pyster

SINUSOIDAL ANALYSIS AND MODELING OF WEAKLY NONLINEAR CIRCUITS, by Donald D. Weiner and John F. Spina

Sinusoidal Analysis and Modeling of Weakly Nonlinear Circuits

With Application to Nonlinear Interference Effects

Donald D. Weiner
Professor of Electrical and Computer Engineering
Syracuse University

John F. Spina
Chief of Compatibility Techniques Section
Rome Air Development Center

Van Nostrand Reinhold Electrical/Computer Science and Engineering Series

VNR **VAN NOSTRAND REINHOLD COMPANY**
NEW YORK   CINCINNATI   ATLANTA   DALLAS   SAN FRANCISCO
LONDON   TORONTO   MELBOURNE

Van Nostrand Reinhold Company Regional Offices:
New York   Cincinnati   Atlanta   Dallas   San Francisco

Van Nostrand Reinhold Company International Offices:
London   Toronto   Melbourne

Library of Congress Catalog Card Number: 79-22229
ISBN: 0-442-26093-8

Manufactured in the United States of America

Published by Van Nostrand Reinhold Company
135 West 50th Street, New York, N.Y. 10020

Published simultaneously in Canada by Van Nostrand Reinhold Ltd.
15 14 13 12 11 10 9 8 7 6 5 4 3 2 1

Library of Congress Cataloging in Publication Data

Weiner, Donald D
  Sinusoidal analysis and modeling of weakly
nonlinear circuits.

  (Van Nostrand Reinhold electrical/computer science
and engineering series)
  Bibliography: p.
  Includes index.
  1. Electric circuits, Nonlinear.   2. Electric
circuit analysis.   3. Electric engineering—
Mathematics.   4. Transfer functions.   5. Volterra
series.   I. Spina, John F., joint author.   II. Title.
III. Series.
TK454.W44   621.319'2   79-22229
ISBN 0-442-26093-8

To Our Wives

Jane and Nancy

and Our Children

Debbie	and	*Lori*
Bill		*Linda*
Kay		*John*
		Jane
		Mike
		Michelle

Preface

Very little attention is traditionally devoted to nonlinear circuits in most engineering curriculums. As a result, practicing engineers tend to approach the analysis of nonlinear problems with considerable uncertainty and apprehension. This is unfortunate because all physical circuits are inherently nonlinear. Therefore, significant desired and undesired nonlinear effects are commonly encountered in practice.

The difficulties associated with analyzing nonlinear problems have long been recognized. However, considerable progress has been made in recent years. Although no single analytical approach is generally applicable, solution procedures do exist for specific classes of nonlinear problems. In this book we deal with the class of "quasi-linear" circuits which we refer to as "weakly nonlinear." The nonlinearities in such circuits depart from linear behavior in a gradual, as opposed to abrupt, manner.

The class of weakly nonlinear circuits is exceptionally large. Such circuits arise in most communication equipments. Small-signal mixers, frequency multipliers, and square-law detectors represent three examples of intentional class members while "linear" amplifier circuits, all of which inadvertently contain weak nonlinearities, are examples of unintentional members. It is important to be able to analyze the effects of weak nonlinearities in all cases. Even in "linear" circuits, where the nonlinearities are negligible as far as the intended mode of operation is concerned, nonlinear effects can be generated that severely limit system performance.

For many years weakly nonlinear circuits have been analyzed with limited success by employing the classical power series approach. Even though excellent results were obtained in some applications, severe discrepancies occurred in others. A newly developed analytical tool, referred to as the nonlinear transfer function approach, provides the basis for this book. It corrects many of the shortcomings associated with the classical technique. The new approach is based upon the Volterra functional series and is capable of determining the response of weakly nonlinear systems to arbitrary inputs.

Unfortunately, when presented in its full generality, the nonlinear transfer function approach is obscured by relatively complicated mathematics. This has greatly impeded its acceptance by practicing engineers. The purpose of this book is to provide a simplified, highly readable presentation that preserves the important concepts and their utility. We accomplish this by limiting our discussion to the sinusoidal steady-state response of weakly nonlinear circuits. Because the significant features of a weakly nonlinear circuit can be extracted from its sinusoidal response and because sinusoidal excitations and sums of sinusoids are frequently used as input signals, our presentation is not overly restrictive.

It has been our experience that the greatest impediment to learning about nonlinear circuit theory is the engineer's fear of the subject matter. The expressions that arise in nonlinear circuit analysis are very lengthy and, although not basically difficult, are certainly confusing. Typically, the engineer automatically assumes that difficult and sophisticated mathematics are involved and is easily discouraged when he fails to understand a particular point. Consequently, we have incorporated a relatively low level of mathematics in our discussions and have derived most equations in considerable detail. The results are deliberately not presented in handbook form. All too frequently, tabulated results are used incorrectly because of lack in understanding of the assumptions employed in their development. We have attempted a simple and coherent presentation that enables the interested engineer to both correctly use the results and understand how they came about. All that is required by the reader is an introductory course in linear circuit theory.

This is the first book to deal in depth with applications of the Volterra series method. The subject matter is quite broad in its application and will be of interest to engineers in many disciplines. However, because of our backgrounds, we have illustrated the material in terms of the specific application of nonlinear distortion and interference effects in communication circuits. The book may also prove to be attractive to certain college professors for use in their undergraduate or beginning graduate courses dealing with nonlinear systems. Although the topics in our book are not those normally included in such courses, they are easier to understand and, therefore, can be taught with greater success. In addition, they deal with significant nonlinear phenomena which the student is likely to encounter once he becomes an engineer. In addition, since the book deals with a state-of-the-art approach that is still in the embryonic stages of development, it will be of interest to graduate students performing research in the nonlinear systems area.

Organization of the text is indicated by the table of contents and is discussed in Section 1.2. The book is distinguished by the fact that much of the material in Chapters 2 through 7 is unavailable in other books. Some of the features of interest are as follows:

(1) Measurements of linear and nonlinear effects are commonly recorded in terms of power. However, an assortment of different powers and power gains are widely used. A thorough discussion of power considerations in the sinusoidal steady-state is given in Chapter 2. Conditions under which the various quantities become equivalent are pointed out.

(2) An attractive feature of the Volterra functional series is that it generalizes upon results already familiar to the engineer through his exposure to the power series approach. To develop, in a familiar context, certain concepts needed later in the text, the power series approach is discussed in considerable detail in Chapter 3. Assumptions and limitations that are often overlooked are clearly explained. Expressions relating output powers to input powers are developed. Insight is provided into the development and application of these widely used expressions.

(3) Strong similarities to the power series method are exploited in Chapter 4 where the nonlinear transfer function approach is introduced. Experimental results are presented for the nonlinear transfer functions of two practical transistor amplifiers.

(4) Other weakly nonlinear circuit analysis techniques are computationally efficient and emphasize numerical solution of the circuit equations. They tend to yield specific numerical solutions to specific problems. The advantage of the nonlinear transfer function approach is that it is algebraic in nature. The nonlinear transfer functions are properties of the circuit and not of the circuit input. They completely characterize the behavior of weakly nonlinear systems and, like linear transfer functions, can be determined without a specific input in mind. Linear circuit techniques for determining the nonlinear transfer functions are developed in Chapter 5. Nodal, loop, and mixed-variable analyses are discussed. Hence, the very same computer-aided techniques used in the numerical solution to linear circuit problems can also be used to evaluate nonlinear transfer functions. This is a decided advantage of the Volterra series method.

(5) Due to a lack of standardization, many different criteria exist in the literature for characterizing such nonlinear effects as intermodulation. To avoid confusion, each criterion must be carefully defined. Several commonly used nonlinear distortion and interference criteria are defined in Chapter 6 and are ex-

pressed in terms of the nonlinear transfer functions. Therefore, criteria originally developed in conjunction with the power series method are generalized to the Volterra approach.

(6) Nonlinearities in active devices are usually responsible for the weakly nonlinear behavior of electronic circuits. To analyze such circuits, it is first necessary to replace each active device with a suitable nonlinear circuit model. In other words, the modeling of weakly nonlinear circuits requires the nonlinear circuit modeling of active devices. Nonlinear incremental equivalent circuits are developed for solid-state devices in Chapter 7. Particular attention is devoted to extraction of the model parameters.

The authors would like to thank Jacob Scherer of Rome Air Development Center whose keen understanding of the need for improved modeling and analysis procedures in the area of nonlinear receiver modeling led to the material reported herein. Thanks are also due the engineers at Signatron, Inc., Lexington, Mass., who conducted for the Rome Air Development Center much of the research on which this work is based. Drs. Phillip A. Bello, James W. Graham, Leonard Ehrman, John Pierce, and Julian J. Bussgang directed various portions of the effort. Valuable contributions were also made by S. Ahmed Meer, Nicholas Johnson, Charles J. Boardman, Steven Richman, and John O'Donnell. John Spina was the RADC project engineer while Donald Weiner served as a consultant to RADC during the project. Signatron's outstanding research is documented in a final contract report that was published under the title, "Nonlinear System Modeling and Analysis with Applications to Communications Receivers," (RADC-TR-73-178, June 1973). Our Figs. 1.8, 4.4 through 4.11, 6.3, 7.9, 7.17, 7.18, and 7.21 through 7.23 have been extracted from this excellent report which is highly recommended and which may be obtained from the National Technical Information Service (AD 766278). The report is not restricted to sinusoidal inputs and covers many topics not presented in our book. Unfortunately, the extended coverage requires additional mathematical complexity. Our book was motivated when it became evident that, although there was significant interest in the subject matter, many engineers were unable to read and understand the Signatron report. Additional insight to the Volterra approach was provided to the authors by Drs. William Whyland, Gerald Naditch, Michael Rudko, Jack Williams, and Edward Ewen who, as graduate students at Syracuse University, completed Ph.D. dissertations related to the analysis of weakly nonlinear systems. We would also like to thank Joseph Naresky and Samuel Zaccari of Rome Air Development Center and Drs. James A. Luker, Wilbur R. LePage, and Bradley J. Strait of Syracuse University for their support and encouragement. Finally, we are pleased to acknowledge the services of Mary Jo Fairbanks and Barbara Howden who did excellent work, respectively, in typing the manuscript and preparing the figures.

DONALD D. WEINER
JOHN F. SPINA

Nomenclature

The following superscripts are used in order to account for those entries that have not been listed in the glossary. In some cases, more than one superscript applies.

(a) A superscript 1 indicates that uppercase letters express the quantity in decibels. For example, IMAGR denotes the intermodulation average gain ratio, imagr, expressed in decibels.

(b) A superscript 2 indicates that the dc component of a variable is denoted by uppercase symbols with uppercase subscripts while the incremental component is denoted by lowercase symbols with lowercase subscripts. For example, V_{DS} and v_{ds} are the dc and incremental components, respectively, of v_{DS}. Note that $v_{DS} = V_{DS} + v_{ds}$.

(c) A superscript 3 indicates that the subscript n and the argument (f_m) are added whenever the quantity applies to the nth-order intermodulation frequency f_m. For example, $\text{srtgr}_n(f_m)$ is the nth-order spurious response transducer gain ratio applicable to the intermodulation frequency f_m.

(d) A superscript 4 indicates that the subscript E is appended whenever the quantity pertains to the BJT emitter-base junction while the subscript C is appended whenever the quantity pertains to the BJT collector-base junction. For example, C_{1E} is the BJT emitter-base junction transition capacitance while v_{JC} is the BJT collector-base junction voltage.

(e) A superscript 5 indicates that the subscript e is appended whenever the quantity pertains to the BJT emitter-base junction while the subscript c is appended whenver the quantity pertains to the BJT collector-base junction. For example, I_{se} denotes the BJT emitter-base junction saturation current while c_{1cl} denotes the power series coefficient in the expansion of the expression for C_{1C}.

ABBREVIATIONS

Some abbreviations used in the text are:

B	base terminal of bipolar junction transistor
BJT	bipolar junction transistor
C	collector terminal of bipolar junction transistor
cmr	cross-modulation ratio[1]
D	drain terminal of field-effect transistor
dB	decibel
dBm	dB relative to a milliwatt
dBv	decibel voltage gain
dr	desensitization ratio[1]
E	emitter terminal of bipolar junction transistor
FET	field-effect transistor
G	gate terminal of field-effect transistor
gcer	gain compression/expansion ratio[1]
imagr	intermodulation average gain ratio[1,3]
imigr	intermodulation input gain ratio[1,3]
imogr	intermodulation output gain ratio[1,3]
imtgi	intermodulation transducer gain intercept[1,3]
imtgr	intermodulation transducer gain ratio[1,3]
JFET	junction field-effect transistor
$k$hd	kth harmonic distortion
MOSFET	metal-oxide-semiconductor field-effect transistor
mf	modulation factor
S	source terminal of field-effect transistor
srtgr	spurious response transducer gain ratio[1,3]
thd	total harmonic distortion
vg	voltage gain

CONVENTIONS

The following conventions have been used:*

j	imaginary unit satisfying the equation $j^2 = -1$

*Vectors are denoted by boldface in the text whereas they are denoted by underbars in the figures.

$\lvert A \rvert$	magnitude of the complex number A is denoted by $\lvert A \rvert$
ang A	angle of the complex number A is denoted by ang A
Re$\{A\}$	real part of the complex number A is denoted by Re$\{A\}$
Im$\{A\}$	imaginary part of the complex number A is denoted by Im$\{A\}$
$\underline{/\theta_A}$	quantity, exp $(j\theta_A)$, is sometimes denoted by $\underline{/\theta_A}$
A^*	conjugate of the complex number A is denoted by A^*
$(f_1 - f_2 + \cdots + f_n)$	as explained in Section 3.3, the frequency mix involving the frequencies $f_1, -f_2, \ldots, f_n$ is denoted by $(f_1 - f_2 + \cdots + f_n)$
$(n; \mathbf{m})$	as defined by (4-14), the multinomial coefficient is denoted by $(n; \mathbf{m})$
$\overline{A(f_1, \ldots, f_n)}$	as illustrated in (4-77), the arithmetic average of the $n!$ terms generated by all possible permutations of the n frequencies $f_1, \ldots, f_n$ is denoted by $\overline{A(f_1, \ldots, f_n)}$
p	time differential operator, d/dt, is denoted by p

SYMBOLS

The principal symbols are defined below. In some cases, a symbol has more than one interpretation. The correct interpretation is obvious from the usage of the symbol in the text.

A	voltage gain expressed in decibels
A	μth root of the capacitance ratio defined in (7-58)
A	a parameter defined in (7-171) to simplify the expression for $i(v_{GJ})$
$A(\alpha)$	a polynomial defined in (5-151) to aid with development of the recursion relation derived in Section 5.6
A_k	complex amplitude of the kth harmonic
$A_n(f_1, \ldots, f_n)$	nth-order transfer function of $a(t)$
$A_n(q_1, \ldots, q_n)$	complex amplitude defined in (3-19)
A_1	complex amplitude of output component at f_S
A_2	complex amplitude of output component at $f_S - f_m$
A_3	complex amplitude of output component at $f_S + f_m$
A_4	complex amplitude of output component at $f_S - 2f_m$
A_5	complex amplitude of output component at $f_S + 2f_m$
a	either a node designation or mesh designation
a	a parameter used in the empirical relationship for h_{FE}
$a(t)$	either a node-to-datum voltage, mesh current, or link current
a_k	circuit parameter defined in (A-9)

$a_{k_1 k_2 \ldots k_n}$	power series coefficient and circuit parameter defined in (A-19)
a_n	power series coefficient in expansion of zero-memory nonlinearity
$a_n(t)$	nth-order component of $a(t)$
α	angle of E
α	BJT common-base short-circuit forward-current transfer ratio
α	single-tuned circuit 3-dB half bandwidth
α	dummy variable used in the development of the recursion relation derived in Section 5.6
$\alpha(\cdot)$	zero-memory functional for current-controlled current source
α_I	BJT common-base inverted-mode dc current gain
α_k	power series coefficient in expansion of $\alpha(\cdot)$
α_k	angle of A_k
α_k	power series coefficient resulting from series reversion of (7-107)
α_N	BJT common-base normal-mode dc current gain
α_1	current amplification factor for current-controlled current source
B	susceptance
B	single-tuned circuit 3-dB bandwidth
B	a parameter defined in (7-171) to simplify the expression for $i(v_{GJ})$
$B(\alpha)$	a polynomial defined in (5-152) to aid with development of the recursion relation derived in Section 5.6
B_i	imaginary part of $(\frac{3}{4}) |E_1|^2 H_3(-f_1,f_1,f_1)/H_1(f_1)$
B_{in}	2-port input susceptance
B_k	complex amplitude of input sinusoid at f_k
B_L	load susceptance
B_N	bound defined in (7-32) on normalized truncation error
$B_n(f_1,\ldots,f_n)$	nth-order transfer function of $b(t)$
$B_n(q_1,\ldots,q_n)$	complex amplitude defined in (3-23)
B_r	real part of $(\frac{3}{4}) |E_1|^2 H_3(-f_1,f_1,f_1)/H_1(f_1)$
b	either a node designation or mesh designation
$b(t)$	either a node-to-datum voltage, mesh current, or link current
β	angle of I
β_i	single-tuned circuit off-frequency rejection factor for interferer at frequency f_i
$C(0)$	semiconductor diode transition capacitance evaluated at $v_J = 0^4$

C_c	BJT linear incremental collector capacitance
C_d'	semiconductor diode diffusion capacitance constant[5]
C_e	BJT linear incremental emitter capacitance
C_{GD}	FET gate-to-drain capacitance
$C_{GD}(0)$	FET gate-to-drain capacitance evaluated at $v_{GD} = 0$
C_{GS}	FET gate-to-source capacitance
$C_{GS}(0)$	FET gate-to-source capacitance evaluated at $v_{GJ} = 0$
C_{gp}	vacuum-tube pentode grid-to-plate capacitance
C_j	linear approximation to C_1 under forward-biased conditions[4]
C_{jD}	linear approximation to C_{GD}
C_K	cathode bypass capacitor
$C_n(f_1, \ldots, f_n)$	nth-order transfer function of $c(t)$
C_s	capacitance in series with a voltage source
C_T	total semiconductor diode junction capacitance[4]
C_x	controlling element capacitance for a controlled source
C_1	semiconductor diode transition capacitance, also known as the barrier capacitance or space-charge capacitance[4]
C_2	semiconductor diode diffusion capacitance[4]
c	either a node designation or mesh designation
$c(\cdot)$	incremental capacitance defined in (5-21)
$c(m)$	intermodulation multiplier defined in (3-59)[1]
$c(t)$	either a node-to-datum voltage, mesh current, or tree branch voltage
c_{gsk}	power series coefficient in expansion of expression for C_{GS}
c_{gso}	linear incremental gate-to-source capacitance
c_k	power series coefficient in expansion of $c(\cdot)$
c_0	linear incremental capacitance at the Q point
c_{1l}	power series coefficient[5] in expansion of expression for C_1
c_{10}	linear incremental capacitance[5] associated with C_1
c_{2l}	power series coefficient[5] in expansion of expression for C_2
c_{20}	linear incremental capacitance associated with C_2
$D_n(f_1, \ldots, f_n)$	nth-order transfer function of $d(t)$
d	node designation
$d(t)$	either a node-to-datum voltage, mesh current, or tree branch voltage
Δf	difference between frequency f and resonant frequency
Δf_i	difference between frequency f_i and resonant frequency
Δi_C	increment in i_C
Δi_D	increment in i_D
$\Delta P_L(f_m)$	decibel difference between $P_L(f_m)$ and P_L^{min}
ΔR_L	difference in load resistance needed for a conjugate match
Δv_{CB}	increment in v_{CB}

Δv_{CE}	increment in v_{CE}
Δv_D	increment in v_D
ΔX_L	difference in load reactance needed for a conjugate match
$\Delta \tau$	width of $p(t)$
$\delta(t)$	unit impulse
E	a complex voltage
E_I	complex amplitude of unmodulated interfering signal
$|E_m|$	amplitude of $\hat{y}_n(t; m)$
$\mathbf{E}_n(f_1, \ldots, f_n)$	unknown vector appearing in the frequency-domain nodal equation of (5-114)
E_q	complex amplitude of input sinusoid at f_q
E_r	complex amplitude of a reference voltage
E_S	complex amplitude of a voltage source
E_S	complex amplitude of sinusoidal desired signal
E_{TH}	complex amplitude of Thévenin voltage source
E_1	complex amplitude of 2-port input voltage
E_1	complex amplitude of input component at f_S
E_2	complex amplitude of 2-port output voltage
E_2	complex amplitude of input component at f_I
E_3	complex amplitude of input component at $f_I - f_m$
E_4	complex amplitude of input component at $f_I + f_m$
e	node designation
e	base of Naperian logarithms
$e(t)$	tree branch voltage
$\mathbf{e}(t)$	unknown vector appearing in the time-domain nodal equation of (5-103)
$e_C(t)$	capacitor total voltage[2]
$e_{CS}(t)$	controlled source total terminal voltage[2]
$e_k(t)$	cathode impedance voltage
$e_L(t)$	inductor total voltage[2]
$e_R(t)$	resistor total voltage[2]
$e_s(t)$	independent source voltage
$\mathbf{e}_s(t)$	independent voltage source vector appearing in the time-domain mesh equation (5-123)
$e_{TH}(t)$	Thévenin equivalent voltage
e_X	controlled source total control voltage[2]
e_0	voltage at quiescent operating point
η	exponent in avalanche multiplication factor, M
f	frequency in hertz
f	node designation
$f(\cdot)$	nonlinear function of $(K + 1)$ input variables
$f(i_C)$	a zero-memory functional defined in (7-99) for α_N nonlinearity

$f(t)$	link current
$f(x)$	a zero-memory functional used in Taylor series discussion
f_I	carrier frequency of amplitude-modulated interfering signal
f_k	power series coefficient in expansion of $f(i_C)$
f_{Lo}	local oscillator frequency
f_m	frequency of modulation
$f_{\mathbf{m}}$	intermodulation frequency generated by frequency mix vector $\mathbf{m}$
f_q	frequency of qth input sinusoid
f_S	desired signal frequency
f_0	receiver-tuned frequency
f_0	single-tuned circuit resonant frequency
G	conductance
G	power gain expressed in decibels
G	input system in cascade connection of Fig. 4.12
G_{in}	input conductance of 2-port network
G_L	load conductance
$G_n(f_1, \ldots, f_n)$	nth-order transfer function of system G
G_p	parasitic conductance
G_r	a reference conductance
G_s	conductance in series with a voltage source
G_x	controlling element conductance for a controlled source
g	node designation
$g(\cdot)$	zero-memory functional, usually relating a current to a voltage
$g(t)$	continuous time function
$g(v_B, v_C - v_A)$	nonlinear controlled current source used to represent the current gain and avalanche nonlinearities in the nonlinear equivalent circuit of Fig. 3.7
g_A	available power gain of a loaded 2-port network[1]
g_a	linear conductance
g_c	linear conductance
g_E	exchangeable power gain of a loaded 2-port network[1]
g_{el}	power series coefficient in expansion of current-voltage relation for R_{JE}
g_I	insertion power gain of a loaded 2-port network[1]
g_i	power series coefficient in expansion of zero-memory non-linearity of system G
g_k	power series coefficient in expansion of $g(\cdot)$
g_{ks}	power series coefficient in expansion of voltage-controlled current source control equation
g_m	FET linear incremental transconductance
g_P	average power gain of a loaded 2-port network[1]

$g_s(t)$	independent source vector appearing in the time-domain mixed-variable analysis equation of (5-137)
g_T	transducer power gain of a loaded 2-port network[1]
g_1	linear incremental conductance
g_{1s}	voltage-controlled current source mutual conductance
$\Gamma(\cdot)$	zero-memory functional relating inductance current to flux linkage
Γ_k	power series coefficient in expansion of $\Gamma(\cdot)$
Γ_s	reciprocal inductance in series with a voltage source
Γ_x	controlling element reciprocal inductance for a controlled source
Γ_1	linear incremental reciprocal inductance at the Q point
$\gamma_c(v_C - v_B)$	nonlinear controlled current source used to represent the varactor capacitance nonlinearity in the nonlinear equivalent circuit of Fig. 3.7
$\gamma_e(v_B)$	nonlinear controlled current source used to represent the emitter capacitance nonlinearity in the nonlinear equivalent circuit of Fig. 3.7
H	output system in cascade connection of Fig. 4.12
$H(f)$	linear filter transfer function
$H_n(f_1, \ldots, f_n)$	nth-order transfer function of system H
$H_n(\mathbf{m})$	nth order transfer function defined in (4-23)
$h(t)$	linear system impulse response
h_{FE}	BJT common emitter dc current gain
$h_{FE_{\max}}$	maximum value of h_{FE}
h_j	power series coefficient in expansion of zero-memory nonlinearity of system H
$h_n(\tau_1, \tau_2, \ldots, \tau_n)$	nth-order Volterra kernel
I	complex current
$I_n(\mathbf{m})$	intermodulation conversion gain factor introduced in Example 3.14
$I_{C_{\max}}$	collector current at which h_{FE} equals $h_{FE_{\max}}$
$I_{cn}(f_1, \ldots, f_n)$	nth-order transfer function of incremental component of $i_C(t)$
I_{DO}	parameter defined in (7-188) for JFET drain current
$I_{g_1 n}(f_1, \ldots, f_n)$	nth-order transfer function of incremental current through g_1
$I_{ln}(f_1, \ldots, f_n)$	nth-order transfer function of incremental component of $i_L(t)$
I_m	parameter defined in (7-168) for JFET controlled source
I_s	semiconductor diode saturation current[5]
I_1	complex amplitude of 2-port input current

$-I_2$	complex amplitude of 2-port output current
$\mathbf{i}(t)$	unknown vector appearing in the time-domain mesh equation of (5-123)
$i(v_{GJ})$	FET voltage-controlled current source
i_B	BJT total base current[2]
i_C	BJT total collector current[2]
$i_C(t)$	capacitor total current[2]
$i_{CS}(t)$	controlled source total terminal current[2]
i_{C_1}	C_1 total current[2]
i_{C_2}	C_2 total current[2]
$i_c(t)$	tree branch current
$i_{cn}(t)$	nth-order component of $i_c(t)$
i_D	semiconductor diode total current[2]
i_D	FET total drain current[2]
$i_d(t)$	tree branch current
i_E	BJT total emitter current[2]
$i_e(t)$	tree branch current
i_G	FET total gate current[2]
$i_{g_1}(t)$	incremental current through g_1
$i_{g_1 n}(t)$	nth-order component of $i_{g_1}(t)$
i_J	semiconductor diode total junction current[2,4]
$i_L(t)$	inductor total current[2]
$i_{ln}(t)$	nth-order component of $i_l(t)$
$i_N(t)$	Norton equivalent current
$i_p(t)$	vacuum-tube pentode controlled source in equivalent circuit of Fig. 3.3
i_R	total current flowing through R_D in vacuum-tube diode global model[2]
$i_R(t)$	resistor total current[2]
i_S	FET total source current[2]
$i_s(t)$	independent current source
$\mathbf{i}_s(t)$	independent current source vector appearing in the time-domain nodal equation of (5-103)
i_X	controlled source total control current[2]
i_0	current at quiescent operating point
$J_{kn}(f_1, \ldots, f_n)$	entry in $\mathbf{J}_n(f_1, \ldots, f_n)$ arising from the nonlinear circuit elements connected to node k
$\mathbf{J}_n(f_1, \ldots, f_n)$	nonlinear excitation current vector appearing in the frequency-domain nodal equation of (5-114)
$\mathbf{j}(t)$	nonlinear excitation current vector appearing in the time-domain nodal equation of (5-103)
K	cascade system of Fig. 4.12

K	either the number of node-to-datum voltages, meshes, or mixed-variable independent variables in a network
$K(f)$	linear filter transfer function
$K(v_B)$	nonlinear controlled current source used to represent the emitter resistance nonlinearity in the nonlinear equivalent circuit of Fig. 3.7
$K_n(f_1, \ldots, f_n)$	nth-order transfer function of system K
$K_n(f_1, \ldots, f_n)$	nth-order transfer function of $k(t)$
k	either a node designation or mesh designation
k	Boltzmann's constant
$k(t)$	either a node-to-datum voltage or mesh current
k_a	constant introduced in (7-121) for use in evaluating η
k_b	constant introduced in (7-121) for use in evaluating η
k_l	power series coefficient in expansion of zero-memory nonlinearity of system K
$k_n(t)$	nth-order component of $k(t)$
k_1	constant assumed for linear transfer function in Example 6.11
k_3	constant assumed for third-order transfer function in Example 6.11
L_p	inductance in parallel with current source
L_x	controlling element inductance for a controlled source
$l(\cdot)$	incremental inductance defined in (5-41)
l_a	constant introduced in (7-132) for use in evaluating η
l_b	constant introduced in (7-132) for use in evaluating η
l_k	power series coefficient in expansion of $l(\cdot)$
l_0	linear incremental inductance at the Q point
$\lambda(\cdot)$	zero-memory functional defined in (5-46) relating inductive flux linkage to current
$\lambda_L(t)$	inductor total flux linkage[2]
λ_k	power series coefficient in expansion of $\lambda(\cdot)$
λ_0	flux linkage at quiescent operating point
M	maximum value in the interval (x_0, x) for the $(N+1)$th derivative of $f(x)$
M	avalanche multiplication factor defined in (7-73)
$[M(\cdot)]$	immittance matrix appearing in the time- and frequency-domain mixed-variable analysis equations of (5-137) and (5-140)
m	frequency mix vector defined in (3-24)
m^0	frequency mix vector resulting in dc intermodulation component
m$'$	frequency mix vector defined in (3-39)

m_I	modulation index of amplitude-modulated interfering signal
m_k	number of times frequency f_k appears in a frequency mix
m_k	power series coefficient in expansion of expression for M, the avalanche multiplication factor
μ	semiconductor diode junction grading constant[4]
$\mu(\cdot)$	zero-memory functional for voltage-controlled voltage source
μ_D	exponent appearing in expression for JFET C_{GD}
μ_k	power series coefficient in expansion of $\mu(\cdot)$
μ_S	exponent appearing in expression for JFET C_{GS}
μ_1	voltage amplification factor for voltage-controlled voltage source
N	degree of highest significance in zero-memory nonlinearity
N	order of highest significant response in weakly nonlinear circuit
n	semiconductor diode ideality factor[4]
n_b	number of branches in a network graph
n_c	number of independent current sources in a network
n_t	number of nodes in a network graph
n_v	number of independent voltage sources in a network
ω	frequency in radians per second
ω_0	single-tuned circuit resonant frequency in radians per second
P	denotes parallel element
$P_L^{\min}$	average power, in decibels, dissipated at the IF output by the minimum detectable signal
P_L^0	average power, in decibels, dissipated in the load for 0 dB available input power
$P_{\lim}$	available input power, in decibels, at which higher order effects become noticeable
$P_S^{\min}(f_0)$	average input power, in decibels, of the minimum detectable signal
$P_S[f_0,f_m]$	assuming linear behavior, the equivalent input power, in decibels, of an on-tune signal which produces the same load power as does the intermodulation component at f_m
$P_T(f_1,f_2)$	parameter, defined in (3-107), to denote "total" input power in Example 3.14
$p(t)$	instantaneous power
$p(t)$	rectangular pulse of width $\Delta\tau$ and height $1/\Delta\tau$
p_A	available power[1]
p_{AO}	2-port available output power[1]

p_{AS}	generator available power[1]
p_{ave}	average power[1]
p_{com}	complex power
p_E	generator exchangeable power[1]
p_{EO}	2-port exchangeable output power[1]
p_{ES}	generator exchangeable power[1]
p_L	2-port average power delivered to a load[1]
$\tilde{p}_L$	generator average power delivered directly to a load[1]
$\tilde{p}_L'$	value of $\tilde{p}_L$ when the load is close to a conjugate match
$p_L(f_{\mathbf{m}})$	load average power due to the intermodulation component generated by the frequency mix vector $\mathbf{m}$[1]
p_r	reference power
p_S	2-port average input power delivered by a source[1]
$p_S(f_q)$	average output power in the input signal at frequency f_q^1
p_1	average output power at the fundamental frequency
ϕ	semiconductor junction contact potential[4]
$\phi(f)$	angle of $K(f)$
$\psi(f)$	angle of $H(f)$
$\psi_{\mathbf{m}}$	angle defined in (3-44) to simplify expression for $\hat{y}_n(t; \mathbf{m})$
$\psi_n(f_1, \ldots, f_n)$	angle of $H_n(f_1, \ldots, f_n)$
$\psi_n(\mathbf{m})$	angle of $H_n(\mathbf{m})$
Q	number of input sinusoids to a weakly nonlinear system
Q	single-tuned circuit quality factor
Q	quiescent operating point
q	electronic charge
$q(\cdot)$	zero-memory functional defined in (5-29) relating capacitive charge to voltage
$q_C(t)$	capacitor total charge
q_h	power series coefficient in expansion of $q(\cdot)$
$q_k(v_{cb}, v_{je})$	terms of kth degree appearing in (7-108)
q_0	charge at quiescent operating point
R	resistance
R_{in}	2-port input resistance
R_J	semiconductor diode resistor[4]
R_K	cathode resistor
R_L	load resistance
R_L	semiconductor diode leakage resistance[4]
R_L'	load resistance near a conjugate match
$R_N(x)$	Taylor series remainder for an N-term approximation
R_p	resistance in parallel with current source
R_S	source resistance
R_S	semiconductor diode bulk resistance[4]

R_S	JFET source bulk resistance
R_{SB}	BJT base bulk resistance
R_{TH}	resistance of Thévenin impedance
R_x	controlling element resistance for a controlled source
$r(\cdot)$	zero-memory functional, usually relating a voltage to a current
r_a	linear resistor resistance
r_b	BJT linear incremental base resistance
r_c	BJT linear incremental collector resistance
r_c	linear resistor resistance
r_d	semiconductor diode linear incremental resistance
r_e	BJT linear incremental emitter resistance
r_k	power series coefficient in expansion of $r(\cdot)$
r_{ks}	power series coefficient in expansion of current-controlled voltage source control equation
r_p	vacuum-tube plate resistance
r_1	linear incremental resistance
r_{1s}	current-controlled voltage source mutual resistance
ρ	constant introduced in (7-168) for the saturation region expression for the JFET controlled source
S	denotes series element
S_p	elastance in parallel with current source
S_x	controlling element elastance for a controlled source
$s(\cdot)$	zero-memory functional defined in (5-36) relating capacitive voltage to charge
$s(t)$	nonlinear portion of either a capacitive voltage or inductive current as defined in (5-189)
s_k	power series coefficient in expansion of $s(\cdot)$
s_1	linear incremental elastance at the Q point
T	absolute temperature of p-n junction in degrees Kelvin
t	time in seconds
θ_I	angle of unmodulated interfering signal
θ_m	angle of modulation
$\theta_{\mathbf{m}}$	angle defined in (3-43) to simplify expression for $\hat{y}_n(t; \mathbf{m})$
θ_q	angle of qth input sinusoid
θ_S	angle of desired signal
U_k	complex amplitude of sinusoidal component of $u(t)$ at f_k
$U_{kn}(f_1,\ldots,f_n)$	entry in $\mathbf{U}_n(f_1,\ldots,f_n)$ arising from the nonlinear circuit elements
$\mathbf{U}_n(f_1,\ldots,f_n)$	nonlinear excitation vector appearing in the frequency-domain mixed-variable equation of (5-140)
$u(t)$	input signal to system G

$\mathbf{u}(t)$	nonlinear excitation vector appearing in the time-domain mixed-variable equation of (5-137)
V_{CBO}	BJT collector-to-base breakdown voltage, also known as the avalanche voltage
V_{CEO}	BJT collector-to-emitter sustaining voltage
V_f	complex amplitude of sinusoidal component of $v(t)$ at f
$V_{hn}(f_1,\ldots,f_n)$	entry in $\mathbf{V}_n(f_1,\ldots,f_n)$ arising from the nonlinear circuit elements in mesh h
$\mathbf{V}_n(f_1,\ldots,f_n)$	nonlinear excitation voltage vector appearing in the frequency-domain mesh equation of (5-126)
V_{po}	JFET pinchoff voltage
$v(t)$	output signal of system G
$v(t)$	nonlinear portion of either a capacitive current or inductive voltage as defined in (5-184)
$\mathbf{v}(t)$	nonlinear excitation voltage vector appearing in the time-domain mesh equation of (5-122)
$v_a(t)$	a cotree branch voltage
v_{BE}	BJT total base-to-emitter voltage[2]
$v_b(t)$	a cotree branch voltage
v_{CB}	BJT total collector-to-base voltage[2]
v_{CE}	BJT total collector-to-emitter voltage[2]
v_D	semiconductor diode total voltage[2]
v_{DS}	FET total drain-to-source voltage[2]
v_{EB}	BJT total emitter-to-base voltage[2]
$v_f(t)$	a cotree branch voltage
v_{GD}	FET total gate-to-drain voltage[2]
v_{GI}	FET total gate-to-internal node voltage[2]
v_{GS}	FET total gate-to-source voltage[2]
v_J	semiconductor diode total junction voltage
v_j	node-to-datum voltage at node j in Figs. 3.6 and 3.7; $j = A, B, C$
$W_n(f_1,\ldots,f_n)$	nth-order transfer function of $w(t)$
$W_{n,r}(f_1,\ldots,f_n)$	nth-order transform which is generated by a term of rth degree as appears in (5-194)
$w(t)$	zero-memory nonlinearity input signal
$w(t)$	output signal of system H
$w_n(t)$	nth-order component of $w(t)$
$w_n(t;f)$	sum of components in $w_n(t)$ at frequency f
$w_{n,r}(t)$	nth-order component that is generated by a term of rth degree as appears in (5-146)
X	reactance
X	denotes control element

X_{in}	2-port input reactance
X_L	load reactance
X_L'	load reactance near a conjugate match
$X_{n,r}(f_1, \ldots, f_n)$	nth-order transform which is generated by a term of rth degree as appears in (5-198)
X_S	source reactance
X_{TH}	reactance of Thévenin impedance
$x(t)$	linear system input signal
$x(t)$	weakly nonlinear system input signal
$\mathbf{x}(t)$	unknown vector appearing in the time-domain mixed-variable equation of (5-137)
x_k	value of $x(t)$ at $t = k\Delta\tau$
x_0	point about which Taylor series expansion of $f(x)$ is made
Y	admittance
$[Y(\cdot)]$	admittance matrix appearing in the time- and frequency-domain nodal equations of (5-103) and (5-114)
Y_L	load admittance
Y_{11}	2-port short-circuit input admittance
Y_{12}	2-port short-circuit reverse-transfer admittance
Y_{21}	2-port short-circuit forward-transfer admittance
Y_{22}	2-port short-circuit output admittance
y	incremental variable defined in (7-106) to simplify analysis of the α_N nonlinearity
$y(t)$	linear system output signal
$y(t)$	weakly nonlinear system output signal
$y(t;f)$	sum of components in $y(t)$ at frequency f
$y(t; f_S, f_S \pm f_m, f_S \pm 2f_m)$	sum of components with frequencies in the vicinity of f_s defined in (6-103) for the cross-modulation discussion
y_{ie}	BJT common-emitter incremental short-circuit input admittance
y_{is}	FET common-source incremental short-circuit input admittance
$y_n(t)$	nth-order component of $y(t)$
$y_n(t; \mathbf{m})$	component of $y_n(t)$ generated by frequency mix vector $\mathbf{m}$
$\hat{y}_n(t; \mathbf{m})$	the sum of $y_n(t; \mathbf{m})$ and its complex conjugate
Z	impedance
$[Z(\cdot)]$	impedance matrix appearing in the time- and frequency-domain mesh equations of (5-123) and (5-126)
Z_{in}	2-port input impedance
Z_L	load impedance
Z_S	source impedance
Z_{TH}	Thévenin impedance

Z_{11}	2-port open-circuit input impedance
Z_{12}	2-port open-circuit reverse-transfer impedance
Z_{21}	2-port open-circuit forward-transfer impedance
Z_{22}	2-port open-circuit output impedance
$z(t)$	zero-memory nonlinearity output signal
$z_n(t)$	nth-order component of $z(t)$

Contents

1

Introduction

All physical components and devices are inherently nonlinear. Nevertheless, the preponderance of books devoted to circuit and system theory deal almost exclusively with linear analysis. This paradox exists because: (1) linear systems are characterized in terms of linear algebraic, differential, integral, and difference equations that are relatively easy to solve; (2) most nonlinear systems can be adequately approximated by equivalent linear systems for suitably small inputs; and (3) closed-form analytical solutions of nonlinear equations are not ordinarily possible. The absence of general nonlinear analysis techniques greatly complicates the treatment of nonlinear problems. As a result, nonlinearities are frequently ignored in the modeling of physical systems. However, linear models are incapable of explaining important nonlinear phenomena. The purpose of this book is to present a powerful and useful analytical tool, known as the nonlinear transfer function approach, that can be used for a large class of nonlinear systems to model, analyze, and predict significant nonlinear effects.

In this text we distinguish between strong and weak nonlinearities. Strong nonlinearities are characterized by abrupt changes within the operating regions of their characteristics. For example, consider the transistor amplifier shown in Fig. 1.1. The transfer characteristic relating the input and output voltages appears in Fig. 1.2. Note that the input signal drives the amplifier into the cutoff and saturation regions. Consequently, the input sinusoidal waveform is converted to

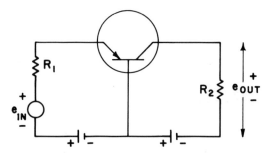

Fig. 1.1. Transistor amplifier circuit.

a clipped sinusoid at the output. For this situation the strong nonlinearities are conveniently modeled in terms of piecewise-linear characteristics. A piecewise-linear approximation to the characteristic of Fig. 1.2 is shown in Fig. 1.3 where the strong nonlinearities in this example are modeled as the abrupt changes in

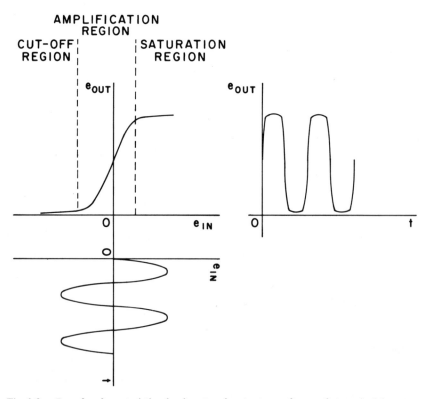

Fig. 1.2. Transfer characteristic plus input and output waveforms of strongly driven transistor amplifier.

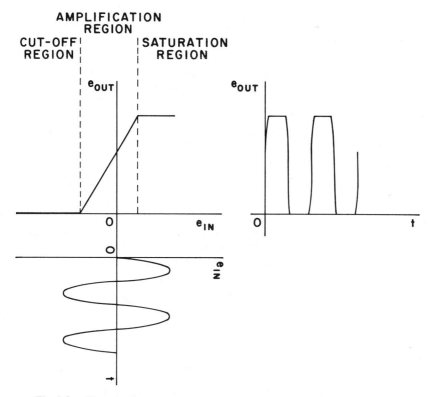

Fig. 1.3.   Piecewise-linear approximation to transfer characteristic of Fig. 1.2.

the transfer characteristic from the cutoff to amplification and from the amplification to saturation regions. Strong nonlinearities are not discussed in this text.

Our attention is directed to problems dealing with weak nonlinearities. The corresponding input–output relationships vary gradually as a function of the input amplitude and lack abrupt changes in slope. By way of illustration, consider, once again, the transistor amplifier of Fig. 1.1. Provided operation is restricted to the amplification region, the transfer characteristic may be classified as a weak nonlinearity. Ideally, the amplifier should behave linearly and the output waveform should be an amplified replica of the input. In fact, the transistor is nonlinear and, as shown in Chapter 7, models for the transistor contain nonlinear resistors, nonlinear capacitors, and nonlinear controlled sources. In the design and analysis of small-signal amplifiers, we usually make use of linear incremental equivalent circuits. This approach is valid only if we are willing to ignore nonlinear behavior. However, in many applications nonlinear effects such as nonlinear distortion and intermodulation can severely degrade amplifier performance. Obviously such effects cannot be accounted for by a linear model.

Some systems are intentionally designed to contain weak nonlinearities. This is the case, for example, in mixer and detector circuits where either diodes or field-effect transistors are employed in order to obtain square-law behavior. On the other hand, weak nonlinearities appear inadvertently in all systems that are intended to be linear. In the past, various approximations have been utilized in the modeling of weakly nonlinear problems in order to make the analyses more tractable. These assumptions led to highly inaccurate solutions in some situations. The nonlinear transfer function approach, developed in this book, enables weakly nonlinear problems to be solved with relative ease and accuracy.

Our emphasis is on the modeling and analysis of weakly nonlinear effects in electronic circuits. The topics discussed are quite general and have application to such diverse areas as automatic control, broadcasting, cable television, communications, electromagnetic compatibility, electronic devices, instrumentation, signal processing, and systems theory. In this text the material is illustrated by analyzing distortion and interference effects in weakly nonlinear communication circuits. A brief introduction to these effects is presented in Section 1.1.

1.1   DISTORTION AND INTERFERENCE EFFECTS IN A COMMUNICATIONS RECEIVER

The basic purpose of a communications system is the transmission of information from one place to another. Because electronic equipments have experienced a "population" explosion in recent years, communications has become increasingly difficult. As the numbers of electronic equipments have grown, the electromagnetic spectrum has become more crowded with a corresponding decrease in the spatial separations between equipments. To further aggravate the problem, improved sensitivities have enabled equipments to respond to signals that previously went undetected. As a result, desired input signals are frequently accompanied by a host of unwanted signals.

To combat interference, equipments are designed to be frequency selective. A typical selectivity curve, assuming a linear system, is shown in Fig. 1.4. Note that the equipment is tuned to f_0, the center frequency of the desired signal spectrum. At frequencies for which the attenuation is large, the rejection of the incoming signal is also large. Therefore, from linear considerations, unwanted signals falling outside the system passband should be severely attenuated. Such interferers are expected to cause only minimal degradation in system performance. Actual experience does not support this conclusion. In practice it is found that signals falling well outside the system passband may, in fact, result in significant interference. This irregular behavior, not predicted by linear systems theory, may be caused by nonlinearities inherent in solid-state devices such as transistors contained within the equipment.

In order to elaborate on this point, we consider the superheterodyne receiver

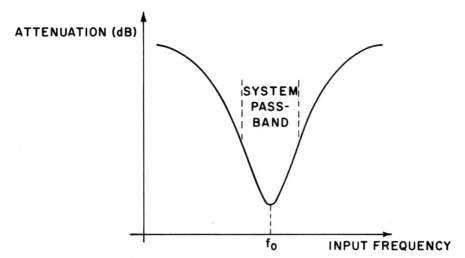

Fig. 1.4.   Linear frequency selectivity curve.

illustrated in Fig. 1.5. The functional blocks preceding the detector provide the frequency selectivity, tuning, and amplification needed for proper receiver operation. When a receiver is tuned across its frequency range, it is desirable that the receiver bandwidth and gain remain constant. This is difficult to achieve with tunable networks incorporated into each of the amplifier stages. Consequently, superheterodyne receivers are used that translate the carrier frequency of the incoming signal to a fixed predetermined intermediate frequency. The frequency conversion (or heterodyning) is carried out by mixing the received RF signal with a local oscillator signal having its frequency separated from that of the incoming signal by an amount equal to the center frequency of the IF amplifiers. Tuning the receiver requires adjusting only the RF amplifier stages and local oscillator.

The advantage of the superheterodyne approach is that the receiver gain and bandwidth are determined by the fixed-frequency IF amplifier stages following the mixer. The constant bandwidth results because the RF stages have much

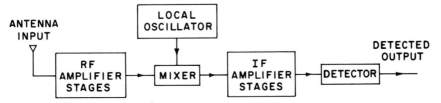

Fig. 1.5.   Simplified block diagram of superheterodyne receiver.

larger bandwidths than do the IF stages. For example, in a VHF communications receiver capable of being tuned over the 30–100 MHz frequency range, typical RF and IF bandwidths are 5 MHz and 500 kHz, respectively. Since the IF stages are much narrower than the RF stages, they determine the overall selectivity. It follows that, even though the bandwidths of the adjustable RF stages vary as the receiver is tuned across its frequency range, the overall bandwidth remains essentially constant.

Conventional operation of a superheterodyne receiver is obtained by tuning the receiver to the desired signal. This results in the carrier frequency of the desired signal being separated from the local oscillator frequency by an amount equal to the IF frequency. Typically, the local oscillator frequency is larger than the tuned receiver frequency. For this situation the image frequency is defined to be the sum of the local oscillator and IF frequencies. A superheterodyne receiver exhibits an image response in addition to its conventional response, because a signal located at the image frequency is also translated by the mixer to the fixed IF passband. A typical selectivity curve for a superheterodyne receiver tuned to 45 MHz appears in Fig. 1.6. Note that the local oscillator frequency is 66 MHz and that the image response occurs at 87 MHz. We conclude from this that the IF frequency is 21 MHz. The image response is not as strong as the conventional response because a signal located at the image frequency is severely attenuated by the RF selectivity. Nevertheless, if an undesired signal at the image frequency is considerably larger than the signal to which the receiver is tuned, significant interference can result. Clearly, the greater the IF frequency, the larger is the image rejection that can be provided by the RF circuitry. The

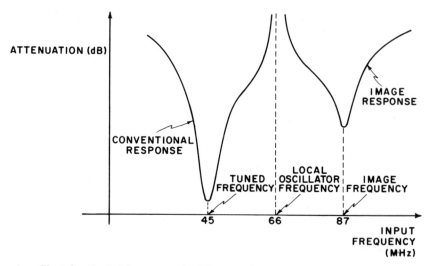

Fig. 1.6.   Typical frequency selectivity curve for a superheterodyne receiver.

infinite attenuation appearing at 66 MHz occurs because the receiver does not transmit the dc component generated when the frequency of the incoming RF signal is identical to the local oscillator frequency.

We now discuss some interference considerations in conjunction with the selectivity curve of Fig. 1.6. Interfering signals are customarily classified as being either: (1) cochannel; (2) adjacent channel; or (3) out of band, depending upon their frequencies relative to the receiver tuned frequency. Uniform agreement does not exist concerning the precise interpretation of these terms. For our purposes, we make the following definitions: A cochannel signal is one whose frequency falls within the 3-dB receiver bandwidth. Signals whose frequencies fall outside the 3-dB receiver bandwidth but inside the RF bandwidth are referred to as adjacent channel interferers. Out of band signals are those whose frequencies appear outside the RF bandwidth.

After frequency conversion, note that cochannel signals are located inside the 3-dB bandwidth of the IF stages whereas adjacent channel and out of band signals, except for those located near the image frequency, are not. Cochannel signals can be a serious source of interference because they appear at the IF output having experienced approximately the same gain as the desired signal. To aid in minimizing cochannel interference, frequency allocation and assignment procedures are commonly used. Such procedures are not completely effective because transmitter emissions are nonideal. In addition to transmitting power at the desired operating frequency, transmitters also emit harmonics of the desired operating frequency, signal components at various other frequencies, and broadband noise. Filters are employed in transmitters to suppress these undesired transmissions. Nevertheless, an unwanted emission 80 dB below the fundamental output of a 100 kW transmitter still has a power level of 1 mW. This is more than enough power to severely degrade the performance of a nearby communications receiver that is capable of responding to signals with powers much less than a milliwatt.

Adjacent channel and out of band signals are also amplified, heterodyned, and detected in the same manner as the desired signal. Because they fall on the skirts of the receiver selectivity curve, they experience considerable attenuation. When processed in this manner, such signals cause significant interference only if they are much stronger than the desired signal. Adjacent channel signals need not be quite as large as out of band signals because they suffer less attenuation. However, out of band signals can be troublesome when centered within the receiver's image response. Also, because of the large frequency range encompassed by such signals, there are apt to be a large number of out of band signals at the receiver input. This increases the likelihood of some out of band signals being present that are strong enough to overcome receiver selectivity so as to cause significant interference.

The preceding discussion of interference has hinged upon the linear response of the frequency selective portions of the receiver. For this reason the interference effects discussed thus far are commonly referred to as linear effects. With respect to adjacent channel and out of band signals, additional interference problems may arise because of nonlinearities inherent in the electronic devices. The major nonlinear interference effects are known as intermodulation, desensitization, cross modulation, and spurious responses. A nonlinear distortion effect, called gain compression, is produced by a strong desired signal in the absence of interferers. This book develops an analytical tool for modeling, analyzing, and predicting each of these effects. In this section we simply give a brief description of what is meant by the various terms. (A more mathematical treatment is found in Chapter 6.)

Intermodulation: In a linear time-invariant system the frequency content at the output is always identical to that of the input. To put it another way, linear time-invariant systems do not generate new frequencies. This is not true with nonlinear systems. The process by which two or more signals combine in a nonlinear manner so as to produce new frequency components is termed intermodulation.

It is the process of intermodulation that enables signals appearing outside a receiver's passband to generate interference that falls inside the IF bandwidth. For example, consider a receiver with an amplifier having both quadratic and cubic nonlinearities. Let the input consist of two sinusoidal signals at frequencies f_1 and f_2, respectively, where f_2 is greater than f_1. Figure 1.7 shows the frequency content at both the amplifier input and output. The terms marked (1) indicate components resulting from the linear behavior of the amplifier. Terms that are generated by the quadratic and cubic nonlinearities are marked by (2) and (3), respectively. Although there are only two frequency components at the input, intermodulation produces thirteen different frequencies at the output. Should any of these frequencies fall into the receiver's IF passband, severe degradation in performance may result. Even in the absence of interfering signals, intermodulation may produce serious distortion when the desired input signal consists of several frequency components.

Desensitization: The nonlinear effect by which an interfering signal reduces the apparent gain of a receiver is known as desensitization. For example, consider an experiment whereby a communications receiver is excited by two sinusoidal signals at the distinct frequencies f_1 and f_2. Assume the amplitude of the input signal at f_1 is maintained constant throughout the experiment. Beginning with extremely small values of the signal at f_2, let its magnitude be gradually increased. From linear considerations the amplitude of the response at f_1 should be unaffected by the presence of the second signal at f_2. However, in actual practice, it is observed that the amplitude of the response at f_1 decreases as the amplitude of the input signal at f_2 exceeds some critical level. Since receiver gain is reduced, we say that the receiver has been desensitized by the interfering signal at f_2.

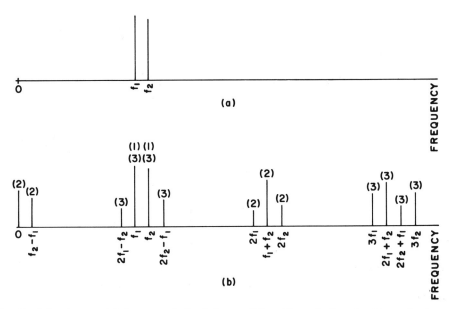

Fig. 1.7. Intermodulation example involving amplifier with quadratic and cubic nonlinearities. (a) Frequency content at input. (b) Frequency content at output.

A typical plot of the desensitization, or gain reduction, in a communications receiver is shown in Fig. 1.8. The data pertain to an experiment in which the sinusoidal signal at f_1 was positioned at the tuned receiver frequency of 19.75 MHz and had an available input power of -55 dBm. The plot shows the amount of desensitization experienced by the signal at f_1 as a function of the available power in the signal at f_2. Separate curves are given for $f_2 = 24.75, 29.75, 34.75, 39.75,$ and 49.75 MHz. Observe that the amount of desensitization depends upon the frequency of the interferer and occurs abruptly as the input power at f_2 exceeds a critical level.

Cross Modulation: Cross modulation is the nonlinear effect whereby modulation from one signal is transferred to another. In this way severe distortion can be generated by an interferer that is widely separated in frequency from the desired signal. The effect is illustrated in Fig. 1.9. As shown in Fig. 1.9(a), the desired signal is assumed to be an unmodulated carrier with frequency f_1. The interferer, which appears in Fig. 1.9(b), is an amplitude-modulated signal at frequency f_2. Note that the modulation consists of a triangular waveform. When these two signals are applied to a receiver having a cubic nonlinearity, a distorted version of the modulation from the signal at f_2 is transferred to the carrier at f_1. This is shown in Fig. 1.9(c). Whereas the envelope of the desired signal was originally constant, it now varies in a periodic manner. Obviously, cross modulation can be extremely troublesome in communication systems.

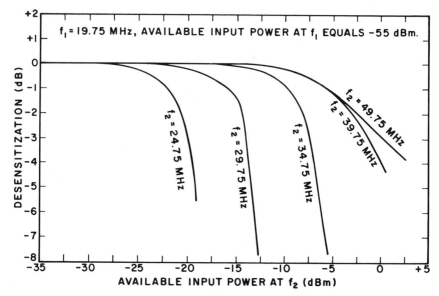

Fig. 1.8. Typical desensitization in a communications receiver.

Spurious Responses: Spurious responses are a special case of intermodulation. The amplitude of an intermodulation component depends upon the power levels of the signals involved. The weaker the signals incident upon a nonlinearity, the smaller are the intermodulation components generated. Since out of band signals are greatly attenuated by a receiver's frequency selectivity, they do not normally combine to produce significant interference via the intermodulation process.

However, many receivers employ strong local oscillator signals. In such instances, the local oscillator signal is much larger than the received signals at the mixer input. For example, the power level of a local oscillator signal at the input to a mixer may be 120 dB greater than that of the desired signal. In practice, a strong local oscillator signal can neutralize the attenuation experienced by an out of band signal. Consequently, out of band signals that combine nonlinearly with the local oscillator signal may produce significant intermodulation components. A receiver's image response is one example of this effect.

Spurious responses are defined to be those responses that are generated by an interfering signal, or one of its harmonics, combining with the local oscillator signal, or one of its harmonics. Of particular concern are those frequency combinations that result in interference components falling within or near the IF passband. It is the mechanism of spurious responses that allows out of band signals to severely disrupt, what would otherwise be, conventional receiver operation.

Gain Compression/Expansion: Thus far our discussion of nonlinear effects has involved interfering signals. Nonlinearities may adversely effect receiver

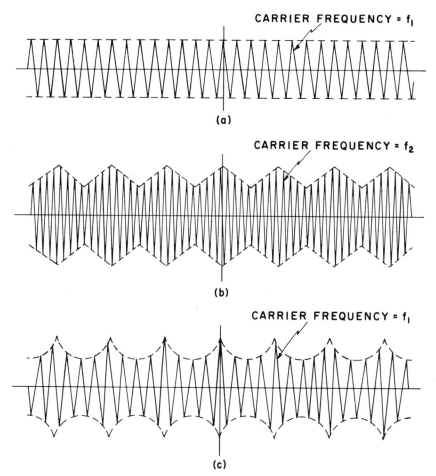

Fig. 1.9. Cross-modulation example. (a) Unmodulated carrier at f_1. (b) Carrier at f_2 amplitude modulated by triangular waveform. (c) Output at f_1 of receiver having a cubic nonlinearity.

performance even when interferers are absent. One such nonlinear effect is called gain compression or expansion. In a linear receiver the receiver gain remains constant as the input signal amplitude is varied. This is not the case with actual receivers that are inherently nonlinear. The terms, gain compression and expansion, are used to describe the variation in gain of a receiver as the input signal is increased in amplitude.

As a demonstration of these effects, an experiment was performed on an IF amplifier with a center frequency of 21.4 MHz, a 3-dB bandwidth of 500 kHz, and a midband power gain of 16 dB. Figure 1.10 shows the amplifier's measured input-output power characteristics for sinusoidal inputs at 21.4 and 20.0 MHz, respectively. When an amplifier performs linearly, the output power in decibels increases by the same amount as the input power in decibels. Thus the straight lines at 45° in Fig. 1.10(a) and (b) correspond to the response of a perfectly linear amplifier. For small enough input signals, we see that the amplifier behaves linearly.

The plot in Fig. 1.10(a) is for the case in which the input signal is applied at the amplifier's tuned frequency of 21.4 MHz. The amplifier has a linear gain of 16 dB at this frequency since an input power of -70 dBm results in an output power of -54 dBm. We say that gain expansion occurs when the amplifier gain is greater than the linear gain. Similarly, gain compression is said to occur when the amplifier gain is less than the linear gain. In our example, gain expansion occurs gradually as the input power level exceeds -60 dBm and transforms abruptly into gain compression at an input level of approximately -29 dBm. The plot is interesting because it illustrates that both expansion and compression can occur at a single frequency for different values of the input level.

The input-output power characteristic shown in Fig. 1.10(b) for an input frequency of 20.0 MHz follows the more conventional type of gain compression curve. Note that an input power of -30 dBm produces an output power of -64 dBm. Therefore, since the 20.0-MHz signal falls outside the IF passband, it experiences a linear attenuation of 34 dB. Gain compression first becomes apparent when the input power level is approximately -15 dBm.

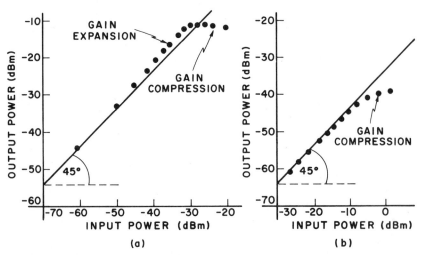

Fig. 1.10. Measured input-output power characteristics of IF amplifier. (a) Input signal at 21.4 MHz. (b) Input signal at 20.0 MHz.

When the gain varies as a function of the input signal level, serious nonlinear distortion can be generated in the desired signal.

Summary: In this section we have discussed the nonlinear effects of intermodulation, desensitization, cross modulation, spurious responses, and gain compression/expansion. These important effects, several of which may occur simultaneously in a communications receiver, cannot be analyzed, predicted, or explained by linear analysis techniques.

1.2 SERIES REPRESENTATION OF WEAK NONLINEARITIES

The Taylor series expansion plays a fundamental role in the modeling of weak nonlinearities. In this section we briefly consider the power series description of nonlinear elements. If the input and output are denoted by x and y, respectively, the power series description assumes the form

$$y = a_1 x + a_2 x^2 + a_3 x^3 + \cdots. \tag{1-1}$$

Equation (1-1) can be used to describe the nonlinear behavior of a variety of nonlinear elements. For example, x could be the voltage and y the current associated with a nonlinear resistor. Alternatively, for a nonlinear capacitor, x could be the voltage and y the charge. The power series representation is most useful when only a few terms are needed for a good characterization. As a result, it is not suited for use with strong nonlinearities.

Single-Frequency Input: Consider the application of a single-frequency sinusoid to a nonlinear element whose representation is given by (1-1). For simplicity, assume only the coefficients a_1, a_2, and a_3 are nonzero (i.e., $a_n = 0$ for $n > 3$). For an input given by

$$x = B_1 \cos 2\pi f_1 t, \tag{1-2}$$

the output is

$$y = a_1 B_1 \cos 2\pi f_1 t + a_2 B_1^2 \cos^2 2\pi f_1 t + a_3 B_1^3 \cos^3 2\pi f_1 t. \tag{1-3}$$

Application of the trigonometric identities

$$\cos^2 2\pi f_1 t = \tfrac{1}{2} + \tfrac{1}{2} \cos 2\pi (2f_1) t$$
$$\cos^3 2\pi f_1 t = \tfrac{3}{4} \cos 2\pi f_1 t + \tfrac{1}{4} \cos 2\pi (3f_1) t \tag{1-4}$$

results in

$$y = \tfrac{1}{2} a_2 B_1^2 + [a_1 B_1 + \tfrac{3}{4} a_3 B_1^3] \cos 2\pi f_1 t + \tfrac{1}{2} a_2 B_1^2 \cos 2\pi (2f_1) t$$
$$+ \tfrac{1}{4} a_3 B_1^3 \cos 2\pi (3f_1) t. \tag{1-5}$$

Observe that, in addition to the response at frequency f_1, the output contains intermodulation components at the frequencies 0, $2f_1$, and $3f_1$. The generation of new frequencies not present in the input signal is an important feature of nonlinear circuits. Also, note that the response at frequency f_1 is generated by both the linear and cubic terms in (1-1). Depending upon the sign of a_3 relative to a_1, the output amplitude at frequency f_1 may be either larger or smaller than that generated by the linear term alone. This is the gain compression/expansion phenomenon mentioned earlier.

Three-Frequency Input: As before, assume the power series of (1-1) contains only the linear, quadratic, and cubic terms. Let the input to the nonlinear element be

$$x = B_1 \cos 2\pi f_1 t + B_2 \cos 2\pi f_2 t + B_3 \cos 2\pi f_3 t. \tag{1-6}$$

Substituting this expression into (1-1) and making use of the appropriate trigonometric identities, it is readily shown that the output $y(t)$ contains components at the various frequencies listed in Table 1.1. The three-frequency input is seen to generate many more intermodulation components than does the single-frequency input. In fact, the number of intermodulation components grows dramatically with further increases in the number of input frequencies. For an element that is intended to be linear, the desired outputs are those due to the term $a_1 x$. All other outputs are intermodulation components that contribute to objectionable noise and interference.

In summary, a nonlinear element produces at its output additional frequencies that are various combinations of sum and difference frequencies of the input signal. Although our discussion has been for nonlinear elements with weak non-

TABLE 1.1.   **Frequencies in Output $y = a_1 x + a_2 x^2 + a_3 x^3$ for Three-Frequency Input.**

Power Series Term	Frequency Content
$a_1 x$	f_1, f_2, f_3
$a_2 x^2$	0 $2f_1, 2f_2, 2f_3$ $f_1 + f_2, f_1 + f_3, f_2 + f_3$ $f_1 - f_2, f_1 - f_3, f_2 - f_3$
$a_3 x^3$	f_1, f_2, f_3 $3f_1, 3f_2, 3f_3$ $2f_1 + f_2, 2f_1 + f_3, 2f_2 + f_1, 2f_2 + f_3, 2f_3 + f_1, 2f_3 + f_2$ $2f_1 - f_2, 2f_1 - f_3, 2f_2 - f_1, 2f_2 - f_3, 2f_3 - f_1, 2f_3 - f_2$ $f_1 + f_2 + f_3, f_1 + f_2 - f_3, f_1 - f_2 + f_3, f_1 - f_2 - f_3$

linearities, it should be pointed out that the same generation of additional frequencies occurs with strong nonlinearities.

1.3 ORGANIZATION OF TEXT

This book is devoted to the sinusoidal analysis and modeling of nonlinear effects in weakly nonlinear circuits. In particular, we develop the nonlinear transfer function approach which is a generalization of the linear transfer function concept. The advantage of this technique is that insight developed from experience with linear systems can be used to understand the behavior of nonlinear systems.

Both sinusoidal and nonsinusoidal inputs can be handled by the nonlinear transfer function approach. However, the mathematics becomes considerably more complicated when nonsinusoidal inputs are involved. In order to keep our presentation as simple as possible, only sinusoidal excitations are considered in the main body of this volume. This is adequate for obtaining considerable insight into the response of weakly nonlinear systems. Nonsinusoidal inputs are treated briefly in Appendix A.

In order to provide a review of the sinusoidal steady-state analysis of linear circuits, power considerations are discussed in Chapter 2. This material has application in later chapters where nonlinear effects are characterized in terms of input and output powers. Various power gains are defined and their equivalence, under special conditions, is pointed out.

The classical power series approach to the analysis of weakly nonlinear systems is presented in Chapter 3. Although this technique has been widely used in the past, it is not generally applicable. Nevertheless, it is a convenient analytical tool where appropriate. Using this model, analytical expressions are developed for the response of weakly nonlinear systems to a sum of sinusoidal inputs.

The more general nonlinear transfer function approach is developed in Chapter 4. It is shown that the amplitudes and phases of the various sinusoidal outputs are determined directly by the nonlinear transfer functions. Examples of some experimentally determined nonlinear transfer functions are presented and discussed.

In Chapter 5 the harmonic input method is presented for determining the nonlinear transfer functions from circuit diagrams. Terminal relationships are discussed for nonlinear resistors, capacitors, inductors, and controlled sources. The harmonic input method is shown to reduce a weakly nonlinear problem to an equivalent linear problem where the nonlinearities manifest themselves as known excitations.

Criteria for characterizing nonlinear distortion and interference effects are discussed in Chapter 6. Analytical expressions utilizing the nonlinear transfer functions are developed for the various criteria.

Before the harmonic input method for determining the nonlinear transfer functions can be applied, it is necessary to model each electronic device appearing in a weakly nonlinear system by a nonlinear incremental equivalent circuit. In Chapter 7 we present such models for the semiconductor diode, bipolar junction transistor, junction field-effect transistor, and metal-oxide semiconductor field-effect transistor. Emphasis is on device behavior as opposed to device physics. Attention is also given to the problem of obtaining model parameters from laboratory measurements.

2

Power Considerations for Linear Circuits in the Sinusoidal Steady State

The study of distortion and interference in weakly nonlinear circuits excited by sinusoidal inputs is inherently concerned with the power generated in various sinusoidal components at the output. Since linear circuit theory plays a major role in determining the nonlinear response, this chapter provides background material by reviewing power considerations associated with linear networks operating in the sinusoidal steady state. In order to avoid complicated and difficult to remember trigonometric formulas, representations involving complex quantities are introduced. These lead to the useful concepts of negative frequency, complex voltage, complex current, and complex power.

The topic of power gain is next developed in detail. The discussion is complicated by the existence of several different definitions for power. In particular, one may speak of either average power, available power, and/or exchangeable power. These, in turn, are used in definitions of average power gain, insertion power gain, available power gain, transducer power gain, and exchangeable power gain. Although the gains become identical in certain specialized situations, the definitions are not equivalent. Consequently, the various power gains cannot be used interchangeably.

This chapter begins by reviewing complex number arithmetic and the complex representation of sinusoids.

2.1 COMPLEX NUMBER ARITHMETIC AND THE COMPLEX REPRESENTATION OF SINUSOIDS

Recall that a complex number is given by

$$A = A_1 + jA_2 \tag{2-1}$$

where A_1 and A_2 are real numbers and j has the property that $j^2 = -1$. A_1 is called the real part of A while A_2 is called its imaginary part. The symbols, $\text{Re}\{\cdot\}$ and $\text{Im}\{\cdot\}$, are used to denote the real and imaginary parts of the quantities inside the brackets. Thus $A_1 = \text{Re}\{A\}$ and $A_2 = \text{Im}\{A\}$. Two complex numbers are equal if and only if their real and imaginary parts are equal, respectively. Specifically, if $A = A_1 + jA_2$ and $B = B_1 + jB_2$, then $A = B$ if and only if $A_1 = B_1$ and $A_2 = B_2$.

A complex number may be represented in vector form as shown in Fig. 2.1. The magnitude of a complex number is defined to be the length of the corresponding vector and is given by

$$\left| A \right| = (A_1^2 + A_2^2)^{1/2} = [\text{Re}^2\{A\} + \text{Im}^2\{A\}]^{1/2}. \tag{2-2}$$

Similarly, the angle of a complex number is the angle of the corresponding vector. We will denote it by ang A. Therefore,

$$\text{ang } A = \theta_A = \tan^{-1}\left(\frac{A_2}{A_1}\right) = \tan^{-1}\left[\frac{\text{Im}\{A\}}{\text{Re}\{A\}}\right]. \tag{2-3}$$

(In some texts the angle is referred to as the argument and is denoted by arg A.) From Fig. 2.1 it is obvious that the real and imaginary parts of A are expressed in terms of its magnitude and angle according to the relations

$$A_1 = \text{Re}\{A\} = \left| A \right| \cos \theta_A$$
$$A_2 = \text{Im}\{A\} = \left| A \right| \sin \theta_A. \tag{2-4}$$

It follows that the complex number A can be written as

$$A = A_1 + jA_2 = \left| A \right| \cos \theta_A + j\left| A \right| \sin \theta_A$$
$$= \left| A \right| (\cos \theta_A + j \sin \theta_A). \tag{2-5}$$

Euler's formula states that

$$e^{\pm j\theta} = \cos \theta \pm j \sin \theta. \tag{2-6}$$

As a result, the complex number A can be expressed in polar form as

$$A = \left| A \right| e^{j\theta_A} = \left| A \right| \underline{/\theta_A}. \tag{2-7}$$

The second form in (2-7) is in every way equivalent to the first and is preferred when the angle is in degrees rather than in radians. We see that a complex num-

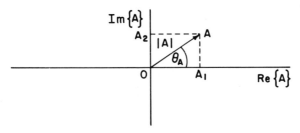

Fig. 2.1. Vector representation of the complex number A.

ber may be written either in rectangular form, in terms of its real and imaginary parts, or in polar form, in terms of its magnitude and angle. Conversion between rectangular and polar forms is accomplished by means of equations (2-2) through (2-4).

The addition of two complex numbers is conveniently carried out in rectangular form. Suppose the two complex numbers are $A = A_1 + jA_2$ and $B = B_1 + jB_2$. Their sum is obtained by adding separately the real and imaginary parts. Therefore,

$$A + B = (A_1 + B_1) + j(A_2 + B_2). \tag{2-8}$$

Note that the real part of the sum is the sum of the real parts and the imaginary part of the sum is the sum of the imaginary parts. With symbols, this is indicated by

$$\text{Re}\{A + B\} = \text{Re}\{A\} + \text{Re}\{B\} = A_1 + B_1$$

$$\text{Im}\{A + B\} = \text{Im}\{A\} + \text{Im}\{B\} = A_2 + B_2. \tag{2-9}$$

When multiplying two complex numbers, it is preferable to first write them in polar form. The polar forms of A and B are

$$A = |A| \, e^{j\theta_A} = |A| \, \underline{/\theta_A}.$$
$$B = |B| \, e^{j\theta_B} = |B| \, \underline{/\theta_B}. \tag{2-10}$$

The product is obtained by multiplying the magnitudes and adding the angles. Thus

$$AB = |A| \, |B| \, e^{j(\theta_A + \theta_B)} = |A| \, |B| \, \underline{/\theta_A + \theta_B}. \tag{2-11}$$

We see that the magnitude of the product equals the product of the magnitudes and the angle of the product equals the sum of the angles. Specifically,

$$|AB| = |A| \, |B|$$

$$\text{ang } AB = \theta_A + \theta_B = \text{ang } A + \text{ang } B. \tag{2-12}$$

Although it is convenient to have the numbers in polar form for multiplication, this is not necessary. In rectangular form they are multiplied by the conventional rules of algebra. Thus the product may also be written as

$$AB = (A_1 + jA_2)(B_1 + jB_2)$$
$$= (A_1 B_1 - A_2 B_2) + j(A_1 B_2 + A_2 B_1). \tag{2-13}$$

Polar form is also preferable for dividing two complex numbers. The quotient of A divided by B is

$$\frac{A}{B} = \frac{|A| e^{j\theta_A}}{|B| e^{j\theta_B}} = \frac{|A|}{|B|} e^{j(\theta_A - \theta_B)} = \frac{|A|}{|B|} \underline{/\theta_A - \theta_B}. \tag{2-14}$$

From (2-14) it follows that the magnitude of the quotient is the quotient of the magnitudes and the angle of the quotient equals the difference between the angles. Symbolically,

$$\left| \frac{A}{B} \right| = \frac{|A|}{|B|}$$

$$\text{ang} \frac{A}{B} = \theta_A - \theta_B = \text{ang } A - \text{ang } B. \tag{2-15}$$

Should A and B be given in rectangular form, their quotient is obtained by multiplying numerator and denominator with $B_1 - jB_2$. This results in

$$\frac{A}{B} = \frac{A_1 + jA_2}{B_1 + jB_2} \frac{B_1 - jB_2}{B_1 - jB_2}$$

$$= \frac{A_1 B_1 + A_2 B_2}{B_1^2 + B_2^2} + j \left(\frac{A_2 B_1 - A_1 B_2}{B_1^2 + B_2^2} \right). \tag{2-16}$$

The above method of division is referred to as the process of rationalization.

Rationalization involves multiplying numerator and denominator with the complex conjugate of the denominator. Given the number $A = A_1 + jA_2$, its complex conjugate is defined to be

$$A^* = A_1 - jA_2. \tag{2-17}$$

In taking the complex conjugate of a number, the real part remains unchanged while the imaginary part is reversed in sign. The complex conjugate is illustrated in Fig. 2.2. Note that conjugate numbers have identical magnitudes but angles that are equal and opposite in sign. From Fig. 2.2 we conclude that

$$|A^*| = |A| = (A_1^2 + A_2^2)^{1/2}$$

$$\text{ang } A^* = -\text{ang } A = -\theta_A. \tag{2-18}$$

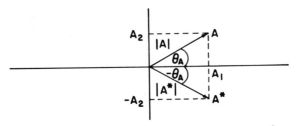

Fig. 2.2. The complex number A and its conjugate A^*.

It follows that

$$A^* = |A| e^{-j\theta_A} = |A| \underline{/-\theta_A}. \tag{2-19}$$

Conjugation has the interesting properties that the conjugate of a sum equals the sum of the conjugates, the conjugate of a product equals the product of the conjugates, and the conjugate of a quotient equals the quotient of the conjugates. These relationships are demonstrated below:

$$(A + B)^* = [(A_1 + jA_2) + (B_1 + jB_2)]^* = [(A_1 + B_1) + j(A_2 + B_2)]^*$$

$$= (A_1 + B_1) - j(A_2 + B_2) = (A_1 - jA_2) + (B_1 - jB_2)$$

$$= A^* + B^* \tag{2-20}$$

$$(AB)^* = [|A| e^{j\theta_A} |B| e^{j\theta_B}]^* = [|A| |B| e^{j(\theta_A + \theta_B)}]^*$$

$$= |A| |B| e^{-j(\theta_A + \theta_B)} = |A| e^{-j\theta_A} |B| e^{-j\theta_B}$$

$$= A^* B^* \tag{2-21}$$

$$\left(\frac{A}{B}\right)^* = \left[\frac{|A| e^{j\theta_A}}{|B| e^{j\theta_B}}\right]^* = \left[\frac{|A|}{|B|} e^{j(\theta_A - \theta_B)}\right]^*$$

$$= \frac{|A|}{|B|} e^{-j(\theta_A - \theta_B)} = \frac{|A| e^{-j\theta_A}}{|B| e^{-j\theta_B}}$$

$$= \frac{A^*}{B^*}. \tag{2-22}$$

It can be shown, no matter how complicated the expression is for a complex quantity, that its conjugate may be obtained by simply changing the sign of j wherever it occurs.

Example 2.1

$$\left[\frac{(A + jB) jC(D - jE)}{(F + jG) e^{j\theta}}\right]^* = \frac{(A - jB) (-jC) (D + jE)}{(F - jG) e^{-j\theta}}. \tag{2-23}$$

Additional useful relationships involving the complex conjugate are:

$$|A|^2 = AA^* \tag{2-24}$$

$$\text{Re}\{A\} = \text{Re}\{A^*\} = \frac{1}{2}(A + A^*) = A_1 \tag{2-25}$$

$$\text{Im}\{A\} = -\text{Im}\{A^*\} = \frac{1}{2j}(A - A^*) = A_2. \tag{2-26}$$

Thus the magnitude squared, the real part, and the imaginary part of a complex number can all be obtained from simple expressions involving the complex number itself and its complex conjugate.

We have introduced complex numbers because they are useful for representing sinusoidal voltages and currents. Consider the sinusoidal voltage

$$e(t) = |E| \cos(2\pi ft + \alpha). \tag{2-27}$$

By means of the Euler formula given in (2-6), we can write

$$e(t) = \text{Re}\{|E| e^{j(2\pi ft + \alpha)}\} = \text{Re}\{|E| e^{j\alpha} e^{j2\pi ft}\}. \tag{2-28}$$

If we define the complex voltage E to be

$$E = |E| e^{j\alpha} \tag{2-29}$$

and make use of (2-25), $e(t)$ becomes

$$e(t) = \text{Re}\{E e^{j2\pi ft}\} = \tfrac{1}{2}(E e^{j2\pi ft} + E^* e^{-j2\pi ft}). \tag{2-30}$$

This is an interesting result because the concept of negative frequency is suggested by the complex conjugate term. Provided the frequency of the voltage is known, it is straightforward to convert between the complex voltage E and the sinusoidal voltage $e(t)$. Therefore, specification of one is equivalent to specification of the other. The complex voltage E is also referred to as a phasor.

Similarly, sinusoidal currents can be expressed in terms of complex currents. Note that

$$i(t) = |I| \cos(2\pi ft + \beta)$$
$$= \text{Re}\{|I| e^{j(2\pi ft + \beta)}\} = \text{Re}\{|I| e^{j\beta} e^{j2\pi ft}\}. \tag{2-31}$$

Defining the complex current to be

$$I = |I| e^{j\beta}, \tag{2-32}$$

we have

$$i(t) = \text{Re}\{I e^{j2\pi ft}\} = \tfrac{1}{2}(I e^{j2\pi ft} + I^* e^{-j2\pi ft}). \tag{2-33}$$

The term, phasor, is also applied to the complex current I.

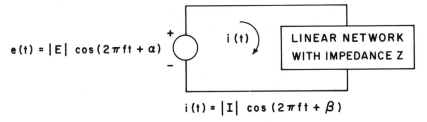

$$i(t) = |I| \cos (2\pi ft + \beta)$$

Fig. 2.3. A linear network in the sinusoidal steady state with impedance Z.

In our work we shall find it convenient to deal with complex voltages and currents. For sinusoidal inputs the principal advantage accruing from this approach is that network differential equations are replaced by linear algebraic equations. These equations are time independent and can be solved using any of the methods of linear algebra.

Example 2.2

Let $e(t)$ and $i(t)$ be the sinusoidal voltage and current associated with the linear network having the impedance Z as shown in Fig. 2.3. The problem is to determine the loop current $i(t)$. The first step in the solution is to replace $e(t)$ by its complex voltage E. From sinusoidal steady-state considerations, the complex voltage and current are related by the linear algebraic equation

$$E = IZ. \tag{2-34}$$

Given E, this equation is easily solved for I. By means of (2-33), the complex current I is then transformed back to the time domain so as to obtain $i(t)$. More complicated networks are handled in a similar manner.

2.2 INSTANTANEOUS, AVERAGE, AND COMPLEX POWERS

By definition, the instantaneous power delivered to the linear network of Fig. 2.3 is the product of the voltage across the input terminal pair multiplied by the current entering the positive terminal. In general, it is a function of time caused by the time variations in the voltage and current. For the case of sinusoidal voltages and currents, the instantaneous power is

$$
\begin{aligned}
p(t) = e(t)\,i(t) &= |E| \cos (2\pi ft + \alpha)\,|I| \cos (2\pi ft + \beta) \\
&= \tfrac{1}{4}(E\,e^{j2\pi ft} + E^*\,e^{-j2\pi ft})\,(I\,e^{j2\pi ft} + I^*\,e^{-j2\pi ft}) \\
&= \tfrac{1}{4}(EI^* + E^*I) + \tfrac{1}{4}(EI\,e^{j2\pi(2f)t} + E^*I^*\,e^{-j2\pi(2f)t}). \tag{2-35}
\end{aligned}
$$

Observe that each bracketed term in the final expression of (2-35) consists of a complex quantity plus its conjugate. From (2-25) it follows that the instantaneous power can be written as

$$p(t) = \tfrac{1}{2} \operatorname{Re}\{EI^*\} + \tfrac{1}{2} \operatorname{Re}\{EI\, e^{j2\pi(2f)t}\}. \qquad (2\text{-}36)$$

Recalling that E has magnitude $|E|$ and angle α while I has magnitude $|I|$ and angle β, we conclude that

$$p(t) = \tfrac{1}{2} \operatorname{Re}\{|E|\,|I|\, e^{j(\alpha-\beta)}\} + \tfrac{1}{2} \operatorname{Re}\{|E|\,|I|\, e^{j[2\pi(2f)t+\alpha+\beta]}\}$$

$$= \tfrac{1}{2}|E|\,|I| \cos(\alpha-\beta) + \tfrac{1}{2}|E|\,|I| \cos[2\pi(2f)\,t + \alpha + \beta]. \qquad (2\text{-}37)$$

Thus the instantaneous power in the sinusoidal steady state consists of a constant plus a sinusoid at twice the frequency of the voltage and current.

In many applications the time average of the instantaneous power delivered to the network is of most interest. Recall that the time average of a sum equals the sum of the time averages. In addition, the time average of a constant equals the constant while the time average of a sinusoid, over an integer number of periods, is zero. Applying a time average to (2-37), it follows that the average power delivered to the network is

$$p_{\text{ave}} = \tfrac{1}{2}|E|\,|I| \cos(\alpha-\beta)$$

$$= \tfrac{1}{2} \operatorname{Re}\{EI^*\} = \tfrac{1}{2} \operatorname{Re}\{E^*I\}. \qquad (2\text{-}38)$$

The average power is seen to depend upon both the magnitudes and angles of the complex voltage and current. Equation (2-38) also points out that p_{ave} may be determined by taking the real part of either of the complex quantities $\tfrac{1}{2}EI^*$ and $\tfrac{1}{2}E^*I$. This approach is frequently more convenient to use. As a result, the complex power is defined to be

$$p_{\text{com}} = \tfrac{1}{2}EI^*. \qquad (2\text{-}39)$$

The concept of complex power is extremely useful in deriving various expressions for the average power delivered to a network.

Example 2.3

Let the impedance of the linear network in Fig. 2.3 have resistance R and reactance X. Hence

$$Z = R + jX. \qquad (2\text{-}40)$$

Since $E = ZI$, p_{ave} can be written as

$$p_{\text{ave}} = \tfrac{1}{2} \operatorname{Re}\{EI^*\} = \tfrac{1}{2} \operatorname{Re}\{ZII^*\}$$

$$= \tfrac{1}{2}|I|^2 \operatorname{Re}\{Z\} = \tfrac{1}{2}|I|^2 R. \qquad (2\text{-}41)$$

Although (2-41) looks very much like the instantaneous power formula for a resistor, we must remember that I is a complex current and that R is the real part of a complex impedance.

Example 2.4

Now assume that the linear network in Fig. 2.3 is characterized by its admittance

$$Y = \frac{1}{Z} = G + jB \qquad (2\text{-}42)$$

where G is the conductance and B is the susceptance. Noting that $I = EY$, the average power becomes

$$P_{ave} = \tfrac{1}{2} \, \text{Re}\{EI^*\} = \tfrac{1}{2} \, \text{Re}\{E^*I\}$$

$$= \tfrac{1}{2} \, \text{Re}\{E^*EY\} = \tfrac{1}{2}|E|^2 \, \text{Re}\{Y\} = \tfrac{1}{2}|E|^2 \, G. \qquad (2\text{-}43)$$

Equation (2-43) is the dual expression to (2-41). In general, the conductance of a network does not equal the reciprocal of its resistance. The true relationship between G and R is obtained by observing that

$$Y = G + jB = \frac{1}{Z} = \frac{1}{R+jX} = \frac{1}{R+jX} \frac{R-jX}{R-jX}$$

$$= \frac{R}{R^2+X^2} - j \frac{X}{R^2+X^2}. \qquad (2\text{-}44)$$

Taking the real part of the admittance, it follows that

$$G = \text{Re}\{Y\} = \frac{R}{R^2+X^2} = \frac{\text{Re}\{Z\}}{|Z|^2}. \qquad (2\text{-}45)$$

Consequently, an alternate expression for the average power is

$$P_{ave} = \frac{1}{2}|E|^2 \frac{R}{R^2+X^2} = \frac{1}{2}|E|^2 \frac{\text{Re}\{Z\}}{|Z|^2}. \qquad (2\text{-}46)$$

For the special case in which the impedance is real, $Z = R$ and $X = 0$. Then the admittance reduces to $Y = G = 1/R$. The average power for this very special situation is

$$P_{ave} = \frac{1}{2} \frac{|E|^2}{R}. \qquad (2\text{-}47)$$

It is important to realize that, although (2-41) is valid for all linear networks, (2-47) is valid only for those networks whose impedances are purely resistive.

As is seen from (2-45), the sign of $\text{Re}\{Y\}$ is the same as that of $\text{Re}\{Z\}$. Also, from (2-41) and (2-46), it is apparent that the sign of p_{ave} is identical to the sign of $\text{Re}\{Z\}$. When $\text{Re}\{Z\}$ is positive, p_{ave} is positive. The delivered average power is dissipated and the network is said to be passive. On the other hand, p_{ave} is negative when $\text{Re}\{Z\}$ is negative. Negative power implies that average power is being generated by the network instead of being absorbed. The network is now said to be active. For example, circuits containing transistors and vacuum tubes are likely to be active. The equivalent circuits for these devices contain controlled (or dependent) sources that account for their active behavior.

2.3 AVERAGE POWER GAIN

A port is defined to be a terminal pair for which current into one terminal equals current out of the other. As shown in Fig. 2.4, 2-port networks are commonly used in the transmission of power from a signal source to a load. Because the network is assumed to be operating in the sinusoidal steady state, circuit variables are indicated as complex voltages and currents. By convention, port currents are positive when entering the positive port terminals. It is assumed that no independent sources are contained within the 2-port. The signal source driving the 2-port is modeled as a Thévenin equivalent circuit. This consists of the ideal voltage source E_S in series with the generator impedance Z_S. The linear load is designated by the impedance Z_L.

Two linear equations between the port variables E_1, E_2, I_1, and I_2 are all that is needed to completely characterize the behavior of a linear 2-port. Of the four variables, any two can be chosen to be independent. The various choices result in six different parameter sets. For our discussion in this chapter we arbitrarily choose the open-circuit impedance parameters given by

$$E_1 = Z_{11}I_1 + Z_{12}I_2$$
$$E_2 = Z_{21}I_1 + Z_{22}I_2. \tag{2-48}$$

Here I_1 and I_2 are the independent variables while E_1 and E_2 are the dependent variables.

Power is often the most significant consideration in applications involving sinusoidal voltages and currents. Of interest is the amount of power delivered to the load as compared to that generated by the source. Although we have assumed that no independent generators are contained within the 2-port, con-

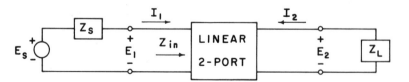

Fig. 2.4.   A loaded linear 2-port in the sinusoidal steady state.

trolled sources are allowed. Therefore, power gains greater than unity are possible. For passive 2-ports, where the gain is less than unity, the power loss is defined to be the reciprocal of the power gain.

As pointed out previously, there exist several different definitions of power gain. In this section we discuss the concept of average power gain. With reference to Fig. 2.4, let

p_L average power delivered to the load from the 2-port;
p_S average power delivered to the 2-port from the source;
g_P average power gain of the loaded 2-port.

The average power gain is then defined as

$$g_P = \frac{p_L}{p_S}. \tag{2-49}$$

To pursue the topic further, let the load and input impedances of the 2-port be given by

$$Z_L = R_L + jX_L$$
$$Z_{in} = R_{in} + jX_{in}. \tag{2-50}$$

Observe that the current into the load is $-I_2$ while that into the 2-port is I_1. It follows from (2-41) that the average power gain is

$$g_P = \frac{\frac{1}{2}|I_2|^2 \; \text{Re}\{Z_L\}}{\frac{1}{2}|I_1|^2 \; \text{Re}\{Z_{in}\}} = \left|\frac{I_2}{I_1}\right|^2 \frac{R_L}{R_{in}}. \tag{2-51}$$

g_P is seen to be the product of the magnitude squared of the 2-port current gain times the ratio of the real part of the load impedance to the real part of the input impedance.

Example 2.5

An interesting relation results when the 2-port and load are entirely resistive. With $X_L = X_{in} = 0$, R_L and R_{in} can be expressed as

$$R_L = -\frac{E_2}{I_2}$$
$$R_{in} = \frac{E_1}{I_1}. \tag{2-52}$$

In addition, the current gain for a resistive network is purely real. Therefore, the expression in (2-51) reduces to

$$g_P = \left(\frac{I_2}{I_1}\right)^2 \left(-\frac{E_2}{I_2}\right)\left(\frac{I_1}{E_1}\right) = \left(\frac{E_2}{E_1}\right)\left(-\frac{I_2}{I_1}\right). \tag{2-53}$$

This is recognized as the product of the loaded 2-port voltage and current gains. The minus sign is associated with I_2 since $-I_2$ represents the current flowing from the 2-port into the load. In general, (2-53) is not valid when the 2-port and/or load contain energy storage elements. The average power is then calculated using (2-51).

Evaluation of (2-51) requires knowledge of both the 2-port current gain and input impedance. These quantities can be expressed in terms of whichever parameters have been chosen to characterize the 2-port. In any event, the output port variables are constrained by the relation

$$E_2 = -I_2 Z_L. \tag{2-54}$$

Example 2.6

For the Z-parameter formulation, the second equation in (2-48) becomes

$$-I_2 Z_L = Z_{21} I_1 + Z_{22} I_2. \tag{2-55}$$

The 2-port current gain, therefore, is

$$-\frac{I_2}{I_1} = \frac{Z_{21}}{Z_{22} + Z_L}. \tag{2-56}$$

The input impedance is the ratio of E_1 to I_1. Elimination of I_2 from the first equation in (2-48) by means of (2-56) yields the equation

$$E_1 = Z_{11} I_1 - \frac{Z_{12} Z_{21}}{Z_{22} + Z_L} I_1. \tag{2-57}$$

It follows that

$$Z_{in} = \frac{E_1}{I_1} = Z_{11} - \frac{Z_{12} Z_{21}}{Z_{22} + Z_L}. \tag{2-58}$$

Substitution of (2-56) and (2-58) into (2-51) results in

$$g_P = \left| \frac{Z_{21}}{Z_{22} + Z_L} \right|^2 \frac{R_L}{\text{Re} \left\{ Z_{11} - \dfrac{Z_{12} Z_{21}}{Z_{22} + Z_L} \right\}}. \tag{2-59}$$

This is an expression for the average power gain in terms of the Z-parameters of the 2-port. Observe that the source impedance Z_S does not appear in the expression. This is a consequence of both the 2-port current gain and the input impedance being independent of Z_S. In this sense, average power gain is a figure of merit for the loaded 2-port irrespective of the generator exciting the network.

An alternate expression for the average power gain is obtained by writing p_L and p_S in the form given by (2-43). p_L and p_S now involve the load and input admittances of the 2-port given by

$$Y_L = \frac{1}{Z_L} = G_L + jB_L$$

$$Y_{in} = \frac{1}{Z_{in}} = G_{in} + jB_{in}. \tag{2-60}$$

Noting that the voltage across the load is E_2 while the voltage at the input port is E_1, the average power gain becomes

$$g_P = \frac{\frac{1}{2}|E_2|^2 \, \text{Re}\{Y_L\}}{\frac{1}{2}|E_1|^2 \, \text{Re}\{Y_{in}\}} = \left|\frac{E_2}{E_1}\right|^2 \frac{G_L}{G_{in}}. \tag{2-61}$$

In this form g_P is the product of the magnitude squared of the 2-port voltage gain times the ratio of the real part of the load admittance to the real part of the input admittance. The expression reduces to (2-59) when the voltage gain and G_{in} are written as functions of the 2-port Z-parameters. In addition, (2-61) reduces to (2-53) when the network is entirely resistive.

Frequently, it is preferable to compare powers on a logarithmic scale where the unit, called the decibel, is abbreviated as dB. The number of decibels by which a power p_2 exceeds the power p_1 is defined to be the decibel gain

$$G = 10 \log_{10} \frac{p_2}{p_1}. \tag{2-62}$$

Example 2.7
If p_2 is 500 mW and p_1 is 5 mW, the power gain in decibels is 20 dB. The gain is 0 dB when $p_2 = p_1$.

A brief tabulation of power ratios along with their approximate values in decibels is given in Table 2.1. Although the decibel is a unit based on the ratio between two powers, it may also be used as an indication of the actual power level provided the reference power p_1 is known. When the reference power is 1 mW, the abbreviation dBm is used. Thus 0 dBm corresponds to $p_2 = 1$ mW of power while 30 dBm corresponds to $p_2 = 1$ W. Since

$$10 \log_{10} A + 10 \log_{10} B = 10 \log_{10} AB, \tag{2-63}$$

the addition of decibels corresponds to multiplication of their ratios.

**TABLE 2.1.   Power Ratios
Expressed in Decibels.**

p_2/p_1	$10 \log_{10} p_2/p_1$ (dB)
1.00	0
1.25	1
1.6	2
2	3
3	4.8
5	7
7	8.5
10	10
100	20
1000	30

Example 2.8

The sum of 3 dB and 2 dB corresponds to multiplication of the approximate ratios 2 and 1.6. Therefore, 5 dB is approximately equivalent to a power ratio of 3.2.

In a similar fashion, subtraction of decibels results in a division of the ratios.

In accordance with the above discussion, the average power gain in decibels is

$$G_P = 10 \log_{10} \frac{p_L}{p_S} = 10 \log_{10} g_P. \qquad (2\text{-}64)$$

Observe that we reserve an uppercase letter for a gain expressed in decibels. Substitution of (2-61) into (2-64) yields

$$G_P = 10 \log_{10} \left| \frac{E_2}{E_1} \right|^2 \frac{G_L}{G_{in}} = 20 \log_{10} \left| \frac{E_2}{E_1} \right| + 10 \log_{10} \frac{G_L}{G_{in}}. \qquad (2\text{-}65)$$

For the special case in which $G_L = G_{in}$, $\log_{10} G_L/G_{in} = 0$ and the average power gain becomes

$$G_P = 20 \log_{10} \left| \frac{E_2}{E_1} \right|. \qquad (2\text{-}66)$$

The logarithm of the voltage gain is a useful measure whether or not $G_L = G_{in}$. With (2-66) in mind, the voltage gain in decibels is defined as

$$A = 20 \log_{10} \left| \frac{E_2}{E_1} \right|. \qquad (2\text{-}67)$$

Therefore, if $G_L = G_{in}$, the voltage gain in decibels equals the power gain in decibels. However, these two quantities are not equal when G_L differs from G_{in}.

Equation (2-65) shows, in general, that the two gains are different by the amount $10 \log_{10} G_L/G_{in}$. When there is the possibility of confusion, the designation dBv is used to indicate decibel voltage gain. Otherwise, the logarithmic unit for both power and voltage gains is denoted by dB.

2.4 INSERTION POWER GAIN

The 2-port of Fig. 2.4 serves as an interface for the power transfer from source to load. Another way of evaluating 2-port performance is to compare average powers delivered to the load with and without the 2-port in place. For this purpose, consider the circuit of Fig. 2.5 in which the load is connected directly to the source. Let $\tilde{p}_L$ be the average power delivered to the load without the 2-port in place. With reference to (2-41) and Fig. 2.5,

$$\tilde{p}_L = \frac{1}{2}|I_L|^2 \operatorname{Re}\{Z_L\} = \frac{1}{2}\left|\frac{E_S}{Z_S + Z_L}\right|^2 R_L. \tag{2-68}$$

On the other hand, with reference to (2-46) and Fig. 2.4, the average power delivered to the load with the 2-port in place is

$$p_L = \frac{1}{2}|E_2|^2 \frac{R_L}{|Z_L|^2}. \tag{2-69}$$

The insertion power gain is defined to be

$$g_I = \frac{p_L}{\tilde{p}_L}. \tag{2-70}$$

Therefore, substitution of (2-68) and (2-69) into (2-70) results in

$$g_I = \left|\frac{E_2}{E_S}\right|^2 \left|\frac{Z_S + Z_L}{Z_L}\right|^2. \tag{2-71}$$

For many systems, such as in microwave applications, it is common to have identical generator and load impedances. The insertion gain then reduces to

$$g_I = 4\left|\frac{E_2}{E_S}\right|^2. \tag{2-72}$$

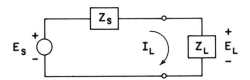

Fig. 2.5.   Circuit of Fig. 2.4 with 2-port removed.

Recall that the voltage E_2 is associated with the circuit of Fig. 2.4 while the voltage E_S is associated with that of Fig. 2.5. Nevertheless, since the generator is assumed to be the same in both circuits, it is possible to interpret the ratio E_2/E_S as a voltage gain involving the loaded 2-port of Fig. 2.4.

Example 2.9

The insertion gain is readily expressed in terms of the open-circuit impedance parameters of (2-48). The input port variables of Fig. 2.4 are related by the equation

$$E_1 = E_S - I_1 Z_S. \tag{2-73}$$

Substitution of this constraint into (2-57) yields

$$E_S = \frac{(Z_{11} + Z_S)(Z_{22} + Z_L) - Z_{12}Z_{21}}{Z_{22} + Z_L} I_1. \tag{2-74}$$

Similarly, the output port variables are constrained by (2-54). In conjunction with (2-56), it follows that

$$E_2 = \frac{Z_{21} Z_L}{Z_{22} + Z_L} I_1. \tag{2-75}$$

Dividing (2-75) by (2-74), we obtain the voltage gain

$$\frac{E_2}{E_S} = \frac{Z_{21} Z_L}{(Z_{11} + Z_S)(Z_{22} + Z_L) - Z_{12}Z_{21}}. \tag{2-76}$$

Finally, substituting (2-76) into (2-71), the insertion power gain in terms of the Z-parameter formulation becomes

$$g_I = \left| \frac{Z_{21}(Z_S + Z_L)}{(Z_{11} + Z_S)(Z_{22} + Z_L) - Z_{12}Z_{21}} \right|^2. \tag{2-77}$$

Note that g_I does depend on the source impedance Z_S whereas the average power gain, g_P, does not [see (2-59)]. Therefore, given the parameters characterizing the 2-port and load, it is possible to maximize the insertion power gain with respect to the generator impedance. This is not true for the average power gain.

The average and insertion power gains differ in a rather significant way. In the definition of average power gain one of the powers is dissipated in the load while the other is dissipated at the input to the 2-port. However, in the definition of insertion power gain both powers are dissipated in the load. From an experimental point of view, measurement of average power gain involves a single circuit. On the other hand, measurement of insertion power gain requires that the load be removed from the 2-port and connected to the source. This may, or may not, be convenient depending upon the application.

2.5 AVAILABLE POWER GAIN

In this section we introduce the concept of available power. Consider the situation shown in Fig. 2.5 in which the generator is connected directly to the load. Let the load and generator impedances be given by

$$Z_S = R_S + jX_S$$

$$Z_L = R_L + jX_L. \tag{2-78}$$

Assume the generator impedance is fixed but that the load impedance is to be chosen so as to maximize the average power delivered to the load. Obviously, the choice

$$Z_L = -R_S - jX_S \tag{2-79}$$

results in infinite power flow. To eliminate this physically unrealizable solution from consideration, we require both the load and generator impedances to correspond to passive networks. This, in turn, implies

$$R_S \geqslant 0$$

$$R_L \geqslant 0. \tag{2-80}$$

The unrealizable solution of (2-79) is no longer acceptable since a positive value of R_S results in a negative value of R_L which is prohibited by the assumptions in (2-80). The maximum power delivered by the generator is now finite and is defined to be the generator's available power, denoted by p_A.

From (2-68) the average power delivered to the load is

$$\tilde{p}_L = \frac{|E_S|^2}{2} \frac{R_L}{(R_S + R_L)^2 + (X_S + X_L)^2}. \tag{2-81}$$

Since the objective is to vary the load so as to maximize $\tilde{p}_L$, we consider R_L and X_L to be variables with R_S, X_S, and E_S treated as constants. Differential calculus tells us that $\tilde{p}_L$ is maximized by selecting R_L and X_L such that

$$\frac{\partial \tilde{p}_L}{\partial R_L} = \frac{\partial \tilde{p}_L}{\partial X_L} = 0. \tag{2-82}$$

Partial differentiation of $\tilde{p}_L$ with respect to X_L yields

$$\frac{\partial \tilde{p}_L}{\partial X_L} = -\frac{|E_S|^2 R_L(X_S + X_L)}{[(R_S + R_L)^2 + (X_S + X_L)^2]^2}. \tag{2-83}$$

The partial derivative is zero either when $R_L = 0$ or $X_L = -X_S$. However, $R_L = 0$ implies $\tilde{p}_L = 0$ and, therefore, is ruled out as an acceptable solution. We conclude that the reactance of the load for maximum power transfer must satisfy the relation

$$X_L = -X_S. \tag{2-84}$$

The partial differentiation of (2-81) with respect to R_L gives

$$\frac{\partial \tilde{p}_L}{\partial R_L} = \frac{|E_S|^2}{2} \left\{ \frac{(R_S + R_L)^2 + (X_S + X_L)^2 - 2R_L(R_S + R_L)}{[(R_S + R_L)^2 + (X_S + X_L)^2]^2} \right\}. \tag{2-85}$$

Making use of (2-84), (2-85) reduces to

$$\frac{\partial \tilde{p}_L}{\partial R_L} = \frac{|E_S|^2}{2} \frac{(R_S - R_L)}{(R_S + R_L)^3}. \tag{2-86}$$

Obviously, this partial derivative is zero for

$$R_L = R_S. \tag{2-87}$$

From (2-84) and (2-87) we conclude that

$$Z_L = R_S - jX_S = Z_S^* \tag{2-88}$$

is the choice of load impedance necessary for $\tilde{p}_L$ to be maximum. The relationship in (2-88) is referred to as a conjugate impedance match. Recall that the available power of a generator is the maximum average power deliverable by the generator. Substituting (2-84) and (2-87) into (2-81), the available power is seen to be

$$p_A = \frac{|E_S|^2}{8R_S}. \tag{2-89}$$

It is important to emphasize that available power is a property of the generator alone. A generator may be characterized in terms of its available power whether or not it is connected to a conjugate-matched load. Of course, when the load is conjugate matched, the average power delivered by the generator is its available power. Otherwise, a lesser amount is delivered. Available power is an extremely convenient concept because it can be determined independent of the network to which the generator is connected.

Strictly speaking, the conditions given in (2-82) apply to a minimum as well as a maximum. Therefore, we now verify that (2-88) does, indeed, result in a maximum. This could be accomplished by examining the various second partial derivatives of $\tilde{p}_L$. However, a more intuitive approach is to examine the behavior of $\tilde{p}_L$ in the vicinity of $Z_L = Z_S^*$. For this approach, let

$$R_L' = R_S + \Delta R_L$$

$$X_L' = -X_S + \Delta X_L. \tag{2-90}$$

Denote the resulting value of $\tilde{p}_L$ by $\tilde{p}_L'$. Substitution of (2-90) into (2-81) results in

$$\tilde{p}_L' = \frac{|E_S|^2}{2} \frac{R_S + \Delta R_L}{(2R_S + \Delta R_L)^2 + (\Delta X_L)^2}. \qquad (2\text{-}91)$$

Subtraction of (2-91) from (2-89) yields the difference

$$p_A - \tilde{p}_L' = \frac{|E_S|^2}{2} \left[\frac{1}{4R_S} - \frac{R_S + \Delta R_L}{(2R_S + \Delta R_L)^2 + (\Delta X_L)^2} \right]$$

$$= \frac{|E_S|^2}{2} \frac{(\Delta R_L)^2 + (\Delta X_L)^2}{4R_S[(2R_S + \Delta R_L)^2 + (\Delta X_L)^2]}. \qquad (2\text{-}92)$$

However, from (2-80), R_S is required to be positive. It follows that the difference in (2-92) is nonnegative for all possible values of the increments ΔR_L and ΔX_L. We conclude that $\tilde{p}_L$ is, in fact, maximized by selecting $Z_L = Z_S^*$. This maximum is the available power given by (2-89).

Additional insight is obtained by assuming both the source and load to be purely resistive. Then

$$X_S = X_L = 0. \qquad (2\text{-}93)$$

A sketch of $\tilde{p}_L$ as a function of R_L is plotted in Fig. 2.6 for this special case. Although R_S is required to be positive, the condition on R_L is relaxed to allow for both positive and negative values. Observe that $\tilde{p}_L > 0$ corresponds to power being dissipated in the load while $\tilde{p}_L < 0$ corresponds to power being generated by the load. Maximum dissipation occurs when $R_L = R_S$ and is the available

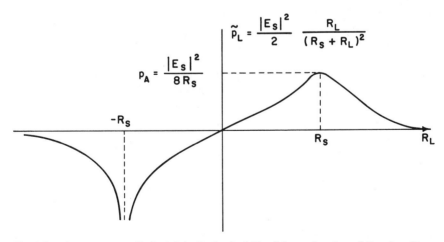

Fig. 2.6.   Average power dissipated in the load of Fig. 2.5 as a function of R_L when $X_L = X_S = 0$ and $R_S > 0$.

power from the generator. $R_L = -R_S$ corresponds to a physically unrealizable situation in which $\tilde{p}_L = -\infty$.

We now introduce the concept of available power gain. With reference to Fig. 2.4, let

p_{AO} available power from the 2-port;
p_{AS} available power from the generator;
g_A available power gain of the loaded 2-port.

The available power gain is then defined as

$$g_A = \frac{p_{AO}}{p_{AS}}. \tag{2-94}$$

p_{AS}, the available power from the generator, is readily obtained from (2-89). Hence the major task in evaluating g_A is to determine p_{AO}, the available power from the 2-port.

To obtain p_{AO}, we represent the linear 2-port at its output terminals by its Thévenin equivalent circuit. This consists of the Thévenin voltage source E_{TH} in series with the Thévenin impedance

$$Z_{TH} = R_{TH} + jX_{TH} \tag{2-95}$$

as illustrated in Fig. 2.7. It follows that the available power from the 2-port can be expressed as

$$p_{AO} = \frac{|E_{TH}|^2}{8R_{TH}}. \tag{2-96}$$

Dividing (2-96) by (2-89), the available power gain is found to be

$$g_A = \left|\frac{E_{TH}}{E_S}\right|^2 \frac{R_S}{R_{TH}}. \tag{2-97}$$

Observe that g_A is independent of Z_L since none of the quantities in (2-97) involve the load impedance. This is a consequence of the available power being independent of the network to which a generator is connected.

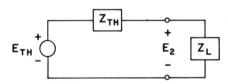

Fig. 2.7. Network of Fig. 2.4 with Thévenin equivalent representation of the linear 2-port at its output terminals.

Example 2.10

To proceed further with the analysis, it is necessary to characterize the 2-port using any convenient set of 2-port parameters. We illustrate the procedure in terms of the Z-parameter formulation of (2-48) so that we may compare the resulting expression for the available power gain with our previous results. We first determine the Thévenin equivalent voltage, E_{TH}. This is the open-circuited voltage seen looking into port 2 with the load removed as shown in Fig. 2.8. The port variables at port 2 are given by $I_2 = 0$ and $E_2 = E_{TH}$. In addition, E_1 and I_1 are constrained by (2-73). As a result, the Z-parameter equations of (2-48) reduce to

$$E_S - Z_S I_1 = Z_{11} I_1$$

$$E_{TH} = Z_{21} I_1. \tag{2-98}$$

Solution for I_1 from the first equation and substitution into the second equation yields

$$E_{TH} = \frac{Z_{21}}{Z_{11} + Z_S} E_S. \tag{2-99}$$

The Thévenin equivalent impedance, Z_{TH}, is determined next. Z_{TH} is the impedance seen looking into port 2 with all independent generators set to zero. This is shown in Fig. 2.9 where $E_S = 0$ is represented as a short circuit. To determine Z_{TH}, we excite port 2 with the current I_2 and calculate the voltage E_2 that is generated. The Thévenin impedance is then given by $Z_{TH} = E_2 / I_2$. From Fig. 2.9,

$$E_1 = -Z_S I_1. \tag{2-100}$$

As a result, the first equation in (2-48) becomes

$$-Z_S I_1 = Z_{11} I_1 + Z_{12} I_2. \tag{2-101}$$

Solution for I_1 results in

$$I_1 = -\frac{Z_{12} I_2}{Z_{11} + Z_S}. \tag{2-102}$$

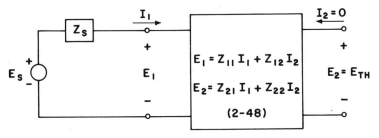

Fig. 2.8. Network used to determine the Thévenin equivalent voltage, E_{TH}.

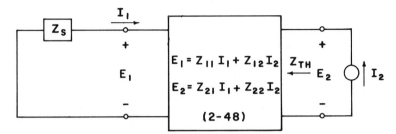

Fig. 2.9.   Network used to determine the Thévenin equivalent impedance, Z_{TH}.

Substitution of this expression into the second Z-parameter equation of (2-48) yields

$$E_2 = \left(Z_{22} - \frac{Z_{12}Z_{21}}{Z_{11} + Z_S}\right)I_2. \qquad (2\text{-}103)$$

It follows that

$$Z_{\text{TH}} = \frac{E_2}{I_2} = Z_{22} - \frac{Z_{12}Z_{21}}{Z_{11} + Z_S}. \qquad (2\text{-}104)$$

Finally, with reference to (2-97), (2-99), and (2-104), the available power gain is expressed as

$$g_A = \left|\frac{Z_{21}}{Z_{11} + Z_S}\right|^2 \frac{R_S}{\text{Re}\left\{Z_{22} - \dfrac{Z_{12}Z_{21}}{Z_{11} + Z_S}\right\}}. \qquad (2\text{-}105)$$

A comparison of this expression with that for the average power gain in (2-59) is enlightening. The two expressions are identical in form. In fact, g_A can be obtained from g_P by interchanging Z_{22} with Z_{11}, Z_{11} with Z_{22}, and Z_L with Z_S. Whereas g_P is independent of Z_S, g_A is independent of Z_L. To put it another way, whereas average power gain depends only on that portion of the network to the right of the generator, available power gain depends only on that portion to the left of the load.

A word of caution is now presented. Available power gain must not be confused with actual power flow in the network. Available power flows from the source only when the 2-port input impedance is conjugate matched to the generator. Similarly, available power flows from the 2-port only when the load is conjugate matched to the 2-port output impedance. In general, these conditions are not met and available powers do not flow. Nevertheless, available power gain may be used to characterize a loaded 2-port even under unmatched conditions.

This is often done when it is desirable to have a figure of merit independent of the load. When conjugate-matched conditions are achieved, then available power gain does equal actual power gain.

2.6 TRANSDUCER POWER GAIN

Yet another power gain is the transducer power gain, denoted by g_T. This is a ratio of the average power delivered to the load to the available power from the source. In terms of our previous notation,

$$g_T = \frac{p_L}{p_{AS}}.$$

(2-106)

Noting that p_L and p_{AS} are given by (2-69) and (2-89), respectively, g_T can be written as

$$g_T = 4 \left| \frac{E_2}{E_S} \right|^2 \frac{R_S R_L}{|Z_L|^2}.$$

(2-107)

Thus g_T involves the same voltage ratio as was encountered in our discussion of insertion power gain.

Example 2.11
   When the 2-port is characterized in terms of Z-parameters, use may be made of (2-76) in (2-107). This results in

$$g_T = \frac{4 |Z_{21}|^2 R_S R_L}{\left| (Z_{11} + Z_S)(Z_{22} + Z_L) - Z_{12} Z_{21} \right|^2}.$$

(2-108)

Interestingly enough, g_T depends upon both Z_s and Z_L.

Transducer power gain is a commonly used figure of merit for transducers, which are devices by means of which power flows from one system to another, because it relates the average power actually dissipated in a load to the maximum power capable of being delivered by the source.

As can be seen from (2-59), (2-77), (2-105), and (2-108), there is no straight-forward relationship between the expressions for the various power gains. Even so, certain observations can be made. For convenience, the various definitions are repeated below

$$g_P = \frac{p_L}{p_S}, \quad g_I = \frac{p_L}{\tilde{p}_L}, \quad g_A = \frac{p_{AO}}{p_{AS}}, \quad g_T = \frac{p_L}{p_{AS}}$$

(2-109)

where

p_L average power delivered to the load from the 2-port;
p_S average power delivered to the 2-port from the source;
$\tilde{p}_L$ average power delivered to the load without the 2-port in place;
p_{AS} available power from the generator;
p_{AO} available power from the 2-port.

From the definition of available power, we conclude

$$p_S \leqslant p_{AS}, \quad \tilde{p}_L \leqslant p_{AS}, \quad p_L \leqslant p_{AO}. \tag{2-110}$$

It follows that

$$g_P \geqslant g_T, \quad g_I \geqslant g_T, \quad g_A \geqslant g_T. \tag{2-111}$$

We see that transducer gain acts as a lower bound for the other gains.

Various gains become equivalent under certain conjugate-matched conditions. These cases are listed in Table 2.2 where

Z_S generator impedance;
Z_L load impedance;
Z_{in} loaded 2-port input impedance;
Z_{TH} Thévenin impedance at the 2-port output with the generator connected to the input.

In microwave systems all generator, load, input, and output impedances are usually chosen to equal the characteristic impedance of the coaxial cable employed in order to eliminate reflections due to mismatch. Since the characteristic impedance is purely real, conjugate-matched conditions exist at all ports. For this special case, all four gains are equivalent, as indicated by the last entry of Table 2.2. In general, equivalence should not be assumed without checking to see whether the appropriate conjugate-matched conditions indicated in the table are satisfied.

TABLE 2.2   Relations Between Power Gains Under Conjugate-Matched Conditions.

Conjugate-Matched Conditions	Equal Powers	Equal Power Gains
$Z_{in} = Z_S^*$	$p_S = p_{AS}$	$g_P = g_T$
$Z_L = Z_S^*$	$\tilde{p}_L = p_{AS}$	$g_I = g_T$
$Z_L = Z_{TH}^*$	$p_L = p_{AO}$	$g_A = g_T$
$Z_{in} = Z_S^*, Z_L = Z_S^*$	$p_S = p_{AS}, \tilde{p}_L = p_{AS}$	$g_P = g_I = g_T$
$Z_{in} = Z_S^*, Z_L = Z_{TH}^*$	$p_S = p_{AS}, p_L = p_{AO}$	$g_P = g_A = g_T$
$Z_S = Z_{TH}, Z_L = Z_S^* = Z_{TH}^*$	$\tilde{p}_L = p_{AS}, p_L = p_{AO}$	$g_I = g_A = g_T$
$Z_S = Z_{TH}, Z_L = Z_S^* = Z_{TH}^*, Z_{in} = Z_S^*$	$p_S = p_{AS} = \tilde{p}_L, p_L = p_{AO}$	$g_P = g_I = g_A = g_T$

2.7 EXCHANGEABLE POWER GAIN

As a final subject, we introduce the topic of exchangeable power which is closely related in concept to available power. In our discussion of available power the generator and load impedances were required to be passive in order to prevent the physically unrealizable situation in which the maximum average power delivered to the load is infinite. This resulted in the inequalities given by (2-80). In many applications the real part of the generator impedance is negative, as occurs with networks employing tunnel diodes. In this section we generalize the concept of available power to accommodate negative resistance sources.

Once again, consider the circuit of Fig. 2.5 with impedances as given by (2-78). Assume the real part of the generator impedance is negative. To avoid the possibility of a physically unrealizable situation, it is necessary that the real part of the load impedance also be negative. Otherwise, $Z_L = -R_S - jX_S$ results in infinite average power being delivered to the load. Note that negative values of R_L imply average power being returned by the load to the source. Assume the generator impedance is fixed and that

$$R_S < 0$$

$$R_L < 0. \qquad (2\text{-}112)$$

The problem is to choose the load impedance so as to maximize the average power returned to the source. This maximum, denoted by p_E, is defined to be the exchangeable power of the generator.

The change in sign on R_S and R_L does not alter any of the equations developed in our discussion of available power gain. For example, (2-81) is still valid as an expression for $\tilde{p}_L$. However, because R_L is negative, $\tilde{p}_L$ is also negative and represents power flow from the load to the generator. Choosing Z_L so as to make $\tilde{p}_L$ as negative as possible amounts to minimizing $\tilde{p}_L$ with respect to R_L and X_L. Since the conditions in (2-82) apply to a minimum as well as a maximum, equations (2-83) through (2-87) are also applicable. We conclude that the maximum power returned from the load to the source, corresponding to a minimum value of $\tilde{p}_L$, results when

$$Z_L = R_S - jX_S = Z_S^*. \qquad (2\text{-}113)$$

The exchangeable power of the generator is then given by

$$p_E = \frac{|E_S|^2}{8R_S}. \qquad (2\text{-}114)$$

Observe that this is negative because of the negative sign associated with R_S. To verify that (2-113) does yield a minimum, we examine $\tilde{p}_L$ in the vicinity of $Z_L = Z_S^*$. Following equations (2-90) through (2-92), the difference $p_E - \tilde{p}_L'$ is given by

$$p_E - \tilde{p}_L' = \frac{|E_S|^2}{2} \frac{(\Delta R_L)^2 + (\Delta X_L)^2}{4R_S[(2R_S + \Delta R_L)^2 + (\Delta X_L)^2]}. \qquad (2\text{-}115)$$

Since R_S is required to be negative, the difference is negative for all possible values of the increments ΔR_L and ΔX_L. We conclude, for this case, that $Z_L = Z_S^*$ does result in a minimum value of $\tilde{p}_L$. As was done for available power, the situation is clarified by sketching in Fig. 2.10 $\tilde{p}_L$ as a function of R_L for the special case in which $X_L = X_S = 0$.

In a sense, $Z_L = Z_S^*$ corresponds to a "maximum" irrespective of the sign of R_S. When $R_S > 0$, maximum power flows from the source to the load. When $R_S < 0$, maximum power flows from the load to the source. The available and exchangeable powers for each case are given by

$$p_A = \frac{|E_S|^2}{8R_S}, \qquad R_S > 0$$

$$p_E = \frac{|E_S|^2}{8R_S}, \qquad R_S < 0. \qquad (2\text{-}116)$$

Sometimes the sign restrictions on R_S are ignored and the term, exchangeable power, is used for both cases. The sign of R_S then determines the direction of

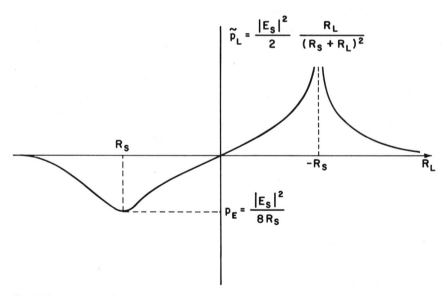

Fig. 2.10. Average power dissipated in the load of Fig. 2.5 as a function of R_L when $X_L = X_S = 0$ and $R_S < 0$.

the power flow. In this way p_E is a measure of the maximum power that may be exchanged between source and load.

The definition of exchangeable power gain is analogous to that for available power gain. With reference to Fig. 2.4, let

p_{EO} exchangeable power from the 2-port;
p_{ES} exchangeable power from the generator;
g_E exchangeable power gain of the loaded 2-port.

The exchangeable power gain is defined to be

$$g_E = \frac{p_{EO}}{p_{ES}}. \tag{2-117}$$

In terms of the Thévenin equivalent representation in Fig. 2.7, g_E is given by

$$g_E = \left|\frac{E_{TH}}{E_S}\right|^2 \frac{R_S}{R_{TH}}. \tag{2-118}$$

This, of course, is identical to (2-97). Observe that g_E may be either positive or negative depending on the sign of the ratio R_S/R_{TH}. As with available power gain, exchangeable power gain corresponds to actual power flow only when the appropriate conjugate-matched conditions are satisfied.

This completes our review of linear circuits operating in the sinusoidal steady state. Our discussion on power has introduced concepts that will be useful later in the text. In the next chapter we begin our study of weakly nonlinear systems.

3

Power Series Approach to the Analysis of Weakly Nonlinear Circuits

Nonlinear considerations play an important role in the operation of communication and electronic circuits. Although no single analytical technique is suitable for handling all types of nonlinear networks, solution procedures do exist for certain classes of nonlinear problems. In this chapter we consider nonlinear circuits that are adequately modeled as isolated blocks consisting of either zero-memory nonlinearities or linear filters. Use of this model is referred to as the power series approach. Because of its extensive application to nonlinear problems, both in the past and present, we also refer to it as the classical approach.

After a general discussion of the power series approach, the response to sinusoidal excitations is derived in detail. The concept of "frequency mix" is introduced to aid in understanding the generation of intermodulation frequencies. Expressions are developed for output powers in the intermodulation components as a function of input powers in the sinusoidal excitations. By way of illustration the power series approach is applied to a frequency assignment problem where operating frequencies for a victim receiver are determined so as to avoid interference from intermodulation effects.

3.1 GENERAL DISCUSSION OF POWER SERIES APPROACH
(The Classical Model)

As a basis for our discussion, consider a weakly nonlinear circuit with input $x(t)$ and output $y(t)$. *The classical model assumes the circuit can be represented, as*

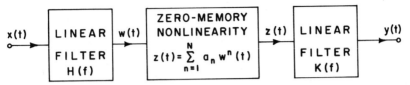

Fig. 3.1. Classical model (power series approach) of a weakly nonlinear system with memory.

shown in Fig. 3.1, in terms of an isolated zero-memory nonlinearity preceded and followed by linear filters. Typically the linear filters are characterized in the frequency domain by their linear transfer functions $H(f)$ and $K(f)$ while the nonlinearity is characterized in the time domain by its power series coefficients $\{a_1, a_2, \ldots, a_N\}$.

A system is said to have memory when the output at time t depends upon the values of the input prior to time t. The nonlinearity in Fig. 3.1 has zero memory because its output at a specific instant of time depends upon its input only at the same instant. Circuits containing energy storage elements have memory while purely resistive circuits have zero memory. Since the linear filters in Fig. 3.1 are intended to be frequency selective, they contain energy storage elements. As a result, their outputs depend upon the past history of their inputs and the non-linear system, as a whole, possesses memory.

The classical model is readily analyzed because the individual blocks shown in Fig. 3.1 can be treated as isolated segments. Specifically, given the input $x(t)$, it is possible to use straightforward linear techniques to obtain the output of the first linear filter $w(t)$. The output of the zero-memory nonlinearity $z(t)$ is then determined by substitution of the expression for $w(t)$ into the power series representation of the nonlinearity. Finally, the circuit output $y(t)$ is easily obtained as the response of the second linear filter to the known input $z(t)$. We see that analysis of the classical model proceeds readily from input to output.

An important question is: "Are there physical systems of interest for which the classical model of Fig. 3.1 is an adequate representation?" The answer is, "Yes. There are many such systems." Historically, the classical model gained prominence in conjunction with vacuum tube circuits. Consequently, for our first example, we discuss a pentode amplifier. The reader should note, however, that the very same points which arise in this example can be made even when the pentode is replaced by a field-effect transistor. Consider the pentode amplifier with tuned input and output as shown in Fig. 3.2. Initially, assume the cathode resistor R_K to be adequately bypassed by the capacitor C_K. Then the cathode terminal of the pentode is effectively grounded as far as alternating signals are concerned. Also, assume the tube to be biased with a negative grid-to-cathode potential such that negligible current flows through the control grid. The non-linear incremental equivalent circuit is shown in Fig. 3.3. The plate circuit of

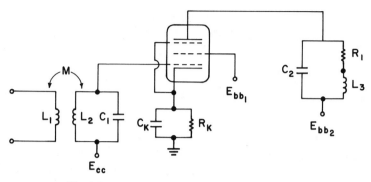

Fig. 3.2. Schematic diagram of pentode amplifier.

the pentode is modeled as a voltage-controlled current source $i_p(t)$ in parallel with the plate resistance r_p. The transconductance of the tube is assumed to be nonlinear and is accounted for by a power series representation for $i_p(t)$ given by

$$i_p(t) = \sum_{n=1}^{N} a_n e_g^n(t). \tag{3-1}$$

In line with our assumptions concerning the bypass capacitor C_K and the grid-to-cathode voltage $e_g(t)$, the cathode impedance is treated as a short circuit while the control grid is modeled as an open circuit.

The nonlinear incremental equivalent circuit of Fig. 3.3 is identical in form to the classical model specified in Fig. 3.1. Because the first linear filter is terminated in an open circuit while the second linear filter is driven by a current source that depends only on the output voltage of the first linear filter, the individual component blocks can be analyzed as isolated segments. Thus given

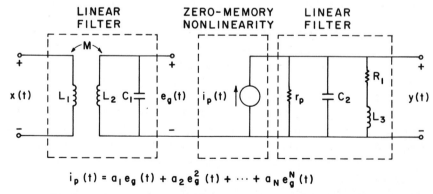

$$i_p(t) = a_1 e_g(t) + a_2 e_g^2(t) + \cdots + a_N e_g^N(t)$$

Fig. 3.3. Nonlinear incremental equivalent circuit of pentode amplifier.

the input voltage $x(t)$, the grid-to-cathode voltage $e_g(t)$ is readily determined as the response of the first linear filter. The current delivered by the controlled source $i_p(t)$ is easily obtained by substitution of the expression for $e_g(t)$ into (3-1). Finally, having determined $i_p(t)$, the output $y(t)$ is simply the voltage generated by the known current source driving the single tuned circuit. We conclude that the pentode amplifier is nicely analyzed using the power series approach.

However, this is not always the case. Suppose for example, the cathode resistor R_K is not adequately bypassed by the capacitor C_K. The appropriate nonlinear incremental equivalent circuit for this situation is shown in Fig. 3.4. Observe that the grid-to-cathode voltage is given by

$$e_g(t) = e_1(t) - e_k(t). \tag{3-2}$$

Because the cathode resistor is not adequately bypassed, the voltage $e_k(t)$ is no longer negligible. Even though knowledge of the input voltage $x(t)$ is adequate to determine the voltage $e_1(t)$, the grid-to-cathode voltage $e_g(t)$ remains undetermined because $e_k(t)$ depends upon the voltage-controlled current source $i_p(t)$ which, in turn, depends nonlinearly upon the unknown voltage $e_g(t)$. The classical approach is seen to break down. Since the individual blocks in Fig. 3.4 are no longer noninteracting, it is not possible to carry out the analysis from input to output as was done with the classical model of Fig. 3.1.

The nonlinear behavior of the pentode complicates the task of providing an adequate bypass for the cathode resistor when interfering signals are present. For example, assume the input voltage $x(t)$ consists of a desired signal at frequency f_1 and an interfering signal at frequency f_2. Let f_1 and f_2 be sufficiently large such that C_K acts effectively as a short circuit at these frequencies. However, because of the nonlinear transconductance of the pentode, an intermodulation component at the difference frequency $f_1 - f_2$ is generated. If f_1 and f_2 are sufficiently close together, the component at $f_1 - f_2$ may cause a significant

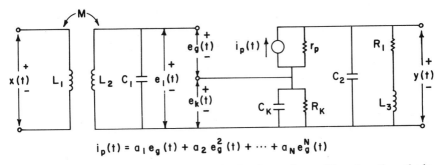

$$i_p(t) = a_1 e_g(t) + a_2 e_g^2(t) + \cdots + a_N e_g^N(t)$$

Fig. 3.4. Nonlinear incremental equivalent circuit of pentode amplifier when the cathode resistor R_K is not adequately bypassed by the capacitor C_K.

voltage to be developed across the parallel combination of R_K and C_K. Therefore, even though the cathode resistor is adequately bypassed for the input signals encountered by the amplifier, an adequate bypass may not exist for intermodulation components that are generated.

The classical model for the pentode amplifier also breaks down when the frequency of excitation is sufficiently high so that interelectrode capacitances are important. For simplicity, assume the cathode resistor is adequately bypassed at all frequencies of interest. The appropriate nonlinear incremental equivalent circuit for this case is shown in Fig. 3.5. C_1' represents the parallel combination of capacitance C_1 and the grid-to-cathode capacitance of the pentode. Similarly, C_2' represents the parallel combination of capacitance C_2 and the plate-to-cathode capacitance of the tube. In addition, C_{gp} denotes the interelectrode capacitance between the control grid and plate.

Note that the voltage-controlled current source $i_p(t)$ now divides its current between $i_1(t)$ and $i_2(t)$. This results in a portion of $i_p(t)$ being fed back to the input tuned circuit. It follows that knowledge of the input voltage $x(t)$ is no longer sufficient for determination of the grid-to-cathode voltage $e_g(t)$. $e_g(t)$ is a function of both $x(t)$ and $i_1(t)$. However, $i_1(t)$ depends upon $i_p(t)$ which, in turn, depends nonlinearly upon the unknown voltage $e_g(t)$. Therefore, it is not possible to model the circuit of Fig. 3.5 in terms of isolated component blocks as required by the classical model.

Depending upon the circumstances, we conclude that the classical model may, or may not, be a useful approach in the nonlinear analysis of a pentode amplifier.

We now investigate the effectiveness of the power series approach in the analysis of a common-emitter amplifier employing a bipolar junction transistor. The circuit is shown in Fig. 3.6(a). Z_S represents the linear generator impedance, R_1 is a biasing resistor, and Z_L is the linear load impedance. The linear incremental equivalent circuit, using a conventional T model for the transistor, is shown in Fig. 3.6(b). $r_b, r_e,$ and r_c are the incremental base, emitter, and collec-

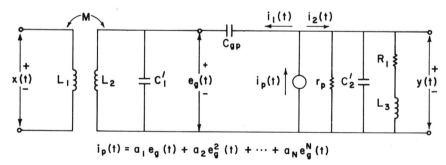

$$i_p(t) = a_1 e_g(t) + a_2 e_g^2(t) + \cdots + a_N e_g^N(t)$$

Fig. 3.5. Nonlinear incremental equivalent circuit of pentode amplifier when the cathode resistor is adequately bypassed and interelectrode capacitances are important.

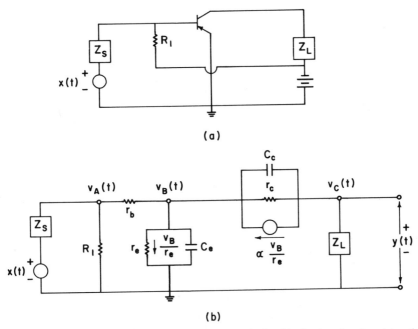

(a)

(b)

Fig. 3.6. Common-emitter transistor amplifier employing bipolar junction transistor. (a) Schematic diagram. (b) Linear incremental equivalent circuit.

tor resistances, respectively. Similarly, C_e and C_c are the incremental emitter and collector capacitances, respectively. Observe that the current source $\alpha v_B / r_e$ is controlled by the current flowing in the resistive portion of the emitter circuit. v_A, v_B, and v_C represent node-to-datum voltages where the emitter terminal has been chosen as the datum node.

In actuality, the circuit components r_e, C_e, C_c and the controlled source $\alpha v_B / r_e$ represent linear approximations to nonlinear elements. Replacement of the nonlinear branches by controlled current sources, whose currents equal the true currents flowing in the branches, results in the nonlinear incremental equivalent circuit shown in Fig. 3.7. Specifically, the current flowing through the base-emitter resistance r_e of the linear model is actually a nonlinear function of the node-to-datum voltage v_B. This is indicated in Fig. 3.7 by the voltage-controlled current source $K(v_B)$. $K(\cdot)$ represents the nonlinear functional relationship between the current in the branch and the voltage v_B. Physically, the emitter capacitance C_e of the linear model is due to the parallel combination of the diffusion and space-charge layer capacitances. These can be shown to be nonlinear functions of the voltage v_B and their behavior is accounted for by the controlled current source $\gamma_e(v_B)$. As with $K(\cdot)$, $\gamma_e(\cdot)$ represents the nonlinear functional relationship between the current in the branch and the voltage v_B.

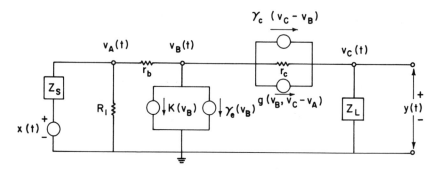

Fig. 3.7. Nonlinear incremental equivalent circuit of common-emitter transistor amplifier.

Similarly, the collector capacitance C_c of the linear model is an approximation to the varactor capacitance which is a nonlinear function of the voltage
$(v_C - v_B)$. This results in the controlled current source $\gamma_c(v_C - v_B)$. Finally, the
nonlinear controlled source $g(v_B, v_C - v_A)$ is inserted in place of $\alpha v_B/r_e$ in order
to model the transistor's nonlinear current gain and avalanche effects. The nonlinear incremental equivalent circuit of a bipolar junction transistor is seen to be
considerably more complicated than the model corresponding to a vacuum-tube
pentode. Note that knowledge of the input voltage $x(t)$ and the node-to-datum
voltages v_A, v_B, and v_C is sufficient to determine the voltages and currents existing everywhere in the network.

It is not possible to reduce the nonlinear incremental equivalent circuit of the
common-emitter transistor amplifier to the form of the classical model in Fig.
3.1. In particular, the circuit cannot be subdivided into isolated blocks, each of
which can be analyzed separately. For example, it would be highly desirable to
be able to work one's way through the network by determining: (1) $v_A(t)$ from
$x(t)$, (2) $v_B(t)$ from $v_A(t)$, and (3) $v_C(t) = y(t)$ from $v_B(t)$. Unfortunately, such
a step-by-step procedure cannot be accomplished no matter which voltages and
currents are chosen to characterize the network. In terms of the node-to-datum
voltages, analysis of the circuit requires the simultaneous solution of three nonlinear coupled equations in the unknowns $v_A(t)$, $v_B(t)$, and $v_C(t)$. A second
difficulty with the circuit of Fig. 3.7 is that the currents flowing through the
nonlinear capacitors involve time derivatives of the voltages. Since time derivatives are defined in terms of two separate instants of time, capacitive nonlinearities contain memory. On the other hand, zero-memory (or resistive) nonlinearities are assumed by the classical model. It follows that the power series approach
is not really suitable for analysis of the nonlinear network shown in Fig. 3.7.

Nevertheless, the classical model is widely used. In some cases it gives excellent
results. In other situations it represents a poorly contrived approximation. Of
course, a crude nonlinear model of a nonlinear system may be better than a
purely linear model that does not predict any nonlinear effects at all! In any

event, whether or not the classical model is a good approximation to the problem at hand, it does provide some insight into the behavior of weakly nonlinear systems. With this in mind, we now study the response of the classical model to sinusoidal inputs.

3.2 RESPONSE OF CLASSICAL MODEL TO SINUSOIDAL EXCITATIONS

System Input: Since we are interested in the effect of interference on weakly nonlinear systems, we assume several sinusoidal signals to be present simultaneously at the input of the classical model shown in Fig. 3.1. In particular, let the excitation be

$$x(t) = \sum_{q=1}^{Q} |E_q| \cos (2\pi f_q t + \theta_q) \qquad (3\text{-}3)$$

where, for discussion purposes, $x(t)$ is arbitrarily chosen to be a voltage. As pointed out in Chapter 2, it is preferable to work with exponential, as opposed to trigonometric, functions. Consequently, we introduce the complex voltage of the qth tone,

$$E_q = |E_q| \exp (j\theta_q), \qquad (3\text{-}4)$$

and define

$$E_{-q} = E_q^*, \qquad E_0 = 0, \qquad f_{-q} = -f_q. \qquad (3\text{-}5)$$

This enables the excitation to be written as

$$x(t) = \frac{1}{2} \sum_{q=1}^{Q} [E_q \exp (j2\pi f_q t) + E_q^* \exp (-j2\pi f_q t)]$$

$$= \frac{1}{2} \sum_{q=-Q}^{Q} E_q \exp (j2\pi f_q t). \qquad (3\text{-}6)$$

Observe that only positive indices are involved in the first equation of (3-6) where terms are grouped into conjugate pairs. On the other hand, the second equation requires the use of both positive and negative indices.

Response of First Linear Filter: The discussion of Section 3.1 demonstrated that the individual blocks of the classical model can be analyzed as isolated segments. Focusing attention on the first block, the transfer function of the linear filter can be expressed in polar form as

$$H(f) = |H(f)| \exp [j\psi(f)]. \qquad (3\text{-}7)$$

Because superposition applies to linear systems, the total response of this filter to a sum of inputs is simply the sum of the individual responses. In particular, from sinusoidal steady-state considerations, the response to the input specified by (3-3) is

$$w(t) = \sum_{q=1}^{Q} |E_q| |H(f_q)| \cos [2\pi f_q t + \theta_q + \psi(f_q)]. \tag{3-8}$$

Equation (3-8) states that the response of the linear filter to a sum of sinusoids is a sum of sinusoids at the same frequencies as those contained in the input. Each output magnitude equals the corresponding input magnitude multiplied by the magnitude of the transfer function while each output phase angle equals the corresponding input angle plus the phase angle of the transfer function. For each term in the sum, the transfer function magnitude and angle are evaluated at the appropriate input frequency.

The linear transfer function of a real system has the property that

$$H(-f) = H^*(f) = |H(f)| \exp [-j\psi(f)]. \tag{3-9}$$

It follows that the magnitude and angle of a linear transfer function are even and odd functions of frequency, respectively. In analogy with (3-6), the response can also be written as

$$w(t) = \frac{1}{2} \sum_{q=1}^{Q} [E_q H(f_q) \exp (j2\pi f_q t) + E_q^* H^*(f_q) \exp (-j2\pi f_q t)]$$

$$= \frac{1}{2} \sum_{q=-Q}^{Q} E_q H(f_q) \exp (j2\pi f_q t). \tag{3-10}$$

The first equation in (3-10) is useful for demonstrating the manner by which negative frequencies arise. However, the form of the second equation in (3-10) is more convenient for nonlinear analysis.

Response of Zero-Memory Nonlinearity: The output of the zero-memory nonlinearity is

$$z(t) = \sum_{n=1}^{N} a_n w^n(t). \tag{3-11}$$

The nth term in the summation is said to be of nth degree because it involves $w(t)$ raised to the nth power. In general, a power series expansion consists of an infinite number of terms. However, the representation is most useful when terms above a certain degree become insignificant. The finite sum in (3-11) indicates that terms above Nth degree contribute negligibly to the response. Obviously, for ease of analysis, it is desirable to have N as small as possible. Use of the second expression from (3-10) results in

$$w^n(t) = \left[\frac{1}{2} \sum_{q=-Q}^{Q} E_q H(f_q) \exp (j2\pi f_q t) \right]^n. \tag{3-12}$$

$w^n(t)$ can be interpreted as an n-fold product involving n separate summations. Therefore,

$$w^n(t) = \left[\frac{1}{2} \sum_{q_1=-Q}^{Q} E_{q_1} H(f_{q_1}) \exp\left(j2\pi f_{q_1} t\right)\right]$$

$$\cdot \left[\frac{1}{2} \sum_{q_2=-Q}^{Q} E_{q_2} H(f_{q_2}) \exp\left(j2\pi f_{q_2} t\right)\right]$$

$$\cdots \left[\frac{1}{2} \sum_{q_n=-Q}^{Q} E_{q_n} H(f_{q_n}) \exp\left(j2\pi f_{q_n} t\right)\right] \qquad (3\text{-}13)$$

where n different indices have been introduced in order to distinguish between the various summations. Interchanging the order of summation and multiplication (3-13) becomes

$$w^n(t) = \frac{1}{2^n} \sum_{q_1=-Q}^{Q} \cdots \sum_{q_n=-Q}^{Q} E_{q_1} \cdots E_{q_n} H(f_{q_1}) \cdots H(f_{q_n})$$

$$\cdot \exp\left[j2\pi(f_{q_1} + \cdots + f_{q_n})t\right]. \qquad (3\text{-}14)$$

In this manner the n-fold product of (3-13) has been converted to an n-fold summation. Observe that $w^n(t)$ contains components at frequencies given by $(f_{q_1} + \cdots + f_{q_n})$ where the indices $q_1, \ldots, q_n$ range from $-Q$ to $+Q$. Since $f_{-q} = -f_q$, frequency differences and negative frequencies arise in (3-14). There are $2Q + 1$ different integers in the range from $-Q$ to $+Q$. However, by definition $E_0 = 0$. Therefore, all terms involving one or more indices of value zero are also zero. Since each summation includes $2Q$ nonzero terms, the total number of nonzero terms in the n-fold sum is $(2Q)^n$. This number grows rapidly with increasing values of Q and n.

Example 3.1

To illustrate (3-14), we evaluate the second-degree term for the two-tone input given by

$$x(t) = \sum_{q=1}^{2} |E_q| \cos\left(2\pi f_q t + \theta_q\right)$$

$$= \frac{1}{2} \sum_{q=-2}^{2} E_q \exp\left(j2\pi f_q t\right). \qquad (3\text{-}15)$$

From (3-14)

$$w^2(t) = \frac{1}{4} \sum_{q_1=-2}^{2} \sum_{q_2=-2}^{2} E_{q_1} E_{q_2} H(f_{q_1}) H(f_{q_2}) \exp\left[j2\pi(f_{q_1} + f_{q_2})t\right].$$

(3-16)

Because $n = Q = 2$, $w^2(t)$ contains 16 nonzero terms. In particular,

$$w^2(t) = \frac{1}{4} \sum_{q_1=-2}^{2} \sum_{q_2=-2}^{2} E_{q_1} E_{q_2} H(f_{q_1}) H(f_{q_2}) \exp\left[j2\pi(f_{q_1} + f_{q_2})t\right]$$

$$= \frac{1}{4} \{ E_2^* E_2^* H^*(f_2) H^*(f_2) \exp\left[j2\pi(-f_2 - f_2)t\right] + E_2^* E_1^* H^*(f_2) H^*(f_1)$$

$$\cdot \exp\left[j2\pi(-f_2 - f_1)t\right] + E_2^* E_1 H^*(f_2) H(f_1) \exp\left[j2\pi(-f_2 + f_1)t\right]$$

$$+ E_2^* E_2 H^*(f_2) H(f_2) \exp\left[j2\pi(-f_2 + f_2)t\right] + E_1^* E_2^* H^*(f_1) H^*(f_2)$$

$$\cdot \exp\left[j2\pi(-f_1 - f_2)t\right] + E_1^* E_1^* H^*(f_1) H^*(f_1) \exp\left[j2\pi(-f_1 - f_1)t\right]$$

$$+ E_1^* E_1 H^*(f_1) H(f_1) \exp\left[j2\pi(-f_1 + f_1)t\right] + E_1^* E_2 H^*(f_1) H(f_2)$$

$$\cdot \exp\left[j2\pi(-f_1 + f_2)t\right] + E_1 E_2^* H(f_1) H^*(f_2) \exp\left[j2\pi(f_1 - f_2)t\right]$$

$$+ E_1 E_1^* H(f_1) H^*(f_1) \exp\left[j2\pi(f_1 - f_1)t\right] + E_1 E_1 H(f_1) H(f_1)$$

$$\cdot \exp\left[j2\pi(f_1 + f_1)t\right] + E_1 E_2 H(f_1) H(f_2) \exp\left[j2\pi(f_1 + f_2)t\right]$$

$$+ E_2 E_2^* H(f_2) H^*(f_2) \exp\left[j2\pi(f_2 - f_2)t\right] + E_2 E_1^* H(f_2) H^*(f_1)$$

$$\cdot \exp\left[j2\pi(f_2 - f_1)t\right] + E_2 E_1 H(f_2) H(f_1) \exp\left[j2\pi(f_2 + f_1)t\right]$$

$$+ E_2 E_2 H(f_2) H(f_2) \exp\left[j2\pi(f_2 + f_2)t\right] \}.$$

(3-17)

This rather simple example illustrates the complexity associated with evaluating (3-14). Reduction of this expression is treated in Section 3.3. At this point, however, we continue the analysis using (3-14) in its present form.

Substituting (3-14) into (3-11) the output of the zero-memory nonlinearity becomes

$$z(t) = \sum_{n=1}^{N} \sum_{q_1=-Q}^{Q} \cdots \sum_{q_n=-Q}^{Q} \frac{a_n}{2^n} E_{q_1} \cdots E_{q_n} H(f_{q_1}) \cdots H(f_{q_n})$$

$$\cdot \exp\left[j2\pi(f_{q_1} + \cdots + f_{q_n})t\right].$$

(3-18)

Introduction of the complex amplitude

$$A_n(q_1, \ldots, q_n) = \frac{a_n}{2^n} E_{q_1} \cdots E_{q_n} H(f_{q_1}) \cdots H(f_{q_n}) \qquad (3\text{-}19)$$

enables $z(t)$ to be written as

$$z(t) = \sum_{n=1}^{N} \sum_{q_1=-Q}^{Q} \cdots \sum_{q_n=-Q}^{Q} A_n(q_1, \ldots, q_n) \exp \left[j2\pi(f_{q_1} + \cdots + f_{q_n}) t \right].$$

$$(3\text{-}20)$$

System Output: Having obtained $z(t)$, the final step in the analysis of the classical model is the determination of $y(t)$, the nonlinear system output. Note that this is also the response of the second linear filter whose transfer function is

$$K(f) = |K(f)| \exp \left[j\phi(f) \right]. \qquad (3\text{-}21)$$

Each term in $z(t)$ is of the form $A \exp \left[j2\pi ft \right]$. From linear circuit theory the response of the filter to $A \exp \left[j2\pi ft \right]$ is simply $AK(f) \exp \left[j2\pi ft \right]$. Therefore, applying superposition, the nonlinear system output is given by

$$y(t) = \sum_{n=1}^{N} \sum_{q_1=-Q}^{Q} \cdots \sum_{q_n=-Q}^{Q} B_n(q_1, \ldots, q_n) \exp \left[j2\pi(f_{q_1} + \cdots + f_{q_n}) t \right]$$

$$(3\text{-}22)$$

where

$$B_n(q_1, \ldots, q_n) = A_n(q_1, \ldots, q_n) K(f_{q_1} + \cdots + f_{q_n})$$

$$= \frac{a_n}{2^n} E_{q_1} \cdots E_{q_n} H(f_{q_1}) \cdots H(f_{q_n}) K(f_{q_1} + \cdots + f_{q_n})$$

$$= \frac{a_n}{2^n} |E_{q_1}| \cdots |E_{q_n}| |H(f_{q_1})| \cdots |H(f_{q_n})| |K(f_{q_1} + \cdots + f_{q_n})|$$

$$\cdot \exp \{ j(\theta_{q_1} + \cdots + \theta_{q_n}) + j[\psi(f_{q_1}) + \cdots + \psi(f_{q_n})]$$

$$+ j\phi(f_{q_1} + \cdots + f_{q_n}) \}. \qquad (3\text{-}23)$$

In the expression for $B_n(q_1, \ldots, q_n)$ the polar form was determined using the phase angles of $E_q, H(f)$, and $K(f)$ which are $\theta_q, \psi(f)$, and $\phi(f)$, respectively. The most striking feature concerning the response of the nonlinear system, as given by (3-22), is the presence of frequencies not contained in the input. Terms involving such frequencies are referred to as intermodulation components. Their complex amplitudes depend upon the complex input voltages, the power series coefficients of the zero-memory nonlinearity, and the linear transfer functions of the two filters, as indicated by (3-23). Evaluation of (3-22) is extremely tedious. Simplification of this expression is discussed in the next section.

3.3 ROLE OF FREQUENCY MIXES IN THE GENERATION OF INTERMODULATION COMPONENTS

Intermodulation components generated in a weakly nonlinear system are of considerable interest because they may fall within the passband of a system even when the original excitations do not.

Example 3.2

Consider a system tuned to 50 MHz with a 1-MHz bandwidth as shown in Fig. 3.8. Assume the input to consist of two sinusoidal tones at f_1 = 46 MHz and f_2 = 48 MHz. Clearly, the two inputs fall outside of the system passband. Yet, if the system contains a nonlinearity, an intermodulation component at $2f_2 - f_1$ = 50 MHz may be generated. This falls at the tuned frequency and may cause significant interference depending upon whether the intermodulation component is sufficiently large relative to the desired signal.

The Frequency Mix Concept: For the remainder of this discussion, we assume the performance of a particular weakly nonlinear system is of interest. From (3-22) it is apparent that, even for relatively small values of Q and N, a very large number of intermodulation components occur in the output. Usually those terms with frequencies falling well outside of the system passband are not troublesome since they are greatly attenuated by the frequency selectivity of the system. It follows that only those intermodulation components with frequencies inside or close to the system passband need be considered. This can enormously simplify evaluation of the response since the significant terms are typically rela-

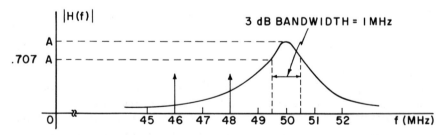

Fig. 3.8. Transfer function magnitude of system tuned to 50 MHz with sinusoidal excitations at 46 and 48 MHz.

tively few in number. Therefore, the first step in evaluating (3-22) is to determine which intermodulation frequencies are of concern.

Having done this, the second step is to determine the manner by which the pertinent intermodulation frequencies are generated. For this purpose, we introduce the concept of a frequency mix. To be specific, reconsider the example associated with Fig. 3.8.

Example 3.3

Assume the system contains a zero-memory nonlinearity of highest degree $N = 5$. Because the inputs are at $f_1 = 46$ MHz and $f_2 = 48$ MHz, the intermodulation component at $2f_2 - f_1 = 50$ MHz falls into the center of the passband and is cause for concern. Terms for which the intermodulation frequency is $2f_2 - f_1$ are produced by the three frequency mixes: (1) $(f_2 + f_2 - f_1)$, (2) $(f_2 + f_2 + f_2 - f_2 - f_1)$, and (3) $(f_2 + f_2 + f_1 - f_1 - f_1)$.

As far as frequency mixes are concerned, the order in which the frequencies appear is unimportant. Therefore $(f_2 - f_1 + f_2)$ represents the same mix as does $(f_2 + f_2 - f_1)$. A mix is characterized by the number of times the various frequencies are involved.

Example 3.4

Consider the three frequency mixes of Example 3.3. $(f_2 + f_2 - f_1)$ involves $-f_1$ once and f_2 twice, $(f_2 + f_2 + f_2 - f_2 - f_1)$ involves $-f_2$ once, $-f_1$ once, and f_2 three times, while $(f_2 + f_2 + f_1 - f_1 - f_1)$ involves $-f_1$ twice, f_1 once, and f_2 twice. With reference to (3-22), observe that the frequency mix $(f_2 + f_2 - f_1)$ occurs in terms for which $n = 3$ while the mixes $(f_2 + f_2 + f_2 - f_2 - f_1)$ and $(f_2 + f_2 + f_1 - f_1 - f_1)$ occur in terms for which $n = 5$.

We have introduced the concept of a frequency mix in order to indicate the manner by which an intermodulation frequency is produced. For clarity of discussion in this text, frequency mixes, such as $(f_2 + f_2 + f_1 - f_1 - f_1)$, are enclosed in parentheses while intermodulation frequencies, such as $2f_2 - f_1$, are not.

To aid in the representation of a frequency mix, let the number of times that the frequency f_k appears be denoted by m_k. Considering negative frequencies, recall that $f_{-k} = -f_k$. Therefore, for an excitation consisting of Q sinusoidal tones, as given by (3-3) and (3-6), the input frequencies are $f_{-Q}, \ldots, f_{-1}, f_1, \ldots, f_Q$. It follows that any possible frequency mix can be represented by the frequency mix vector

$$\mathbf{m} = (m_{-Q}, \ldots, m_{-1}, m_1, \ldots, m_Q). \tag{3-24}$$

Example 3.5

Once again, consider the three frequency mixes of Example 3.3. Since the input frequencies are f_{-2}, f_{-1}, f_1, f_2, the frequency mix vector $\mathbf{m}$ becomes

$$\mathbf{m} = (m_{-2}, m_{-1}, m_1, m_2). \tag{3-25}$$

The frequency mix $(f_2 + f_2 - f_1)$ is represented by $\mathbf{m} = (0, 1, 0, 2)$ while $\mathbf{m} = (1, 1, 0, 3)$ represents the frequency mix $(f_2 + f_2 + f_2 - f_2 - f_1)$ and $\mathbf{m} = (0, 2, 1, 2)$ represents the frequency mix $(f_2 + f_2 + f_1 - f_1 - f_1)$.

Total Response for a Particular Frequency Mix: Having enumerated all of the frequency mixes resulting in a particular intermodulation frequency, the next step is to determine the complex amplitude of the intermodulation component. From (3-22) the total response of the nonlinear system can be written as

$$y(t) = \sum_{n=1}^{N} y_n(t) \tag{3-26}$$

where

$$y_n(t) = \sum_{q_1=-Q}^{Q} \cdots \sum_{q_n=-Q}^{Q} B_n(q_1, \ldots, q_n) \exp \left[j2\pi(f_{q_1} + \cdots + f_{q_n}) t \right]. \tag{3-27}$$

The n-fold sum in (3-27) results in the output frequencies of $y_n(t)$ being generated by adding together all possible combinations of the input frequencies $f_{-Q}, \ldots, f_{-1}, f_1, \ldots, f_Q$ taken n at a time. For the frequency mix corresponding to the frequency mix vector $\mathbf{m}$, the corresponding intermodulation frequency can be expressed as

$$f_{\mathbf{m}} = \sum_{\substack{k=-Q \\ k \neq 0}}^{Q} m_k f_k = m_{-Q} f_{-Q} + \cdots + m_{-1} f_{-1} + m_1 f_1 + \cdots + m_Q f_Q. \tag{3-28}$$

Since exactly n frequencies are involved in each frequency mix of $y_n(t)$, the m_k's obey the constraint

$$\sum_{\substack{k=-Q \\ k \neq 0}}^{Q} m_k = m_{-Q} + \cdots + m_{-1} + m_1 + \cdots + m_Q = n. \tag{3-29}$$

The output frequencies in (3-27) can then be interpreted as those intermodulation frequencies that can be generated by all possible choices of the m_k's such that (3-29) is satisfied.

Many of the terms in the n-fold summation of $y_n(t)$ are equal. Repeating (3-23),

$$B_n(q_1,\ldots,q_n) = \frac{a_n}{2^n} E_{q_1} \cdots E_{q_n} H(f_{q_1}) \cdots H(f_{q_n}) K(f_{q_1} + \cdots + f_{q_n}).$$

(3-30)

Observe that $B_n(q_1,\ldots,q_n)$ is unchanged by a permutation of the indices $q_1,\ldots,q_n$.

Example 3.6

For $n = 3$,

$$B_3(q_1,q_2,q_3) = B_3(q_1,q_3,q_2) = B_3(q_2,q_1,q_3) = B_3(q_2,q_3,q_1)$$

$$= B_3(q_3,q_1,q_2) = B_3(q_3,q_2,q_1). \tag{3-31}$$

In addition, permuting the indices does not affect $\exp\left[j2\pi(f_{q_1} + \cdots + f_{q_n})t\right]$. We conclude that terms in (3-27) are identical whenever they are generated by sets of indices which differ only by the order in which the indices occur.

In terms of the n indices $q_1,\ldots,q_n$, a frequency mix corresponding to the frequency mix vector m results whenever the values of the indices are such that m_{-Q} indices assume the value $-Q,\ldots,m_{-1}$ assume the value -1, m_1 assume the value $1,\ldots$, and m_Q assume the value Q. The order in which the indices assume these values is immaterial. As a result, the number of terms in (3-27) contributing to a frequency mix represented by the vector m equals the number of different ways the n indices can be permuted such that $-Q$ appears m_{-Q} times, $\ldots,-1$ appears m_{-1} times, 1 appears m_1 times, $\ldots$, and Q appears m_Q times. This number is given by the multinomial coefficient

$$(n; \text{m}) = \frac{(n!)}{(m_{-Q}!) \cdots (m_{-1}!)(m_1!) \cdots (m_Q!)}. \tag{3-32}$$

Example 3.7

Consider the frequency mix $(f_2 + f_2 + f_2 - f_2 - f_1)$ that corresponds to $n = 5$ and m $= (1, 1, 0, 3)$. Using (3-32) the number of terms in (3-27) contributing to this mix is

$$(5; 1, 1, 0, 3) = \frac{(5!)}{(1!)(1!)(0!)(3!)} = 20. \tag{3-33}$$

TABLE 3.1. Indices Resulting in the Frequency Mix $(f_2 + f_2 + f_2 - f_2 - f_1)$.

	q_1	q_2	q_3	q_4	q_5		q_1	q_2	q_3	q_4	q_5
(1)	-2	2	2	2	-1	(11)	2	2	-2	-1	2
(2)	-2	2	2	-1	2	(12)	2	2	-1	-2	2
(3)	-2	2	-1	2	2	(13)	2	2	-1	2	-2
(4)	-2	-1	2	2	2	(14)	2	2	-2	2	-1
(5)	-1	2	2	2	-2	(15)	2	-1	2	2	-2
(6)	-1	2	2	-2	2	(16)	2	-1	2	-2	2
(7)	-1	2	-2	2	2	(17)	2	-1	-2	2	2
(8)	-1	-2	2	2	2	(18)	2	-2	2	2	-1
(9)	2	2	2	-1	-2	(19)	2	-2	2	-1	2
(10)	2	2	2	-2	-1	(20)	2	-2	-1	2	2

The 20 sets of indices yielding $\mathbf{m} = (1, 1, 0, 3)$ are tabulated in Table 3.1. Each of the 20 sets results in the identical term

$$\frac{a_5}{32} E_2^* E_1^* E_2^3 H^*(f_2) H^*(f_1) H^3(f_2) K(2f_2 - f_1) \exp [j2\pi(2f_2 - f_1) t].$$

$$(3\text{-}34)$$

Note that the output frequency obtained from this mix is the intermodulation frequency $2f_2 - f_1$.

In general, there are many different frequency mixes generated by the n-fold summation in (3-27). As specified by (3-32), a particular frequency mix, represented by the vector $\mathbf{m}$, includes $(n; \mathbf{m})$ identical terms. Combining identical terms, we denote their sum by $y_n(t; \mathbf{m})$.

Example 3.8

From Example 3.7 the mix $\mathbf{m} = (1, 1, 0, 3)$ yields 20 identical terms each given by (3-34). Adding these terms together, we obtain

$$y_5(t; 1, 1, 0, 3) = \frac{20a_5}{32} E_2^* E_1^* E_2^3 H^*(f_2) H^*(f_1) H^3(f_2) K(2f_2 - f_1)$$

$$\cdot \exp [j2\pi(2f_2 - f_1) t]. \quad (3\text{-}35)$$

In general, $y_n(t; \mathbf{m})$ is given by

$$y_n(t; \mathbf{m}) = (n; \mathbf{m}) \, B_n(q_1, \ldots, q_n) \exp \left[j2\pi(f_{q_1} + \cdots + f_{q_n}) \, t \right]$$

$$= \frac{(n; \mathbf{m}) \, a_n}{2^n} (E_Q^*)^{m_{-Q}} \cdots (E_1^*)^{m_{-1}} (E_1)^{m_1} \cdots (E_Q)^{m_Q} \, [H^*(f_Q)]^{m_{-Q}}$$

$$\cdots [H^*(f_1)]^{m_{-1}} [H(f_1)]^{m_1} \cdots [H(f_Q)]^{m_Q} \, K(f_\mathbf{m}) \exp \left[j2\pi f_\mathbf{m} t \right]$$

$$(3\text{-}36)$$

where the latter expression is obtained by using (3-30) for $B_n(q_1, \ldots, q_n)$, (3-5) to express E_{-q} as E_q^*, (3-9) to express $H(f_{-q}) = H(-f_q)$ as $H^*(f_q)$, and (3-28) for the definition of $f_\mathbf{m}$. Since $y_n(t)$ consists of all possible frequency mixes allowed by the constraint in (3-29), (3-27) can be rewritten as

$$y_n(t) = \sum_\mathbf{m} y_n(t; \mathbf{m}) \qquad (3\text{-}37)$$

where the summation over $\mathbf{m}$ is defined to be

$$\sum_\mathbf{m} = \sum_{m_{-Q}=0}^{n} \cdots \sum_{m_{-1}=0}^{n} \sum_{m_1=0}^{n} \cdots \sum_{m_Q=0}^{n} \qquad (3\text{-}38)$$

$$(3\text{-}29)$$

and the equation number, (3-29), appended below the summation signs indicates that only terms for which the indices sum to n are included in the $2Q$-fold summation.

Although $y_n(t; \mathbf{m})$ is complex, $y_n(t)$ is real. Therefore, terms in (3-37) exist in conjugate pairs. Let $\mathbf{m}'$ denote the frequency mix resulting in the intermodulation frequency $-f_\mathbf{m}$. Given $\mathbf{m}$, $\mathbf{m}'$ is obtained by replacing each value of m_k by m_{-k}.

Example 3.9

If $\mathbf{m} = (1, 1, 0, 3)$, $\mathbf{m}' = (3, 0, 1, 1)$. Whereas $\mathbf{m} = (1, 1, 0, 3)$ represents the frequency mix $(-f_2 - f_1 + f_2 + f_2 + f_2)$, $\mathbf{m}' = (3, 0, 1, 1)$ represents the frequency mix $(-f_2 - f_2 - f_2 + f_1 + f_2)$. Note that $f_\mathbf{m} = f_{(1,1,0,3)} = 2f_2 - f_1$ while $f_{\mathbf{m}'} = f_{(3,0,1,1)} = -2f_2 + f_1 = -f_{(1,1,0,3)} = -f_\mathbf{m}$.

In general, given $\mathbf{m} = (m_{-Q}, \ldots, m_{-1}, m_1, \ldots, m_Q)$, it follows that

$$\mathbf{m}' = (m_Q, \ldots, m_1, m_{-1}, \ldots, m_{-Q}). \qquad (3\text{-}39)$$

Also, since the order of the factors in (3-32) is immaterial,

$$(n; \mathbf{m}) = (n; \mathbf{m}'). \qquad (3\text{-}40)$$

We conclude that

$$y_n(t; \mathbf{m}') = \frac{(n; \mathbf{m})\, a_n}{2^n} (E_Q)^{m-Q} \cdots (E_1)^{m-1} (E_1^*)^{m_1} \cdots (E_Q^*)^{m_Q} [H(f_Q)]^{m-Q}$$

$$\cdots [H(f_1)]^{m-1} [H^*(f_1)]^{m_1} \cdots [H^*(f_Q)]^{m_Q} K^*(f_m)$$

$$\cdot \exp[-j2\pi f_m t].$$

(3-41)

Comparison of (3-41) with (3-36) reveals $y_n(t; \mathbf{m}')$ to be the complex conjugate of $y_n(t; \mathbf{m})$. Let their sum be denoted by $\hat{y}_n(t; \mathbf{m})$. Recalling that θ_q, $\psi(f)$, and $\phi(f)$ are the phase angles of E_q, $H(f)$, and $K(f)$, respectively, $\hat{y}_n(t; \mathbf{m})$ becomes

$$\hat{y}_n(t; \mathbf{m}) = y_n(t; \mathbf{m}) + y_n(t; \mathbf{m}')$$

$$= y_n(t; \mathbf{m}) + y_n^*(t; \mathbf{m}) = 2\operatorname{Re}\{y_n(t; \mathbf{m})\}$$

$$= \frac{(n; \mathbf{m})\, a_n}{2^{n-1}} |E_1|^{(m_1+m_{-1})} \cdots |E_Q|^{(m_Q+m_{-Q})} |H(f_1)|^{(m_1+m_{-1})}$$

$$\cdots |H(f_Q)|^{(m_Q+m_{-Q})} |K(f_m)|$$

$$\cdot \cos[2\pi f_m t + \theta_m + \psi_m + \phi(f_m)]$$

(3-42)

where

$$\theta_m = \sum_{\substack{k=-Q \\ k \neq 0}}^{Q} m_k \theta_k = m_{-Q}\theta_{-Q} + \cdots + m_{-1}\theta_{-1} + m_1\theta_1 + \cdots + m_Q\theta_Q$$

$$= (m_1 - m_{-1})\theta_1 + \cdots + (m_Q - m_{-Q})\theta_Q$$

(3-43)

and

$$\psi_m = \sum_{\substack{k=-Q \\ k \neq 0}}^{Q} m_k \psi(f_k) = m_{-Q}\psi(f_{-Q}) + \cdots + m_{-1}\psi(f_{-1}) + m_1\psi(f_1)$$

$$+ \cdots + m_Q\psi(f_Q)$$

$$= (m_1 - m_{-1})\psi(f_1) + \cdots + (m_Q - m_{-Q})\psi(f_Q).$$

(3-44)

Given a particular frequency mix, (3-42) enables the corresponding sinusoidal component of the response to be readily evaluated.

However, a second type of output component is possible in the response. This is the constant, or dc, output that arises when

$$f_m = \sum_{\substack{k=-Q \\ k \neq 0}}^{Q} m_k f_k = (m_1 - m_{-1})f_1 + \cdots + (m_Q - m_{-Q})f_Q = 0.$$

(3-45)

By assumption, f_k is positive for k positive. Therefore, (3-45) is satisfied provided $m_k = m_{-k}$ for each value of k. Since this represents a special case, we denote by $\mathbf{m}^0$ any frequency mix resulting in dc. From (3-36)

$$y_n(t; \mathbf{m}^0) = \frac{(n; \mathbf{m}^0) a_n}{2^n} [|E_1|^2]^{m_1} \cdots [|E_Q|^2]^{m_Q} [|H(f_1)|^2]^{m_1}$$

$$\cdots [|H(f_Q)|^2]^{m_Q} K(0). \quad (3\text{-}46)$$

Because the phase angle of a linear transfer function is an odd function of frequency, $\phi(0) = 0$ and $K(0)$ is a real quantity. Therefore, it is unnecessary to add a complex conjugate term, as was done in (3-42), in order for the output to be real. In fact, for $f_m = 0$, $\mathbf{m}$ and $\mathbf{m}'$ are the same vector. As a result, only one of these vectors is included in the summation of (3-37). We conclude that (3-42) is used for sinusoidal outputs while dc outputs are determined from (3-46).

Example 3.10

Assume the weakly nonlinear system of Fig. 3.1 is excited by a two-tone input. The vector $\mathbf{m} = (0, 1, 0, 1)$ yields the sinusoidal response

$$\hat{y}_2(t; 0, 1, 0, 1) = a_2 |E_1| |E_2| |H(f_1)| |H(f_2)| |K(f_2 - f_1)|$$

$$\cdot \cos [2\pi(f_2 - f_1) t + \theta_2 - \theta_1 + \psi(f_2) - \psi(f_1) + \phi(f_2 - f_1)] \quad (3\text{-}47)$$

while the vector $\mathbf{m} = (1, 0, 0, 1)$ yields the dc output

$$y_2(t; 1, 0, 0, 1) = 0.5 a_2 |E_2|^2 |H(f_2)|^2 K(0). \quad (3\text{-}48)$$

Total Response at a Particular Intermodulation Frequency: As pointed out earlier, the key to the simplification of $y(t)$ rests with the evaluation of only those components at the frequencies of interest. Let $y(t; f)$ denote that portion of the response at frequency f. Recall that several different frequency mixes may contribute to a particular intermodulation frequency. Therefore, to obtain the total response at f, it is necessary to add together each of the various contributions.

Example 3.11

If a weakly nonlinear system is excited by a two-tone input and if the highest degree of the zero-memory nonlinearity is $N = 5$, then the output at $2f_2 - f_1$ is the sum of the contributions resulting from the frequency mixes given by $\mathbf{m} = (0, 1, 0, 2)$, $\mathbf{m} = (1, 1, 0, 3)$, and $\mathbf{m} = (0, 2, 1, 2)$. In particular, by separating out the common factors, $y(t; 2f_2 - f_1)$ becomes

$$y(t; 2f_2 - f_1) = \hat{y}_3(t; 0, 1, 0, 2) + \hat{y}_5(t; 1, 1, 0, 3) + \hat{y}_5(t; 0, 2, 1, 2)$$

$$= [0.75a_3 + 1.25a_5|E_2|^2 \, |H(f_2)|^2 + 1.875a_5|E_1|^2 \, |H(f_1)|^2]$$

$$\cdot \, |E_1| \, |E_2|^2 \, |H(f_1)| \, |H(f_2)|^2 \, K(2f_2 - f_1) \cos [2\pi(2f_2 - f_1) t$$

$$+ 2\theta_2 - \theta_1 + 2\psi(f_2) - \psi(f_1) + \phi(2f_2 - f_1)]. \tag{3-49}$$

In general, all three contributions should be included when evaluating the total response at $2f_2 - f_1$. However, when the input amplitudes are small enough, those contributions resulting from the fifth-degree terms of the zero-memory nonlinearity may be negligible. For purposes of illustration, assume $a_3 = 5$, $a_5 = 200$, $|E_1| = |E_2| = 10^{-3}$, $|H(f_1)| = 20$, and $|H(f_2)| = 10$. It follows that

$$0.75a_3 = 3.75$$

$$1.25a_5|E_2|^2 \, |H(f_2)|^2 = 2.5 \times 10^{-2}$$

$$1.875a_5|E_1|^2 \, |H(f_1)|^2 = 15.0 \times 10^{-2}. \tag{3-50}$$

Hence even though the fifth-degree coefficient a_5 is 40 times larger than the third-degree coefficient a_3, the contributions involving a_5 are negligible with respect to those involving a_3.

In many applications the total response at a particular frequency is nicely approximated by ignoring those terms corresponding to frequency mixes resulting from higher degree terms of the nonlinearity. However, input amplitudes and system parameters should always be checked to justify the approximation.

Accuracy Considerations: Characterization of a nonlinear system by the classical model requires specification of the prefilter transfer function $H(f)$, the postfilter transfer function $K(f)$, and the power series coefficients $\{a_1, a_2, \ldots, a_N\}$. When accurate predictions of the nonlinear responses are desired, one might expect that accurate modeling of the power series coefficients is the most critical task in modeling the various system parameters. Equation (3-42) reveals that this is not the case. In the expression for $\hat{y}_n(t; \mathbf{m})$ the prefilter transfer function $H(\cdot)$ appears as a factor n times whereas the power series coefficient a_n and the postfilter transfer function $K(\cdot)$ appear only once. As a result, errors in the modeling of $H(f)$ may be much more serious than similar errors in the modeling of a_n and $K(f)$. Because of the accuracy to which $H(f)$ must be known in order to stay within a prescribed output error criterion, it may be exceedingly difficult to make accurate predictions of nonlinear responses when n is large.

3.4   OUTPUT POWERS AS A FUNCTION OF INPUT POWERS

Typically an intermodulation component at the output of a nonlinear system is described in terms of its average power delivered to the load. In this section we develop an expression for this power as a function of the average powers in the input signals that generate the frequency mix.

In general, the signals in Fig. 3.1 may be either voltages or currents. For convenience, we arbitrarily choose $x(t)$, $w(t)$, $z(t)$, and $y(t)$ to be voltages. Thus $H(f)$ and $K(f)$ are dimensionless voltage ratios and a_n has the dimensions of volts$^{(1-n)}$. The sinusoidal response associated with f_m is given by (3-42). If the amplitude of this response is denoted by $|E_m|$, it follows that

$$|E_m| = \frac{(n; m)\, a_n}{2^{n-1}} |E_1|^{(m_1+m_{-1})} \cdots |E_Q|^{(m_Q+m_{-Q})} |H(f_1)|^{(m_1+m_{-1})}$$

$$\cdots |H(f_Q)|^{(m_Q+m_{-Q})} |K(f_m)|. \qquad (3\text{-}51)$$

To emphasize the frequency dependence of the load admittance, we write it as

$$Y_L(f) = G_L(f) + jB_L(f). \qquad (3\text{-}52)$$

Substituting (3-51) into (2-43), the average power dissipated in the load by the intermodulation component corresponding to the frequency mix vector $\mathbf{m}$ is

$$p_L(f_m) = \frac{1}{2} |E_m|^2\, G_L(f_m) = \frac{(n; m)^2\, a_n^2}{2^{2n-1}} \, [|E_1|^2]^{(m_1+m_{-1})}$$

$$\cdots [|E_Q|^2]^{(m_Q+m_{-Q})} \, [|H(f_1)|^2]^{(m_1+m_{-1})}$$

$$\cdots [|H(f_Q)|^2]^{(m_Q+m_{-Q})} |K(f_m)|^2\, G_L(f_m). \qquad (3\text{-}53)$$

This equation for $p_L(f_m)$ is expressed next as a function of the average powers in the input tones by focusing attention on the squared magnitudes of the input voltages.

Assume the input admittance of the nonlinear system is given by

$$Y_{in}(f) = G_{in}(f) + jB_{in}(f). \qquad (3\text{-}54)$$

Since the average power associated with the input signal at frequency f_q is

$$p_S(f_q) = \tfrac{1}{2} |E_q|^2\, G_{in}(f_q), \qquad (3\text{-}55)$$

we have

$$|E_q|^2 = \frac{2 p_S(f_q)}{G_{in}(f_q)}. \qquad (3\text{-}56)$$

Use of (3-56) in (3-53) results in

$$p_L(f_m) = \frac{(n;\, \mathbf{m})^2\, a_n^2}{2^{2n-1}} \left[\frac{2p_S(f_1)}{G_{in}(f_1)} \right]^{(m_1+m_{-1})} \cdots \left[\frac{2p_S(f_Q)}{G_{in}(f_Q)} \right]^{(m_Q+m_{-Q})}$$

$$[|H(f_1)|^2]^{(m_1+m_{-1})} \cdots [|H(f_Q)|^2]^{(m_Q+m_{-Q})}\, |K(f_m)|^2\, G_L(f_m). \quad (3\text{-}57)$$

However, with reference to (3-29),

$$2^{(m_1+m_{-1})} \cdots 2^{(m_Q+m_{-Q})} = 2^{(m_{-Q}+\cdots+m_{-1}+m_1+\cdots+m_Q)} = 2^n. \quad (3\text{-}58)$$

Therefore, by defining the intermodulation multiplier to be

$$c(\mathbf{m}) = \frac{(n;\, \mathbf{m})^2\, a_n^2}{2^{n-1}} \left[\frac{|H(f_1)|^2}{G_{in}(f_1)} \right]^{(m_1+m_{-1})}$$

$$\cdots \left[\frac{|H(f_Q)|^2}{G_{in}(f_Q)} \right]^{(m_Q+m_{-Q})}\, |K(f_m)|^2\, G_L(f_m), \quad (3\text{-}59)$$

the expression for $p_L(f_m)$ reduces to

$$p_L(f_m) = [p_S(f_1)]^{(m_1+m_{-1})} \cdots [p_S(f_Q)]^{(m_Q+m_{-Q})}\, c(\mathbf{m}). \quad (3\text{-}60)$$

This is the desired result. From (3-60) it is obvious that the dimensions of $c(\mathbf{m})$ are watts$^{(1-n)}$. This can also be verified by observing in (3-59) that $(n;\, \mathbf{m})$, $H(f_q)$, and $K(f_m)$ are dimensionless quantities while a_n has the dimensions of volts$^{(1-n)}$ and $G_{in}(f_q)$ and $G_L(f_m)$ have the dimensions of ampere/volt.

Decibel Form of $p_L(f_m)$: The next step is to express $p_L(f_m)$ in decibels. For this purpose, we introduce into (3-60) the reference power p_r in order to form power ratios as indicated by the following equation:

$$\left[\frac{p_L(f_m)}{p_r} \right] = \left[\frac{p_S(f_1)}{p_r} \right]^{(m_1+m_{-1})} \cdots \left[\frac{p_S(f_Q)}{p_r} \right]^{(m_Q+m_{-Q})}\, [c(\mathbf{m})\, p_r^{n-1}].$$

$$(3\text{-}61)$$

An important feature of (3-61) is that all quantities enclosed by rectangular brackets are dimensionless. Converting to decibels, as defined by (2-62), (3-61) becomes

$$P_L(f_m) = (m_1 + m_{-1}) P_S(f_1) + \cdots + (m_Q + m_{-Q}) P_S(f_Q) + C(\mathbf{m})$$

$$(3\text{-}62)$$

where

$$P_L(f_m) = 10 \log_{10} \left[\frac{p_L(f_m)}{p_r} \right]$$

$$P_S(f_q) = 10 \log_{10} \left[\frac{p_S(f_q)}{p_r} \right]; \qquad q = 1, 2, \ldots, Q$$

$$C(m) = 10 \log_{10} [c(m) p_r^{n-1}]. \tag{3-63}$$

When the reference power is 1 mW, $P_L(f_m)$ and $P_S(f_q)$ are expressed in dBm (i.e., decibels relative to 1 mW). However, regardless of the choice for p_r, $C(m)$ is always expressed in decibels.

Example 3.12

Consider a two-tone input for which $p_S(f_1) = 0.01$ mW and $p_S(f_2) = 0.1$ mW. Assume the intermodulation component of interest is represented by the vector $m = (m_{-2}, m_{-1}, m_1, m_2) = (1, 1, 0, 3)$. Hence, $f_m = (m_2 - m_{-2})f_2 + (m_1 - m_{-1})f_1 = 2f_2 - f_1$ and $n = m_{-2} + m_{-1} + m_1 + m_2 = 5$. Assume further, that $|H(f_1)| = 10.0, |H(f_2)| = 1.0, |K(2f_2 - f_1)| = 100, a_5 = 100 \text{ V}^{-4}, G_{in}(f_1) = G_{in}(f_2) = G_L(2f_2 - f_1) = 0.01$ mho, and $p_r = 1$ mW. Using (3-59), the intermodulation multiplier becomes

$$c(1, 1, 0, 3) = \frac{(5; 1, 1, 0, 3)^2 \, a_5^2}{2^4} \left[\frac{|H(f_1)|^2}{G_{in}(f_1)} \right] \left[\frac{|H(f_2)|^2}{G_{in}(f_2)} \right]^4$$

$$\cdot |K(2f_2 - f_1)|^2 \, G_L(2f_2 - f_1)$$

$$= \frac{(20)^2 (100)^2}{16} \left[\frac{(10)^2}{(0.01)} \right] \left[\frac{(1.0)^2}{(0.01)} \right]^4 (100)^2 (0.01)$$

$$= \frac{1}{4} \times 10^{20} \text{ W}^{-4}. \tag{3-64}$$

It follows from (3-60) that the average power dissipated in the load by this intermodulation component is

$$p_L [f_{(1,1,0,3)}] = p_S(f_1)[p_S(f_2)]^4 \, c(1, 1, 0, 3)$$

$$= (10^{-5}) (10^{-4})^4 \left(\tfrac{1}{4} \times 10^{20} \right)$$

$$= \tfrac{1}{4} \times 10^{-1} \text{ W} = 25 \text{ mW}. \tag{3-65}$$

On the other hand, in terms of decibels,

$$P_S(f_1) = 10 \log_{10} \left[\frac{p_S(f_1)}{p_r} \right] = 10 \log_{10} \left[\frac{10^{-5}}{10^{-3}} \right] = -20 \text{ dBm}$$

$$P_S(f_2) = 10 \log_{10} \left[\frac{p_S(f_2)}{p_r} \right] = 10 \log_{10} \left[\frac{10^{-4}}{10^{-3}} \right] = -10 \text{ dBm}$$

$$C(1, 1, 0, 3) = 10 \log_{10} \left[c(1, 1, 0, 3) p_r^4 \right] = 10 \log_{10} \left[\frac{1}{4} \times 10^{20} \times 10^{-12} \right]$$

$$= 74 \text{ dB}. \tag{3-66}$$

Substituting (3-66) into (3-62), we have

$$P_L[f_{(1,1,0,3)}] = P_S(f_1) + 4P_S(f_2) + C(1, 1, 0, 3)$$

$$= -20 - 40 + 74 = 14 \text{ dBm}. \tag{3-67}$$

This agrees with the result obtained in (3-65).

At this point in our discussion, a word of caution is in order. We have chosen to use the term "intermodulation component," in a rather broad way, to refer to any response resulting from a frequency mix. However, different frequency mixes may produce different types of output distortion. In particular, we may have such different nonlinear effects as harmonic distortion, gain compression, cross modulation, desensitization, etc. In this latter context, the term "intermodulation" is used in a much more restricted sense. Nevertheless, we shall continue with our original usage. The more restricted interpretation is discussed in detail in Chapter 6.

We now illustrate an application of the classical model by investigating the frequency assignment problem associated with incorporating a communications receiver into an existing electromagnetic environment.

3.5 APPLICATION OF THE CLASSICAL MODEL TO A FREQUENCY ASSIGNMENT PROBLEM

Installation of a communications receiver is complicated by the presence of other users of the electromagnetic spectrum. Among the possible users are those involved with the radio and television broadcast industries, air and marine traffic, telecommunications, law enforcement, fire services, civil defense, the military, public utilities, the railroads, trucking, buses, taxis, and amateur radio. Unintentional electromagnetic transmissions may also arise from sources such as tools and machines, switches, power lines, fluorescent lights, automobiles, arc welders, lightning, and galactic noise. These sources of potential interference must be dealt with if the receiver is to perform properly after installation.

We assume the receiver is capable of being tuned over a wide range of frequencies. Our objective is to assign frequencies of operation that are free from interference. Obviously, knowledge is required of the electromagnetic spectrum to which the receiver is exposed. Time and money permitting, this information can be obtained by a site survey. Otherwise, the spectrum can be deduced from spectrum data files maintained by such organizations as the Electromagnetic Compatibility Analysis Center and the Federal Communications Commission.

Receiver Model for Frequency Assignment Problem: In order to solve the frequency assignment problem, the predetector portion of communication receivers is commonly modeled in the form of the classical model of Fig. 3.1. In such applications $x(t)$ denotes the antenna input while $y(t)$ denotes the IF output or, equivalently, the detector input. For simplicity in this section, we assume the predetector portion of the receiver under consideration is adequately characterized by the model shown in Fig. 3.9. Observe that the output of the zero-memory nonlinearity is given by

$$z(t) = a_1 w(t) + a_3 w^3(t). \tag{3-68}$$

Therefore, the nonlinear behavior of the predetector portion is represented by the cubic term, $a_3 w^3(t)$. In addition, the frequency selective filters are all assumed to occur prior to the zero-memory nonlinearity. Hence

$$K(f) = 1. \tag{3-69}$$

Finally, we assume $H(f)$ to be a passive filter and account for the receiver's voltage gain by including it in the linear coefficient a_1.

Frequency Assignment Based Upon Linear Considerations: The receiver output is given by (3-26). Since the only nonzero coefficients in the zero-memory nonlinearity power series expansion are a_1 and a_3, the total response reduces to

$$y(t) = y_1(t) + y_3(t). \tag{3-70}$$

Initially, we concern ourselves with the effect of a single interferer given by

$$x(t) = |E_1| \cos(2\pi f_1 t + \theta_1). \tag{3-71}$$

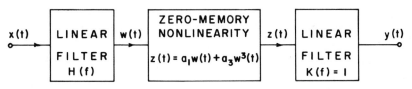

Fig. 3.9. Model of predetector portion of receiver to be used in frequency assignment problem.

In addition to the linear response at f_1, there is also a nonlinear response at f_1 generated by the frequency mix $(f_1 + f_1 - f_1)$. While the nonlinear response arises from the mix represented by the vector $\mathbf{m} = (1, 2)$, the linear response can be interpreted as though it arises from the degenerate frequency mix (f_1) represented by the vector $\mathbf{m} = (0, 1)$. Using (3-69) in (3-42), $\hat{y}_1(t; 0, 1)$ and $\hat{y}_3(t; 1, 2)$ become

$$\hat{y}_1(t; 0, 1) = a_1 |E_1| \, |H(f_1)| \, \cos\, [2\pi f_1 t + \theta_1 + \psi(f_1)] \qquad (3\text{-}72)$$

and

$$\hat{y}_3(t; 1, 2) = \frac{3a_3}{4} |E_1|^3 \, |H(f_1)|^3 \, \cos\, [2\pi f_1 t + \theta_1 + \psi(f_1)]. \qquad (3\text{-}73)$$

Typically, $|E_1|$ is sufficiently small such that $\hat{y}_3(t; 1, 2)$ is negligible with respect to $\hat{y}_1(t; 0, 1)$. As a result, we focus our attention only on the linear portion of the response.

From (3-62), (3-63), and (3-59) the average power dissipated at the IF output due to $\hat{y}_1(t; 0, 1)$ is

$$P_L[f_{(0,1)}] = P_S(f_1) + 10 \log_{10} \, [a_1^2 |H(f_1)|^2 \, G_L(f_1) \, G_{\text{in}}^{-1}(f_1)]. \qquad (3\text{-}74)$$

As is characteristic of linear responses, (3-74) reveals that an increment in the input power by a specified number of decibels produces the same decibel increment in the output power. Assume the receiver is tuned to f_0. Obviously, the minimum signal capable of being detected by the receiver is an on-tune signal since the receiver's sensitivity is greatest at f_0. Let the average input power of the minimum detectable signal be denoted in decibels by $P_S^{\min}(f_0)$. Following the same reasoning as the reasoning that led to (3-74), the average power dissipated at the IF output by the minimum detectable signal is

$$P_L^{\min} = P_S^{\min}(f_0) + 10 \log_{10} \, [a_1^2 |H(f_0)|^2 \, G_L(f_0) \, G_{\text{in}}^{-1}(f_0)]. \qquad (3\text{-}75)$$

Recall that the IF output corresponds to the detector input. The receiver is assumed to contain an ideal detector. Therefore, we evaluate powers at the detector input because ideal detector performance is independent of carrier frequency. For example, an ideal envelope detector demodulates AM signals irrespective of their carrier frequencies. Since $P_L^{\min}$ represents the minimum detectable power at the detector input, extraneous signals that produce detector input powers less than $P_L^{\min}$ can be safely ignored. On the other hand, undesired signals that result in detector input power greater than $P_L^{\min}$ constitute real sources of interference.

To compare $P_L[f_{(0,1)}]$ against $P_L^{\min}$, we subtract (3-75) from (3-74). This results in

$$\Delta P_L[f_{(0,1)}] = P_L[f_{(0,1)}] - P_L^{\min}$$

$$= P_S(f_1) - P_S^{\min}(f_0) + 10 \log_{10}\left[\frac{|H(f_1)|^2}{|H(f_0)|^2}\frac{G_L(f_1)}{G_L(f_0)}\frac{G_{in}(f_0)}{G_{in}(f_1)}\right].$$

$$(3\text{-}76)$$

Also, by analogy with (3-66),

$$\Delta P_L[f_{(0,1)}] = 10 \log_{10}\left\{\frac{p_L[f_{(0,1)}]}{p_r}\right\} - 10 \log_{10}\left\{\frac{p_L^{\min}}{p_r}\right\}$$

$$= 10 \log_{10}\left\{\frac{p_L[f_{(0,1)}]}{p_L^{\min}}\right\}. \qquad (3\text{-}77)$$

Therefore, $\Delta P_L[f_{(0,1)}]$ is nothing more than the ratio, expressed in decibels, of $p_L[f_{(0,1)}]$ to $p_L^{\min}$. Extraneous signals can be safely ignored provided that this ratio is less than unity or, equivalently, that the difference $\Delta P_L[f_{(0,1)}]$ is negative. With the aid of (3-76), this latter condition translates into the inequality

$$P_S(f_1) + 10 \log_{10}\left[\frac{|H(f_1)|^2}{|H(f_0)|^2}\frac{G_L(f_1)}{G_L(f_0)}\frac{G_{in}(f_0)}{G_{in}(f_1)}\right] < P_S^{\min}(f_0). \qquad (3\text{-}78)$$

The inequality in (3-78) is the desired result. As long as this inequality is satisfied, undesired signals will not be sources of interference. By means of (3-78), receiver operating frequencies can be assigned that are free from interference as far as linear effects are concerned.

Let the left-hand side of (3-78) be denoted by $P_S[f_0; f_{(0,1)}]$. Therefore,

$$P_S[f_0; f_{(0,1)}] = P_S(f_1) + 10 \log_{10}\left[\frac{|H(f_1)|^2}{|H(f_0)|^2}\frac{G_L(f_1)}{G_L(f_0)}\frac{G_{in}(f_0)}{G_{in}(f_1)}\right]. \qquad (3\text{-}79)$$

$P_S[f_0; f_{(0,1)}]$ has an interesting interpretation. An on-tune input signal with average power numerically equal to $P_S[f_0; f_{(0,1)}]$ will generate an average power at the IF output given by

$$P_S[f_0; f_{(0,1)}] + 10 \log_{10}[a_1^2|H(f_0)|^2 G_L(f_0) G_{in}^{-1}(f_0)]$$

$$= P_S(f_1) + 10 \log_{10}[a_1^2|H(f_1)|^2 G_L(f_1) G_{in}^{-1}(f_1)]$$

$$= P_L[f_{(0,1)}]. \qquad (3\text{-}80)$$

Therefore, assuming linear receiver operation, $P_S[f_0; f_{(0,1)}]$ is the equivalent input power needed by an on-tune signal to produce the same numerical output power as does the original signal at f_1 with power $P_S(f_1)$. The inequality in (3-78) can be rewritten as

$$P_S[f_0; f_{(0,1)}] < P_S^{\min}(f_0). \qquad (3\text{-}81)$$

Consequently, an undesired signal at frequency f_1 will not cause interference provided its equivalent input power at the tuned frequency f_0 is less than the input power of the minimum detectable signal.

Example 3.13

To gain additional insight into the application of (3-78), let the linear filter with transfer function $H(f)$ be the single-tuned circuit shown in Fig. 3.10(a). This can be analyzed as a voltage divider as indicated by Fig. 3.10(b). By inspection, the linear transfer function of the filter is

$$H(j\omega) = \frac{E_w}{E_x} = \frac{Z_2(j\omega)}{Z_1(j\omega) + Z_2(j\omega)} = \frac{\dfrac{j\omega}{RC}}{\dfrac{1}{LC} - \omega^2 + \dfrac{j\omega}{RC}}. \qquad (3\text{-}82)$$

Introducing into (3-82) the definitions

$$2\alpha = \frac{1}{RC} \quad \text{and} \quad \omega_0^2 = \frac{1}{LC}, \qquad (3\text{-}83)$$

we obtain

$$H(j\omega) = \frac{j\omega 2\alpha}{\omega_0^2 - \omega^2 + j\omega 2\alpha} = \frac{1}{1 + j\dfrac{\omega^2 - \omega_0^2}{\omega 2\alpha}} = \frac{1}{1 + j\dfrac{(\omega + \omega_0)(\omega - \omega_0)}{\omega 2\alpha}}. \qquad (3\text{-}84)$$

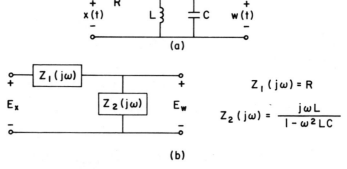

Fig. 3.10. (a) Circuit diagram of linear filter with transfer function $H(f)$. (b) Interpretation of linear filter as a voltage divider.

At resonance $\omega = \omega_0$ and the transfer function assumes its maximum value of unity. Since the frequencies of interest in a typical application are usually reasonably close to the resonant frequency of the tuned circuit, $\omega \approx \omega_0$ and the transfer function can be approximated by

$$H(j\omega) \approx \frac{1}{1 + j\left(\dfrac{\omega - \omega_0}{\alpha}\right)}. \tag{3-85}$$

The 3-dB points occur at those frequencies for which

$$|H(j\omega)| = (1/\sqrt{2})|H(j\omega_0)| = 1/\sqrt{2}.$$

By inspection, the magnitude of the transfer function equals $1/\sqrt{2}$ when $H(j\omega) = 1/(1 \pm j)$. This occurs when $(\omega - \omega_0)/\alpha = \pm 1$ or when $\omega = \omega_0 \pm \alpha$. We conclude that the 3-dB bandwidth equals 2α rad/s. In terms of hertz, the 3-dB bandwidth is denoted by

$$B = \frac{2\alpha}{2\pi} = \frac{\alpha}{\pi} \text{ Hz.} \tag{3-86}$$

Substituting (3-86) into (3-85) and noting that $\omega = 2\pi f$ while $\omega_0 = 2\pi f_0$ yields

$$H(f) \approx \frac{1}{1 + j\left(\dfrac{f - f_0}{B/2}\right)} = \left[1 + j\left(\frac{\Delta f}{B/2}\right)\right]^{-1} = \left[1 + j2Q\left(\frac{\Delta f}{f_0}\right)\right]^{-1} \tag{3-87}$$

where

$$\Delta f = f - f_0 \qquad \text{and} \qquad Q = \frac{f_0}{B}. \tag{3-88}$$

To proceed further with the analysis, let the average input power of the minimum detectable signal be

$$P_S^{\min}(f_0) = -100 \text{ dBm.} \tag{3-89}$$

Also, assume a 50-Ω system in which the input and load impedances are given by resistors of 50 Ω. Thus

$$G_{\text{in}}(f) = G_L(f) = \frac{1}{50} \text{ mho.} \tag{3-90}$$

Recalling that $|H(f_0)| = 1$ and using (3-87), (3-89), and (3-90), the inequality in (3-78) becomes

$$P_S[f_0; f_{(0,1)}] = P_S(f_1) + 10 \log_{10}\left[|H(f_1)|^2\right]$$

$$= P_S(f_1) - 10 \log_{10}\left[1 + \left(\frac{\Delta f}{B/2}\right)^2\right] < -100 \text{ dBm.} \tag{3-91}$$

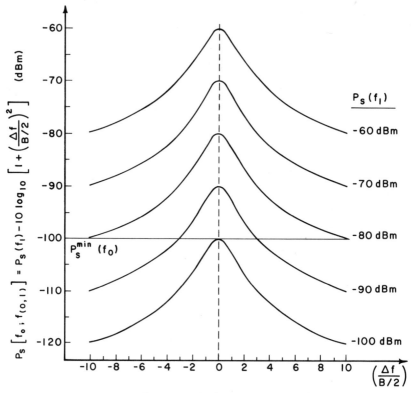

Fig. 3.11. Plot of $P_S[f_0; f_{(0,1)}]$ as used in (3-91).

$P_S[f_0; f_{(0,1)}]$ is plotted in Fig. 3.11 with $P_S(f_1)$ treated as a parameter. Note that the abscissa corresponds to the frequency separation between the undesired signal and the tuned frequency of the receiver normalized by the receiver half bandwidth. The average power of the minimum detectable signal is indicated by the horizontal line drawn at -100 dBm. Given an undesired signal with average input power $P_S(f_1)$, the minimum frequency separation required for interference-free operation can be obtained from the figure. For example, when $P_S(f_1) = -80$ dBm, the receiver-tuned frequency must be separated from the undesired signal by an amount greater than or equal to approximately ten receiver half bandwidths. The larger is $P_S(f_1)$, the greater is the frequency separation required. Because of the assumption leading to (3-85), the plot is symmetrical about the vertical line corresponding to $\Delta f = 0$. As a result, receiver performance is degraded equally by interferers above and below the tuned frequency.

Frequency Assignment Based Upon Intermodulation Considerations: The previous analysis treated the frequency assignment problem based upon linear considerations. However, this is not sufficient because troublesome intermodulation components can be generated by the cubic nonlinearity within the receiver. They arise when two or more sinusoidal tones are present at the receiver input. For simplicity in this part of the analysis, we assume the excitation to be

$$x(t) = |E_1| \cos(2\pi f_1 t + \theta_1) + |E_2| \cos(2\pi f_2 t + \theta_2). \qquad (3\text{-}92)$$

From (3-27), (3-23), and (3-69), that portion of the response due to the cubic nonlinearity is given by

$$y_3(t) = \sum_{q_1=-2}^{2} \sum_{q_2=-2}^{2} \sum_{q_3=-2}^{2} \frac{a_3}{8} E_{q_1} E_{q_2} E_{q_3} H(f_{q_1}) H(f_{q_2}) H(f_{q_3})$$

$$\cdot \exp\left[j2\pi(f_{q_1} + f_{q_2} + f_{q_3}) t\right]. \qquad (3\text{-}93)$$

Taking into consideration all possible frequency mixes and assuming the input frequencies to be reasonably close with $f_2 > f_1$, $y_3(t)$ contains sinusoidal components at $2f_1 - f_2, f_1, f_2, 2f_2 - f_1, 3f_1, 2f_1 + f_2, 2f_2 + f_1$, and $3f_2$ as indicated in Fig. 3.12. Due to the frequency selectivity of the receiver, the most troublesome situation arises when both f_1 and f_2 are near the receiver-tuned frequency of f_0. For this situation it is obvious, from Fig. 3.12, that the only components capable of falling within the receiver passband are those located at $2f_1 - f_2, f_1, f_2$, and $2f_2 - f_1$. However, for small input signals, it has already been shown that the components of $y_3(t)$ at f_1 and f_2 are negligible compared to the linear responses. As a result, the intermodulation components most likely to produce significant interference are those located at $2f_1 - f_2$ and $2f_2 - f_1$. Both components are analyzed in an identical manner. We first consider the component at $2f_1 - f_2$.

The response at $2f_1 - f_2$ is generated by the frequency mix $(f_1 + f_1 - f_2)$

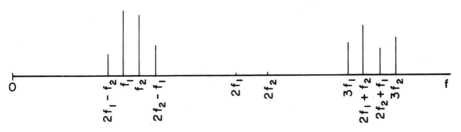

Fig. 3.12. Frequency content of $y_3(t)$.

which is represented by the vector $\mathbf{m} = (1, 0, 2, 0)$. Using (3-69) in (3-42)

$$\hat{y}_3(t; 1, 0, 2, 0) = \frac{3a_3}{4} |E_1|^2 |E_2| |H(f_1)|^2 |H(f_2)|$$

$$\cdot \cos\left[2\pi(2f_1 - f_2)t + 2\theta_1 - \theta_2 + 2\psi(f_1) - \psi(f_2)\right]. \quad (3\text{-}94)$$

From (3-62), (3-63), and (3-59) the average power dissipated at the detector input by $\hat{y}_3(t; 1, 0, 2, 0)$ is

$$P_L[f_{(1,0,2,0)}] = 2P_S(f_1) + P_S(f_2) + C(1, 0, 2, 0) \quad (3\text{-}95)$$

where

$$C(1, 0, 2, 0) = 10 \log_{10}\left\{\frac{9a_3^2}{4}\left[\frac{|H(f_1)|^2}{G_{\text{in}}(f_1)}\right]^2 \frac{|H(f_2)|^2}{G_{\text{in}}(f_2)} G_L(2f_1 - f_2) p_r^2\right\}.$$

$$(3\text{-}96)$$

Recall that the average power dissipated at the detector input by the minimum detectable signal is $P_L^{\min}$. To compare $P_L[f_{(1,0,2,0)}]$ against $P_L^{\min}$, we subtract (3-75) from (3-95). This yields

$$\Delta P_L[f_{(1,0,2,0)}]$$

$$= P_L[f_{(1,0,2,0)}] - P_L^{\min}$$

$$= 10 \log_{10}\left\{\frac{P_L[f_{(1,0,2,0)}]}{P_L^{\min}}\right\}$$

$$= 2P_S(f_1) + P_S(f_2) - P_S^{\min}(f_0)$$

$$+ 10 \log_{10}\left\{\frac{9}{4}\frac{a_3^2}{a_1^2}\frac{|H(f_1)|^4 |H(f_2)|^2}{|H(f_0)|^2}\frac{G_{\text{in}}(f_0) G_L(2f_1 - f_2)}{[G_{\text{in}}(f_1)]^2 G_{\text{in}}(f_2) G_L(f_0)}p_r^2\right\}.$$

$$(3\text{-}97)$$

Obviously, the intermodulation component at $2f_1 - f_2$ can be safely ignored provided either the ratio $p_L[f_{(1,0,2,0)}]/p_L^{\min}$ is less than unity or, equivalently, $\Delta P_L[f_{(1,0,2,0)}]$ is negative. From (3-97) the latter condition leads to the inequality

$$P_S[f_0; f_{(1,0,2,0)}]$$

$$= 2P_S(f_1) + P_S(f_2) + 10 \log_{10}\left\{\frac{9}{4}\frac{a_3^2}{a_1^2}\frac{|H(f_1)|^4 |H(f_2)|^2}{|H(f_0)|^2}\right.$$

$$\cdot \left.\frac{G_{\text{in}}(f_0) G_L(2f_1 - f_2)}{[G_{\text{in}}(f_1)]^2 G_{\text{in}}(f_2) G_L(f_0)}p_r^2\right\} < P_S^{\min}(f_0). \quad (3\text{-}98)$$

As long as this inequality is satisfied, intermodulation components resulting from the frequency mix $(f_1 + f_1 - f_2)$ will not be sources of interference. By interchanging f_2 with f_1 everywhere they appear in (3-98), an analogous result is obtained for the frequency mix $(f_2 + f_2 - f_1)$. As with $P_S[f_0; f_{(0,1)}]$, $P_S[f_0; f_{(1,0,2,0)}]$ has an interesting interpretation. Assuming linear receiver operation, an on-tune input signal with average power numerically equal to $P_S[f_0; f_{(1,0,2,0)}]$ generates an average power at the detector input given by

$$P_S[f_0; f_{(1,0,2,0)}] + 10 \log_{10} [a_1^2 |H(f_0)|^2 G_L(f_0) G_{in}^{-1}(f_0)]$$

$$= 2P_S(f_1) + P_S(f_2) + 10$$

$$\cdot \log_{10} \left\{ \frac{9a_3^2}{4} \left[\frac{|H(f_1)|^2}{G_{in}(f_1)} \right]^2 \frac{|H(f_2)|^2}{G_{in}(f_2)} G_L(2f_1 - f_2) p_r^2 \right\}$$

$$= P_L[f_{(1,0,2,0)}]. \tag{3-99}$$

Therefore, assuming linear behavior, $P_S[f_0; f_{(1,0,2,0)}]$ is the equivalent input power needed by an on-tune signal to produce the same output power as in the intermodulation component at $2f_1 - f_2$ which is generated by the two tones at f_1 and f_2 having input powers $P_S(f_1)$ and $P_S(f_2)$, respectively. The inequality in (3-98) states that the intermodulation component at $2f_1 - f_2$ will not cause interference provided its equivalent input power at the tuned frequency f_0 is less than the input power of the minimum detectable signal. Finally, we point out that, if interference due to both linear and intermodulation effects is to be avoided, then operating frequencies assigned to the receiver must satisfy the two inequalities given by (3-78) and (3-98) for all extraneous signals and all pairs of undesired signals present at the receiver input.

Example 3.14

To demonstrate an application of (3-98), we continue with Example 3.13 which was used to illustrate (3-78). By defining

$$\Delta f_i = f_i - f_0; \qquad i = 1, 2 \tag{3-100}$$

and with reference to (3-87) and (3-90), $P_S[f_0; f_{(1,0,2,0)}]$ becomes

$$P_S[f_0; f_{(1,0,2,0)}]$$

$$= 2P_S(f_1) + P_S(f_2) + 10 \log_{10} \left[\frac{9}{4} \frac{a_3^2}{a_1^2} |H(f_1)|^4 |H(f_2)|^2 (50)^2 p_r^2 \right]$$

$$= 2P_S(f_1) + P_S(f_2) + 20 \log_{10} [|H(f_1)|^2]$$

$$+ 10 \log_{10} [|H(f_2)|^2] + 20 \log_{10} \left[\frac{3}{2} \left| \frac{a_3}{a_1} \right| 50 p_r \right]$$

$$= 2P_S(f_1) + P_S(f_2) - 20 \log_{10} \left[1 + \left(\frac{\Delta f_1}{B/2} \right)^2 \right]$$

$$- 10 \log_{10} \left[1 + \left(\frac{\Delta f_2}{B/2} \right)^2 \right] + 20 \log_{10} \left[\frac{3}{2} \left| \frac{a_3}{a_1} \right| 50 p_r \right]$$

$$= 2[P_S(f_1) - \beta_1] + [P_S(f_2) - \beta_2] + I_3(1, 0, 2, 0) \tag{3-101}$$

where

$$\beta_i = 10 \log_{10} \left[1 + \left(\frac{\Delta f_i}{B/2} \right)^2 \right]; \qquad i = 1, 2 \tag{3-102}$$

and

$$I_3(1, 0, 2, 0) = 20 \log_{10} \left[\frac{3}{2} \left| \frac{a_3}{a_1} \right| 50 p_r \right]. \tag{3-103}$$

β_i is referred to as the off-frequency rejection factor for the interferer at frequency f_i while $I_3(1,0,2,0)$ is called the conversion gain factor of the intermodulation component associated with the frequency mix vector $\mathbf{m} = (1,0,2,0)$.

A plot of (3-101) is complicated by the fact that $P_S[f_0; f_{(1,0,2,0)}]$ is a function of the two frequency variables f_1 and f_2. $P_S[f_0; f_{(1,0,2,0)}]$ can be reduced to a function of a single-frequency variable by imposing the constraint

$$2f_1 - f_2 = f_0. \tag{3-104}$$

This corresponds to a worst-case situation because the frequency of the intermodulation component coincides with the receiver-tuned frequency. Let

$$f_1 = f_0 + \Delta f_1. \tag{3-105}$$

Since $2f_1 - f_2 = f_0$, it follows that

$$f_2 = f_0 + 2\Delta f_1. \tag{3-106}$$

Also define

$$P_T(f_1, f_2) = 2P_S(f_1) + P_S(f_2). \tag{3-107}$$

In terms of $P_T(f_1, f_2)$ and the frequency separation Δf_1, (3-101) reduces to

$$P_S[f_0; f_{(1,0,2,0)}] = P_T(f_1, f_2) - 20 \log_{10} \left[1 + \left(\frac{\Delta f_1}{B/2} \right)^2 \right]$$

$$- 10 \log_{10} \left[1 + \left(\frac{2\Delta f_1}{B/2} \right)^2 \right] + I_3(1, 0, 2, 0). \tag{3-108}$$

Typical numerical values for the parameters in the conversion gain factor are

$$a_1 = 10^3, \quad a_3 = -\tfrac{1}{3} \times 10^8, \quad p_r = 10^{-3}. \tag{3-109}$$

For these values,

$$I_3(1, 0, 2, 0) = 20 \log_{10} \left[\tfrac{3}{2} \times \tfrac{1}{3} \times 10^8 \times 10^{-3} \times 50 \times 10^{-3}\right]$$

$$= 20 \log_{10} \left[2.5 \times 10^3\right] = 68 \text{ dB}. \tag{3-110}$$

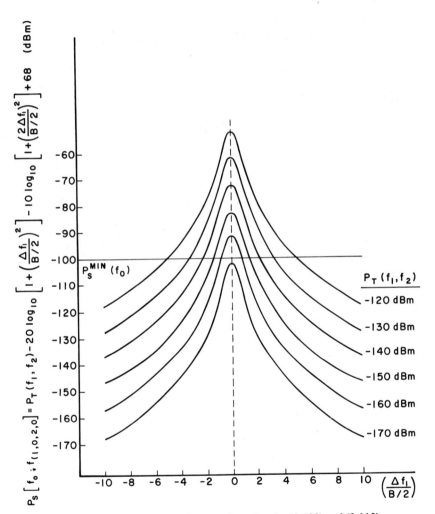

Fig. 3.13. Plot of $P_S[f_0; f_{(1,0,2,0)}]$ as given by (3-108) and (3-110).

With $P_T(f_1, f_2)$ as a parameter, the equivalent input power $P_S[f_0; f_{(1,0,2,0)}]$ is plotted in Fig. 3.13 as a function of the frequency separation Δf_1 normalized by the receiver half bandwidth. As was done in Fig. 3.11, the average power of the minimum detectable signal is indicated by the horizontal line drawn at -100 dBm. Observe that the manner in which the input powers are divided between the interfering signals at f_1 and f_2 is immaterial as long as the "total" power $P_T(f_1, f_2)$ remains constant. Whenever $P_T(f_1, f_2)$ is less than -168 dBm, interference due to the intermodulation component can be ignored. For example, the curve corresponding to $P_T(f_1, f_2) = -170$ dBm falls entirely below the horizontal line at -100 dBm. Otherwise, Fig. 3.13 can be used to determine allowable operating frequencies for the receiver. Figure 3.13 represents a worst-case approach because of the constraint introduced in (3-104). If a more accurate analysis is desirable, $P_S[f_0; f_{(1,0,2,0)}]$ should be evaluated by means of (3-101).

With reference to (3-107), $P_T(f_1, f_2)$ may be converted from decibels to yield

$$p_T(f_1, f_2) = p_S^2(f_1) p_S(f_2). \tag{3-111}$$

Since $p_S(f_1)$ and $p_S(f_2)$ are each assumed to be less than unity, the numerical values used for $P_T(f_1, f_2)$ in Fig. 3.13 are smaller than those used for $P_S(f_1)$ in Fig. 3.11.

This completes our discussion of the classical model. We have tried to demonstrate its utility while, at the same time, emphasizing the assumptions on which the model is based. In the next chapter we consider a more general approach for analyzing weakly nonlinear systems.

4

Nonlinear Transfer Function Approach to the Analysis of Weakly Nonlinear Circuits

The nonlinear transfer function approach is introduced in this chapter. Prior to its development, the primary tool used in the analysis of weakly nonlinear circuits was the classical model discussed in Chapter 3. Unfortunately, results generated by the classical model may be in error by several orders of magnitude when it constitutes an oversimplification of the actual problem.

The nonlinear transfer function approach, which corrects many deficiencies inherent in the classical model, is based upon the Volterra series. Volterra first proposed this series in some of his work dated around 1910. In 1942 Norbert Wiener pointed out that the input/output relationship of certain nonlinear systems involving memory could be conveniently represented by means of the Volterra series expansion. During the late 1950's and early 1960's a sizeable effort was expended by M.I.T. investigators who developed much of the basic groundwork that eventually led to future applications and generalizations. This theoretical framework is documented in reports and publications by Wiener, Bose, Brilliant, George, Zames, Schetzen, Bush, Chesler, and Parente. Other researchers who also pioneered in the initial development of the theory were Barrett, Flake, McFee, Ku, Bedrosian, and Rice. The first practical application of this work was made in 1967 by Narayanan who used the Volterra series to predict nonlinear distortions in transistor amplifiers. Other applications have been made by Van Trees, Maurer, Poon, Kuo, Meyer, and Goldman. Of particular note is the monumental piece of work by Bello, Graham, Ehrman, Meer, O'Donnell, and Bussgang, in which they demonstrated, for the first time, that the nonlinear behavior of communication receivers could be successfully modeled using the Volterra technique. They also introduced the term "nonlinear transfer functions" to describe the frequency-domain kernels that arise in Volterra analysis. Because this term suggests a meaningful interpretation for the Volterra characterization of nonlinear systems, we refer to Volterra analysis as the nonlinear transfer function approach.

In this chapter a brief introduction to the nonlinear transfer function approach is followed by the sinusoidal steady-state analysis of weakly nonlinear systems. The classical model is shown to be a special case of the nonlinear transfer function approach. Expressions are developed for average powers contained in the intermodulation components corresponding to specific frequency mixes. To illustrate the characterization of nonlinear systems, the nonlinear transfer functions of two transistor amplifiers are presented and discussed. Finally, as an introduction to the material of Chapter 5, the cascade connection of two noninteracting nonlinear systems is considered.

4.1 GENERAL DISCUSSION OF NONLINEAR TRANSFER FUNCTION APPROACH

Consider a weakly nonlinear circuit with input $x(t)$ and output $y(t)$. Based upon Volterra analysis, the nonlinear transfer function approach models the circuit as shown in Fig. 4.1. This model consists of the parallel combination of N blocks with each block having, as a common input, the circuit excitation $x(t)$. The output of the nth block is denoted by $y_n(t); n = 1, 2, \ldots, N$. The total response is obtained by summing the outputs of the individual blocks so as to yield

$$y(t) = y_1(t) + y_2(t) + \cdots + y_N(t) = \sum_{n=1}^{N} y_n(t). \qquad (4-1)$$

The nth block, characterized by the nth-order nonlinear transfer function $H_n(f_1, f_2, \ldots, f_n)$, is of nth order in the sense that multiplication of the input $x(t)$ by a constant A results in multiplication of the output $y_n(t)$ by the constant A^n.

The nonlinear transfer function approach is seen to represent the total response of a weakly nonlinear circuit as a sum of N individual responses. The linear portion of the circuit, characterized by the linear transfer function $H_1(f_1)$, generates the first-order component of the response. The quadratic portion of the circuit, characterized by the second-order nonlinear transfer function $H_2(f_1, f_2)$, generates the second-order component of the response. Additional responses are generated in a similar manner. Blocks above Nth order are not included in the model because it is assumed they contribute negligibly to the output.

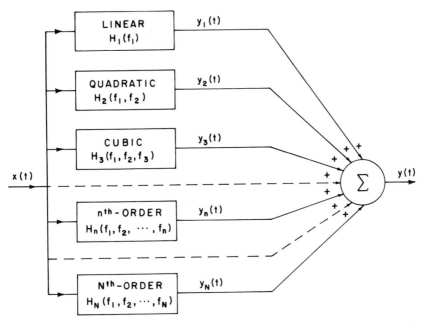

Fig. 4.1. Model of weakly nonlinear circuit used by the nonlinear transfer function approach.

The nonlinear transfer functions completely characterize a weakly nonlinear system and are independent of the excitation. In Appendix A it is explained how knowledge of the nonlinear transfer functions is sufficient for determining the system response to arbitrary inputs. The invariance of the nonlinear transfer functions to system input is a highly desirable feature of the approach. It enables characterization of weakly nonlinear systems without a priori knowledge of inputs that may arise.

In the material that follows it is important to distinguish between the concepts of the degree of a power series term and the order of a nonlinear system response.

Example 4.1

Consider a weakly nonlinear system that is adequately characterized by the model of Fig. 4.1. Assume blocks above third order contribute negligibly to the output. Hence $N = 3$ and the system response is given by

$$y(t) = y_1(t) + y_2(t) + y_3(t). \qquad (4\text{-}2)$$

Let $y(t)$ be a voltage applied across a resistor whose current depends nonlinearly upon its voltage according to the relation

$$i(t) = a_1 y(t) + a_2 y^2(t). \qquad (4\text{-}3)$$

TABLE 4.1.   Various Order Components of $i(t)$.

Order	Components
1	$a_1 y_1(t)$
2	$a_1 y_2(t) + a_2 y_1^2(t)$
3	$a_1 y_3(t) + 2a_2 y_1(t) y_2(t)$
4	$a_2 [y_2^2(t) + 2y_1(t) y_3(t)]$
5	$2a_2 y_2(t) y_3(t)$
6	$a_2 y_3^2(t)$

Since degree refers to the exponent of a power series term, the current-voltage relationship for the resistor involves terms of first and second degree. Substituting (4-2) into (4-3) and carrying out the squaring operation, the resistor current becomes

$$i(t) = a_1 [y_1(t) + y_2(t) + y_3(t)] + a_2 [y_1^2(t) + y_2^2(t) + y_3^2(t)$$
$$+ 2y_1(t)y_2(t) + 2y_1(t)y_3(t) + 2y_2(t)y_3(t)]. \quad (4\text{-}4)$$

Recall that $y_n(t)$ is of nth order in the sense that multiplication of the input $x(t)$ by the constant A results in multiplication of $y_n(t)$ by A^n. It follows that the first-degree term in (4-3) generates components of first, second, and third order. In addition, since multiplication of $x(t)$ by A results in multiplication of the product $y_j(t)y_k(t)$ by A^{j+k}, the order of $y_j(t)y_k(t)$ is $(j + k)$. Consequently, the second-degree term in (4-3) generates components of second through sixth order. Each of the components in the current is tabulated by order in Table 4.1.

The important point is that a power series term of degree n may generate components of order n and above. Use is made of this property in Chapter 5 where we present a method for determining the nonlinear transfer functions of a weakly nonlinear system from its circuit diagram.

4.2   SINUSOIDAL STEADY-STATE RESPONSE OF WEAKLY NONLINEAR SYSTEMS

System Input: To study the sinusoidal steady-state response of weakly nonlinear systems, let the input to the model shown in Fig. 4.1 be the sum of Q sinusoidal signals. With reference to equations (3-3) through (3-6) the excitation can be expressed as

$$x(t) = \sum_{q=1}^{Q} |E_q| \cos (2\pi f_q t + \theta_q) = \frac{1}{2} \sum_{q=-Q}^{Q} E_q \exp (j2\pi f_q t) \quad (4\text{-}5)$$

where $x(t)$ has arbitrarily been chosen to be a voltage and E_q is the complex voltage of the qth tone. The mathematics needed for derivation of the system response is more complicated than we care to introduce in the main body of this volume. Consequently, some of the major results of this chapter are stated without proof. Our objective is to familiarize the reader with the utility of nonlinear transfer functions without obscuring matters by unnecessary mathematics.

System Output: In the nonlinear transfer function approach the total response of the nonlinear system is expressed as

$$y(t) = \sum_{n=1}^{N} y_n(t) \qquad (4\text{-}6)$$

where $y_n(t)$ is the nth-order component of the response generated by the nth-order portion of the system. In Appendix A it is shown that

$$y_n(t) = \frac{1}{2^n} \sum_{q_1=-Q}^{Q} \cdots \sum_{q_n=-Q}^{Q} E_{q_1} \cdots E_{q_n} H_n(f_{q_1}, \cdots, f_{q_n})$$
$$\cdot \exp\left[j2\pi(f_{q_1} + \cdots + f_{q_n})t\right]. \qquad (4\text{-}7)$$

This is a mathematical statement of the fact that, when a sum of Q sinusoids is applied to a nonlinear circuit, additional output frequencies are generated by the nth-order portion of the circuit consisting of all possible combinations of the input frequencies $-f_Q, \ldots, -f_1, f_1, \ldots, f_Q$ taken n at a time. As revealed in (4-7), $H_n(f_{q_1}, \ldots, f_{q_n})$ is the nonlinear transfer function relating the output at frequency $(f_{q_1} + \cdots + f_{q_n})$ to the inputs at frequencies $f_{q_1}, f_{q_2}, \ldots, f_{q_n}$.

All nonlinear transfer functions that differ only by a permutation of their arguments can be shown to be equivalent. Therefore, for ease of discussion and without loss of generality, all transfer functions in the remainder of our work are considered to be symmetrical functions of their arguments unless stated otherwise.

Example 4.2

Assuming symmetry, the third-order transfer functions satisfy the relations

$$H_3(f_1, f_2, f_3) = H_3(f_1, f_3, f_2) = H_3(f_2, f_1, f_3)$$
$$= H_3(f_2, f_3, f_1) = H_3(f_3, f_1, f_2) = H_3(f_3, f_2, f_1). \qquad (4\text{-}8)$$

It can also be shown that transfer functions of real systems are conjugated when the signs of all arguments are changed. Thus

$$H_n(-f_1, \ldots, -f_n) = H_n^*(f_1, \ldots, f_n) \qquad (4\text{-}9)$$

where, once again, the asterisk is used to denote complex conjugate. The symmetry and conjugation properties are now used in (4-7) to express $y_n(t)$ as a sum of sinusoids.

Allowing for the fact that $E_0 = 0$, the n-fold summation in (4-7) contains $(2Q)^n$ individual terms. Many of these terms are identical. Because of the symmetry of the nonlinear transfer functions, interchanging the order of the indices $q_1, \ldots, q_n$ does not alter the general term

$$\frac{1}{2^n} E_{q_1} \cdots E_{q_n} H_n(f_{q_1}, \ldots, f_{q_n}) \exp \left[j2\pi(f_{q_1} + \cdots + f_{q_n})t \right].$$

Example 4.3

When $n = 2$, the two choices $q_1 = -1$, $q_2 = 2$ and $q_1 = 2$, $q_2 = -1$ yield the identical terms

$$\tfrac{1}{4} E_1^* E_2 H_2(-f_1, f_2) \exp \left[j2\pi(-f_1 + f_2)t \right]$$
$$= \tfrac{1}{4} E_2 E_1^* H_2(f_2, -f_1) \exp \left[j2\pi(f_2 - f_1)t \right]. \quad (4\text{-}10)$$

Total Response for a Particular Frequency Mix: At this point it is convenient to consider specific frequency mixes. As was done in our analysis of the classical model, let m_k denote the number of times the frequency f_k appears in a particular frequency mix. All possible frequency mixes are then represented by the frequency mix vector

$$\mathbf{m} = (m_{-Q}, \ldots, m_{-1}, m_1, \ldots, m_Q). \quad (4\text{-}11)$$

The corresponding intermodulation frequency is

$$f_{\mathbf{m}} = \sum_{\substack{k=-Q \\ k \neq 0}}^{Q} m_k f_k = (m_1 - m_{-1})f_1 + \cdots + (m_Q - m_{-Q})f_Q. \quad (4\text{-}12)$$

For an nth-order component of the response, the m_k's obey the constraint

$$\sum_{\substack{k=-Q \\ k \neq 0}}^{Q} m_k = m_{-Q} + \cdots + m_{-1} + m_1 + \cdots + m_Q = n. \quad (4\text{-}13)$$

Therefore, the output frequencies in (4-7) can be interpreted as those intermodulation frequencies that can be generated by all possible choices of the m_k's such that (4-13) is satisfied.

Given a particular vector $\mathbf{m}$, the number of different ways the n indices $q_1, \ldots, q_n$ can be partitioned such that $-Q$ appears m_{-Q} times, $\ldots$, -1 appears m_{-1} times, 1 appears m_1 times, $\ldots$, and Q appears m_Q times is given by the multinomial coefficient

$$(n; \mathbf{m}) = \frac{(n!)}{(m_{-Q}!) \cdots (m_{-1}!)(m_1!) \cdots (m_Q!)}. \quad (4\text{-}14)$$

With respect to the general term from (4-7), that is given by

$$\frac{1}{2^n} E_{q_1} \cdots E_{q_n} H_n(f_{q_1}, \ldots, f_{q_n}) \exp\left[j2\pi(f_{q_1} + \cdots + f_{q_n})t\right],$$

each of the $(n; \mathbf{m})$ realizations yields the identical response

$$\frac{1}{2^n} (E_Q^*)^{m_{-Q}} \cdots (E_1^*)^{m_{-1}} (E_1)^{m_1} \cdots (E_Q)^{m_Q}$$

$$\cdot H_n(\underbrace{f_{-Q}, \ldots, f_{-Q}}_{m_{-Q}}, \ldots, \underbrace{f_{-1}, \ldots, f_{-1}}_{m_{-1}}, \underbrace{f_1, \ldots, f_1}_{m_1}, \ldots, \underbrace{f_Q, \ldots, f_Q}_{m_Q})$$

$$\cdot \exp\left\{j2\pi[(m_1 - m_{-1})f_1 + \cdots + (m_Q - m_{-Q})f_Q] t\right\}.$$

Let their sum be denoted by $y_n(t; \mathbf{m})$. Therefore,

$$y_n(t; \mathbf{m}) = \frac{(n; \mathbf{m})}{2^n} (E_Q^*)^{m_{-Q}} \cdots (E_1^*)^{m_{-1}} (E_1)^{m_1} \cdots (E_Q)^{m_Q}$$

$$\cdot H_n(\underbrace{f_{-Q}, \ldots, f_{-Q}}_{m_{-Q}}, \ldots, \underbrace{f_{-1}, \ldots, f_{-1}}_{m_{-1}}, \underbrace{f_1, \ldots, f_1}_{m_1}, \ldots, \underbrace{f_Q, \ldots, f_Q}_{m_Q})$$

$$\cdot \exp\left[j2\pi f_{\mathbf{m}} t\right]. \tag{4-15}$$

The nth-order portion of $y(t)$ can now be written as

$$y_n(t) = \sum_{\mathbf{m}} y_n(t; \mathbf{m}) \tag{4-16}$$

where the summation over $\mathbf{m}$ is defined to be

$$\sum_{\mathbf{m}} = \sum_{m_{-Q}=0}^{n} \cdots \sum_{m_{-1}=0}^{n} \sum_{m_1=0}^{n} \cdots \sum_{m_Q=0}^{n}, \tag{4-17}$$

$$\tag{4-13}$$

and the equation number, (4-13), appended below the summation signs indicates that only terms for which the indices sum to n are included in the $2Q$-fold summation.

When the excitation consists of Q sinusoidal inputs, as given by (4-5), it can be shown that the summation in (4-16) for obtaining $y_n(t)$ extends over

$$M = \frac{(2Q + n - 1)!}{n!(2Q - 1)!} \tag{4-18}$$

distinct vectors $\mathbf{m}$. Each distinct vector corresponds to a different frequency mix.

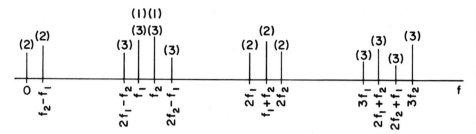

Fig. 4.2.   Line spectra of positive frequency terms listed in Table 4.2

Example 4.4

For a two-tone input (i.e., $Q = 2$), there are four, ten, and twenty different frequency mixes, respectively, contained in $y_1(t)$, $y_2(t)$, and $y_3(t)$. The corresponding responses that are summed in (4-16) are tabulated in Table 4.2 for $n = 1$, 2, and 3. Equation (4-15) was used to evaluate the individual responses appearing in the table. Note that both negative and positive intermodulation frequencies are present and that the complex amplitude of each negative frequency term is the complex conjugate of the corresponding positive frequency term. Also, note the presence of dc terms in the second-order portion of the response.

A pictorial representation of the terms tabulated in Table 4.2, as would be observed in a spectrum analyzer, is shown in Fig. 4.2. The order associated with the frequency mixes leading to each intermodulation frequency is indicated above each line. Only lines corresponding to positive frequency terms are included in the figure.

Because the output of a real system is real for real inputs, $y_n(t)$ is real and the complex terms in (4-16) appear in conjugate pairs. Let $\mathbf{m}'$ denote the frequency mix resulting in the output frequency $-f_{\mathbf{m}}$. As was shown in Chapter 3, $\mathbf{m}'$ is obtained from $\mathbf{m}$ by replacing each value of m_k by m_{-k}. Hence

$$\mathbf{m}' = (m_Q, \dots, m_1, m_{-1}, \dots, m_{-Q}). \tag{4-19}$$

Since $(n; \mathbf{m}') = (n; \mathbf{m})$, it follows that

$$y_n(t; \mathbf{m}') = \frac{(n; \mathbf{m})}{2^n} (E_Q)^{m_{-Q}} \cdots (E_1)^{m_{-1}} (E_1^*)^{m_1} \cdots (E_Q^*)^{m_Q}$$

$$\cdot H_n^*(\underbrace{f_{-Q}, \dots, f_{-Q}}_{m_{-Q}}, \dots, \underbrace{f_{-1}, \dots, f_{-1}}_{m_{-1}}, \underbrace{f_1, \dots, f_1}_{m_1}, \dots, \underbrace{f_Q, \dots, f_Q}_{m_Q})$$

$$\cdot \exp\left[-j2\pi f_{\mathbf{m}} t\right]. \tag{4-20}$$

TABLE 4.2.   Responses Summed in (4-16) to Obtain $y_n(t)$.

m $(m_{-2}, m_{-1}, m_1, m_2)$	Frequency Mix	Response
$n = 1$		
$(1, 0, 0, 0)$	$(-f_2)$	$0.5 E_2^* H_1(-f_2) \exp(-j2\pi f_2 t)$
$(0, 1, 0, 0)$	$(-f_1)$	$0.5 E_1^* H_1(-f_1) \exp(-j2\pi f_1 t)$
$(0, 0, 1, 0)$	(f_1)	$0.5 E_1 H_1(f_1) \exp(j2\pi f_1 t)$
$(0, 0, 0, 1)$	(f_2)	$0.5 E_2 H_1(f_2) \exp(j2\pi f_2 t)$
$n = 2$		
$(1, 1, 0, 0)$	$(-f_2 - f_1)$	$0.5 E_1^* E_2^* H_2(-f_2, -f_1) \exp[-j2\pi(f_1 + f_2)t]$
$(0, 1, 1, 0)$	$(-f_1 + f_1)$	$0.5 |E_1|^2 H_2(-f_1, f_1)$
$(0, 0, 1, 1)$	$(f_1 + f_2)$	$0.5 E_1 E_2 H_2(f_1, f_2) \exp[j2\pi(f_1 + f_2)t]$
$(1, 0, 0, 1)$	$(-f_2 + f_2)$	$0.5 |E_2|^2 H_2(-f_2, f_2)$
$(1, 0, 1, 0)$	$(-f_2 + f_1)$	$0.5 E_1 E_2^* H_2(-f_2, f_1) \exp[j2\pi(-f_2 + f_1)t]$
$(0, 1, 0, 1)$	$(-f_1 + f_2)$	$0.5 E_1^* E_2 H_2(-f_1, f_2) \exp[j2\pi(-f_1 + f_2)t]$
$(2, 0, 0, 0)$	$(-f_2 - f_2)$	$0.25 (E_2^*)^2 H_2(-f_2, -f_2) \exp[-j2\pi(2f_2)t]$
$(0, 2, 0, 0)$	$(-f_1 - f_1)$	$0.25 (E_1^*)^2 H_2(-f_1, -f_1) \exp[-j2\pi(2f_1)t]$
$(0, 0, 2, 0)$	$(f_1 + f_1)$	$0.25 E_1^2 H_2(f_1, f_1) \exp[j2\pi(2f_1)t]$
$(0, 0, 0, 2)$	$(f_2 + f_2)$	$0.25 E_2^2 H_2(f_2, f_2) \exp[j2\pi(2f_2)t]$
$n = 3$		
$(1, 1, 1, 0)$	$(-f_2 - f_1 + f_1)$	$0.75 E_2^* |E_1|^2 H_3(-f_2, -f_1, f_1) \exp[-j2\pi f_2 t]$
$(0, 1, 1, 1)$	$(-f_1 + f_1 + f_2)$	$0.75 |E_1|^2 E_2 H_3(-f_1, f_1, f_2) \exp[j2\pi f_2 t]$
$(1, 0, 1, 1)$	$(-f_2 + f_1 + f_2)$	$0.75 E_1 |E_2|^2 H_3(-f_2, f_1, f_2) \exp[j2\pi f_1 t]$
$(1, 1, 0, 1)$	$(-f_2 - f_1 + f_2)$	$0.75 E_1^* |E_2|^2 H_3(-f_2, -f_1, f_2) \exp[-j2\pi f_1 t]$
$(2, 1, 0, 0)$	$(-f_2 - f_2 - f_1)$	$0.375 (E_2^*)^2 E_1^* H_3(-f_2, -f_2, -f_1) \exp[-j2\pi(2f_2 + f_1)t]$
$(0, 2, 1, 0)$	$(-f_1 - f_1 + f_1)$	$0.375 |E_1|^2 E_1^* H_3(-f_1, -f_1, f_1) \exp[-j2\pi f_1 t]$
$(0, 0, 2, 1)$	$(f_1 + f_1 + f_2)$	$0.375 E_1^2 E_2 H_3(f_1, f_1, f_2) \exp[j2\pi(2f_1 + f_2)t]$
$(1, 0, 0, 2)$	$(-f_2 + f_2 + f_2)$	$0.375 |E_2|^2 E_2 H_3(-f_2, f_2, f_2) \exp[j2\pi f_2 t]$
$(2, 0, 1, 0)$	$(-f_2 - f_2 + f_1)$	$0.375 (E_2^*)^2 E_1 H_3(-f_2, -f_2, f_1) \exp[j2\pi(-2f_2 + f_1)t]$
$(0, 2, 0, 1)$	$(-f_1 - f_1 + f_2)$	$0.375 (E_1^*)^2 E_2 H_3(-f_1, -f_1, f_2) \exp[j2\pi(-2f_1 + f_2)t]$
$(1, 0, 2, 0)$	$(-f_2 + f_1 + f_1)$	$0.375 E_2^* E_1^2 H_3(-f_2, f_1, f_1) \exp[j2\pi(2f_1 - f_2)t]$
$(0, 1, 0, 2)$	$(-f_1 + f_2 + f_2)$	$0.375 E_1^* E_2^2 H_3(-f_1, f_2, f_2) \exp[j2\pi(2f_2 - f_1)t]$
$(2, 0, 0, 1)$	$(-f_2 - f_2 + f_2)$	$0.375 |E_2|^2 E_2^* H_3(-f_2, -f_2, f_2) \exp[-j2\pi f_2 t]$
$(1, 2, 0, 0)$	$(-f_2 - f_1 - f_1)$	$0.375 E_2^* (E_1^*)^2 H_3(-f_2, -f_1, -f_1) \exp[-j2\pi(2f_1 + f_2)t]$
$(0, 1, 2, 0)$	$(-f_1 + f_1 + f_1)$	$0.375 |E_1|^2 E_1 H_3(-f_1, f_1, f_1) \exp[j2\pi f_1 t]$
$(0, 0, 1, 2)$	$(f_1 + f_2 + f_2)$	$0.375 E_1 E_2^2 H_3(f_1, f_2, f_2) \exp[j2\pi(f_1 + 2f_2)t]$
$(3, 0, 0, 0)$	$(-f_2 - f_2 - f_2)$	$0.125 (E_2^*)^3 H_3(-f_2, -f_2, -f_2) \exp[-j2\pi(3f_2)t]$
$(0, 3, 0, 0)$	$(-f_1 - f_1 - f_1)$	$0.125 (E_1^*)^3 H_3(-f_1, -f_1, -f_1) \exp[-j2\pi(3f_1)t]$
$(0, 0, 3, 0)$	$(f_1 + f_1 + f_1)$	$0.125 E_1^3 H_3(f_1, f_1, f_1) \exp[j2\pi(3f_1)t]$
$(0, 0, 0, 3)$	$(f_2 + f_2 + f_2)$	$0.125 E_2^3 H_3(f_2, f_2, f_2) \exp[j2\pi(3f_2)t]$

Comparison of (4-20) with (4-15) reveals $y_n(t; m')$ to be the complex conjugate of $y_n(t; m)$. Let their sum be denoted by $\hat{y}_n(t; m)$. In addition, let the nth-order nonlinear transfer function be expressed in polar form as

$$H_n(f_1, \ldots, f_n) = |H_n(f_1, \ldots, f_n)| \exp [j\psi_n(f_1, \ldots, f_n)]. \quad (4\text{-}21)$$

Recalling that θ_q is the phase angle of E_q, $\hat{y}_n(t; m)$ becomes

$$\hat{y}_n(t; m) = y_n(t; m) + y_n(t; m')$$

$$= y_n(t; m) + y_n^*(t; m) = 2 \operatorname{Re} \{y_n(t; m)\}$$

$$= \frac{(n; m)}{2^{n-1}} |E_1|^{(m_1 + m_{-1})} \cdots |E_Q|^{(m_Q + m_{-Q})} |H_n(m)|$$

$$\cdot \cos [2\pi f_m t + \theta_m + \psi_n(m)] \quad (4\text{-}22)$$

where, by definition,

$$H_n(m) = |H_n(m)| \exp [j\psi_n(m)]$$

$$= H_n(\underbrace{f_{-Q}, \ldots, f_{-Q}}_{m_{-Q}}, \ldots, \underbrace{f_{-1}, \ldots, f_{-1}}_{m_{-1}}, \underbrace{f_1, \ldots, f_1}_{m_1}, \ldots, \underbrace{f_Q, \ldots, f_Q}_{m_Q})$$

$$(4\text{-}23)$$

and

$$\theta_m = \sum_{\substack{k=-Q \\ k \neq 0}}^{Q} m_k \theta_k = (m_1 - m_{-1})\theta_1 + \cdots + (m_Q - m_{-Q})\theta_Q. \quad (4\text{-}24)$$

Example 4.5

For a two-tone input,

$$\hat{y}_3(t; 0, 0, 1, 2) = \tfrac{3}{4} |E_1| |E_2|^2 |H_3(f_1, f_2, f_2)|$$

$$\cdot \cos [2\pi(f_1 + 2f_2)t + \theta_1 + 2\theta_2 + \psi_3(f_1, f_2, f_2)].$$

$$(4\text{-}25)$$

In general, given any frequency mix that does not result in dc, (4-22) is readily used to evaluate the corresponding sinusoidal component at the system output.

Observe that (4-22) is not appropriate for evaluation of dc components in the response. Such terms arise when $f_m = 0$ or, equivalently, when $m_k = m_{-k}$ for each value of k. Since the summation in (4-16) extends over distinct frequency mixes, only one of the two identical vectors $\mathbf{m}$ and $\mathbf{m}'$ is included. Consequently, (4-22) is not applicable for dc. As in Chapter 3, let $\mathbf{m}^0$ represent a frequency mix for which $f_\mathbf{m} = 0$. From (4-15) it follows that

$$y_n(t; \mathbf{m}^0) = \frac{(n; \mathbf{m}^0)}{2^n} \, [|E_1|^2]^{m_1} \cdots [|E_Q|^2]^{m_Q} H_n(\mathbf{m}^0). \qquad (4\text{-}26)$$

Because $\mathbf{m}^0$ equals $(\mathbf{m}^0)'$ and, therefore,

$$H_n(\mathbf{m}^0) = H_n[(\mathbf{m}^0)'] = H_n^*(\mathbf{m}^0), \qquad (4\text{-}27)$$

the nonlinear transfer functions associated with dc outputs are real quantities. Therefore, it is unnecessary to add a complex conjugate term as was done in (4-22). We conclude that (4-22) is used for sinusoidal outputs while dc outputs are determined from (4-26).

Example 4.6

Assume the weakly nonlinear system of Fig. 4.1 is excited by a two-tone input. The vector $\mathbf{m} = (0, 1, 0, 1)$ yields the sinusoidal response

$$\hat{y}_2(t; 0, 1, 0, 1) = |E_1| \, |E_2| \, |H_2(-f_1, f_2)|$$
$$\cdot \cos[2\pi(f_2 - f_1)t + \theta_2 - \theta_1 + \psi_2(-f_1, f_2)] \qquad (4\text{-}28)$$

while the vector $\mathbf{m} = (1, 0, 0, 1)$ yields the dc output

$$y_2(t; 1, 0, 0, 1) = \tfrac{1}{2}|E_2|^2 H_2(-f_2, f_2). \qquad (4\text{-}29)$$

Total Response at a Particular Intermodulation Frequency: When using (4-22) and (4-26), it is important to realize that several different frequency mixes are capable of contributing to the response at a particular output frequency. For very small input amplitudes, the response is dominated by the lowest order terms. However, with larger input amplitudes, higher order terms become important. In general, we denote the total response at frequency f by $y(t; f)$.

Example 4.7

To illustrate the above remarks, consider a weakly nonlinear circuit for which the highest order response is of fifth order (i.e., $N = 5$ in Fig. 4.1). Let the input consist of two sinusoids at frequencies f_1 and f_2 (i.e., $Q = 2$). The output contains a frequency component at $2f_2 - f_1$ that is generated by the three frequency mixes represented by the vectors $\mathbf{m} = (0, 1, 0, 2)$, $\mathbf{m} = (1, 1, 0, 3)$, and

$\mathbf{m} = (0, 2, 1, 2)$. Specifically,

$$y(t; 2f_2 - f_1) = \hat{y}_3(t; 0, 1, 0, 2) + \hat{y}_5(t; 1, 1, 0, 3) + \hat{y}_5(t; 0, 2, 1, 2)$$

$$= 0.75 \, |E_1| \, |E_2|^2 \, |H_3(-f_1, f_2, f_2)|$$

$$\cdot \cos \left[2\pi(2f_2 - f_1)t + 2\theta_2 - \theta_1 + \psi_3(-f_1, f_2, f_2) \right]$$

$$+ 1.25 \, |E_1| \, |E_2|^4 \, |H_5(-f_2, -f_1, f_2, f_2, f_2)|$$

$$\cdot \cos \left[2\pi(2f_2 - f_1)t + 2\theta_2 - \theta_1 + \psi_5(-f_2, -f_1, f_2, f_2, f_2) \right]$$

$$+ 1.875 \, |E_1|^3 \, |E_2|^2 \, |H_5(-f_1, -f_1, f_1, f_2, f_2)|$$

$$\cdot \cos \left[2\pi(2f_2 - f_1)t + 2\theta_2 - \theta_1 + \psi_5(-f_1, -f_1, f_1, f_2, f_2) \right].$$

$$(4\text{-}30)$$

In general, the angles $\psi_3(-f_1, f_2, f_2)$, $\psi_5(-f_2, -f_1, f_2, f_2, f_2)$, and $\psi_5(-f_1, -f_1, f_1, f_2, f_2)$ are not equal. The various terms in (4-30) combine as shown in the phasor diagram of Fig. 4.3. Note that both magnitudes and phase angles of the nonlinear transfer functions are important in determining the total response at a particular output frequency. For very small values of $|E_1|$ and $|E_2|$, the

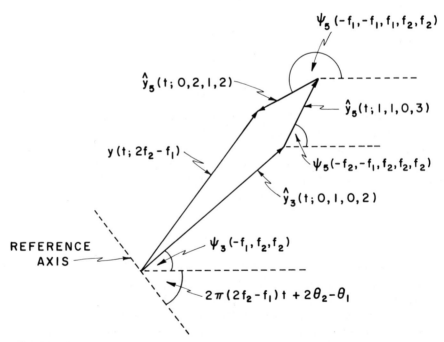

Fig. 4.3.   Phasor diagram illustrating total response of frequency component at $(2f_2 - f_1)$.

response is given mainly by the third-order term. However, as the inputs are increased in magnitude, the fifth-order terms become significant.

The Classical Model–A Special Case: The similarity between the developments of Chapters 3 and 4 is more than coincidental. The classical model is, in fact, a special case of the nonlinear transfer function approach. Comparison of (4-15) with (3-36) reveals that, when a weakly nonlinear system is structured in the form of the classical model, the nth-order nonlinear transfer function corresponding to the frequency mix represented by the vector $\mathbf{m}$ is related to a_n, $H(f)$, and $K(f)$ of the classical model by the relation

$$H_n(\mathbf{m}) = H_n(\underbrace{f_{-Q}, \ldots, f_{-Q}}_{m_{-Q}}, \ldots, \underbrace{f_{-1}, \ldots, f_{-1}}_{m_{-1}}, \underbrace{f_1, \ldots, f_1}_{m_1}, \ldots, \underbrace{f_Q, \ldots, f_Q}_{m_Q})$$

$$= a_n [H^*(f_Q)]^{m_{-Q}} \cdots [H^*(f_1)]^{m_{-1}} [H(f_1)]^{m_1} \cdots [H(f_Q)]^{m_Q} K(f_\mathbf{m}).$$

$$(4\text{-}31)$$

However, in Appendix A it is shown that (4-7) is derived without the need for many of the assumptions required by the classical model. As a result, (4-22) and (4-26) are applicable in situations for which (3-42) and (3-46) are not. One important difference between the two approaches is illustrated by (3-49) and (4-30). In (3-49) the three terms that combine to yield $y(t; 2f_2 - f_1)$ are either in phase or $180°$ out of phase depending upon the signs of the coefficients a_3 and a_5. On the other hand, from (4-30), the three terms may combine at various angles relative to each other, as shown in Fig. 4.3.

Output Powers as a Function of Input Powers: As was the case in Chapter 3, the output of a nonlinear system is frequently expressed in terms of power. Assuming both $x(t)$ and $y(t)$ to be voltages, it is apparent from (4-7) that the nth-order nonlinear transfer function has the dimensions of volts$^{(1-n)}$. From (4-22) the amplitude of the sinusoidal response corresponding to the vector $\mathbf{m}$ is

$$|E_\mathbf{m}| = \frac{(n; \mathbf{m})}{2^{n-1}} |E_1|^{(m_1 + m_{-1})} \cdots |E_Q|^{(m_Q + m_{-Q})} |H_n(\mathbf{m})| \qquad (4\text{-}32)$$

where $H_n(\mathbf{m})$ is given by (4-23). The average power dissipated by this component in a load having conductance $G_L(f)$ is

$$P_L(f_\mathbf{m}) = \frac{1}{2} |E_\mathbf{m}|^2 G_L(f_\mathbf{m})$$

$$= \frac{(n; \mathbf{m})^2}{2^{2n-1}} [|E_1|^2]^{(m_1 + m_{-1})} \cdots [|E_Q|^2]^{(m_Q + m_{-Q})} |H_n(\mathbf{m})|^2 G_L(f_\mathbf{m}).$$

$$(4\text{-}33)$$

Similarly, if the input conductance of the nonlinear system is $G_{in}(f)$, the average power of the input tone at frequency f_q is

$$p_S(f_q) = \tfrac{1}{2} |E_q|^2 G_{in}(f_q) \tag{4-34}$$

from which it follows that

$$|E_q|^2 = \frac{2p_S(f_q)}{G_{in}(f_q)}. \tag{4-35}$$

Finally, with the aid of (4-13), use of (4-35) in (4-33) results in

$$p_L(f_m) = [p_S(f_1)]^{(m_1 + m_{-1})} \cdots [p_S(f_Q)]^{(m_Q + m_{-Q})} c(m) \tag{4-36}$$

where

$$c(m) = \frac{(n;m)^2}{2^{n-1}} \frac{|H_n(m)|^2 G_L(f_m)}{[G_{in}(f_1)]^{(m_1 + m_{-1})} \cdots [G_{in}(f_Q)]^{(m_Q + m_{-Q})}}. \tag{4-37}$$

As in Chapter 3, the dimensions of the intermodulation multiplier $c(m)$ are watts$^{(1-n)}$. In fact, (4-37) reduces to (3-59) for the special case in which the nth-order nonlinear transfer function is given by (4-31). $p_L(f_m)$ is expressed in decibels by introducing the reference power p_r, as was done in (3-61). Then

$$P_L(f_m) = (m_1 + m_{-1})P_S(f_1) + \cdots + (m_Q + m_{-Q})P_S(f_Q) + C(m) \tag{4-38}$$

where $P_L(f_m)$, $P_S(f_q)$, and $C(m)$ are defined in (3-63). Note that (3-60) and (4-36) are identical in form as are (3-62) and (4-38). This serves, once again, to emphasize the similarity between the classical and nonlinear transfer function approaches.

4.3 NONLINEAR TRANSFER FUNCTIONS FOR TWO TRANSISTOR AMPLIFIERS

The previous discussion was carried out in rather general terms. At this point, it is reasonable to ask, "Do nonlinear transfer functions really exist and, if so, what do they look like?" The first step towards answering these questions is provided in this section. As shown in Chapter 5, an analytical technique for determining nonlinear functions is available. However, closed-form analytical expressions are difficult to achieve. Consequently, numerical values are usually obtained by programming the analytical procedure on a digital computer. In this section we present theoretical and experimental results for two different transistor amplifiers. The data presented are taken from the book by Graham and Ehrman (RADC-TR-73-178). Our objective is to demonstrate the close relationship between the linear frequency response of a circuit and the frequency behavior of its nonlinear transfer functions.

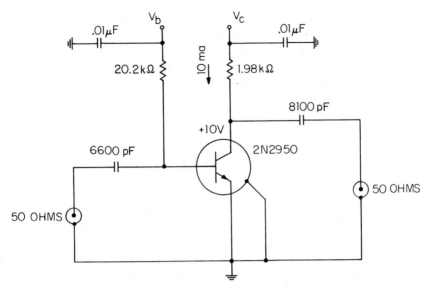

Fig. 4.4.   Schematic diagram of single-stage common-emitter amplifier.

Single-Stage Common-Emitter Amplifier: Figure 4.4 shows the schematic diagram of a relatively simple single-stage common-emitter amplifier. As indicated in the diagram, the amplifier was tested with both the source and load impedances equal to 50 Ω.

Linear Transfer Function: The magnitude of the linear transfer function is plotted in Fig. 4.5. The linear frequency response is seen to be fairly broadband with the lower and higher 3-dB cutoff frequencies at 300 kHz and 9 MHz,

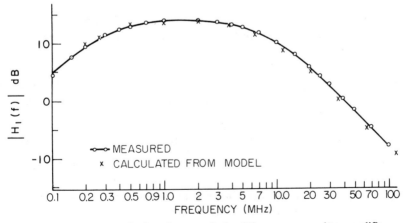

Fig. 4.5.   Linear transfer function magnitude of the common-emitter amplifier.

respectively. Excellent agreement is found to exist between experimental and theoretical values over the three decade frequency span extending from 100 kHz to 100 MHz. In Chapter 5 it is shown that accurate modeling of the linear portion of a circuit is necessary if the higher order nonlinear transfer functions are to be successfully predicted. When the behavior of a linear transfer function is approximated over a frequency range by

$$H_1(f) \approx \frac{K_1}{f^p}, \qquad (4\text{-}39)$$

where K_1 is a constant and p is a positive integer, it follows that

$$20 \log_{10} |H_1(f)| \approx 20 \log_{10} |K_1| - 20p \log_{10} |f|. \qquad (4\text{-}40)$$

The magnitude of the linear transfer function then falls off at a rate of $20p$ dB/decade. From Fig. 4.5 it is seen that the high-frequency linear response falls off at 20 dB/decade. This suggests that $H_1(f)$ behaves inversely proportional to f for high frequencies.

Second-Order Nonlinear Transfer Function: In general, the second-order nonlinear transfer function is a function of the two frequency variables f_1 and f_2. A convenient method for measuring the transfer function magnitude is to excite the amplifier with two tones of known power at frequencies f_1 and f_2. The output power at one of the four intermodulation frequencies $\pm f_1 \pm f_2$ is then measured. Plus and minus signs are used to indicate the possible intermodulation frequencies since the input signal contains frequency components at $-f_2, -f_1, f_1,$ and f_2. Knowledge of the input and output powers along with the input and load impedances is sufficient to determine $|H_2(\pm f_1, \pm f_2)|$, as can be seen from (4-36) and (4-37). In order to treat $|H_2(\pm f_1, \pm f_2)|$ as though it were a function of a single variable, the constraint

$$\pm f_1 \pm f_2 = \text{constant} \qquad (4\text{-}41)$$

may be imposed. Then, as f_2 is swept in frequency, f_1 is determined from (4-41). This enables $|H_2(\pm f_1, \pm f_2)|$ to be plotted as a function of the single variable f_2.

Such a plot is presented in Fig. 4.6 for the magnitude of the second-order nonlinear transfer function of the common-emitter amplifier where the intermodulation component at $f_2 - f_1$ is of interest. The frequency variable f_2 is swept from 3 to 50 MHz under the constraint

$$f_2 - f_1 = 0.5 \text{ MHz}. \qquad (4\text{-}42)$$

Therefore, given f_2, f_1 is always chosen to be 0.5 MHz less than f_2. Also, the intermodulation frequency remains fixed as the frequencies are varied. For this case, $|H_2(-f_1, f_2)|$ has the same shape as the linear response above 3 MHz, decreasing monotonically with increasing frequency. However, at the upper cutoff

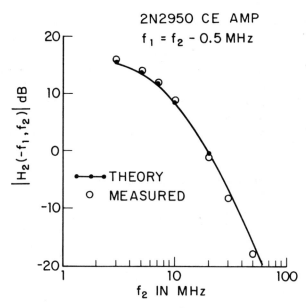

Fig. 4.6. Second-order nonlinear transfer function magnitude of the common-emitter amplifier.

frequency of 9 MHz, $|H_2(-f_1, f_2)|$ is down by 6 dB from its maximum value whereas the linear response is down by 3 dB. In addition, in the high-frequency region, $|H_2(-f_1, f_2)|$ falls off with a slope of 40 dB/decade whereas the linear response decreases by 20 dB/decade. This suggests that, under the constraint of (4-42), the high-frequency behavior of $H_2(-f_1, f_2)$ is approximated by

$$H_2(-f_1, f_2) \approx \frac{K_2}{f_1 f_2} \qquad (4\text{-}43)$$

where K_2 is a constant. Then

$$20 \log_{10} |H_2(-f_1, f_2)| \approx 20 \log_{10} |K_2| - 20 \log_{10} |f_1| - 20 \log_{10} |f_2|.$$

$$(4\text{-}44)$$

Since f_1 approximately equals f_2 during the frequency sweep, it follows that $|H_2(-f_1, f_2)|$ falls off at a rate of 40 dB/decade. This is not too surprising because the linear frequency response causes each of the input tones at f_1 and f_2 to be equally attenuated as the two closely spaced frequencies are varied. The predicted and measured responses are seen to be in good agreement over the entire frequency range. To obtain the predicted response, observe that it was necessary to model the nonlinearities inherent with the transistor. This important topic is discussed in Chapter 7.

Third-Order Nonlinear Transfer Function: The third-order nonlinear transfer function is also conveniently examined using a two-tone test. Assuming sinusoidal inputs of known powers at f_1 and f_2, a measurement is made of the output power in an intermodulation component involving one of the possible third-order frequency mixes, $(\pm f_1 \pm f_2 \pm f_2)$. The corresponding nonlinear transfer function is $H_3(\pm f_1, \pm f_2, \pm f_2)$. Its magnitude is determined by inserting the known powers, input conductances, and load conductance into (4-36) and (4-37). Finally, introducing a constraint between f_1 and f_2 enables $|H_3(\pm f_1, \pm f_2, \pm f_2)|$ to be plotted as a function of f_2 with f_1 being determined by the constraint.

Focusing attention on the intermodulation component at $2f_2 - f_1$, the measured and predicted magnitudes of the third-order nonlinear transfer function for the common-emitter amplifier are presented in Fig. 4.7. The input frequencies are again constrained by (4-42). Therefore, the measured response is at the

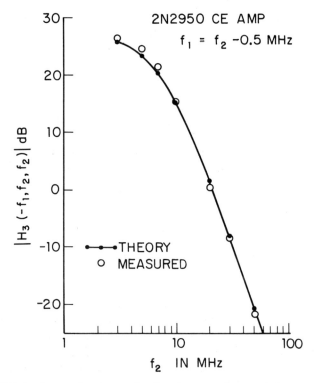

Fig. 4.7.   Third-order nonlinear transfer function magnitude of the common-emitter amplifier.

intermodulation frequency given by

$$2f_2 - f_1 = f_2 + (f_2 - f_1) = f_2 + 0.5 \text{ MHz.} \tag{4-45}$$

In contrast to the second-order case, where the intermodulation frequency remained constant, the intermodulation frequency now varies linearly with f_2. In fact, the input tones and the intermodulation component are all at essentially the same frequency since their frequencies are $f_2 - 0.5, f_2$, and $f_2 + 0.5$ MHz. Although $|H_3(-f_1, f_2, f_2)|$ has the same general shape as the linear response, it is down by 9 dB at 9 MHz, the upper cutoff frequency of the linear transfer function, and falls off at a rate of 60 dB/decade in the high-frequency region. This suggests, for this particular intermodulation component and for the constraint of (4-42), that the high-frequency behavior of $H_3(-f_1, f_2, f_2)$ may be approximated by

$$H_3(-f_1, f_2, f_2) \approx \frac{K_3}{f_1 (f_2)^2} \tag{4-46}$$

where K_3 is a constant. It follows that

$$20 \log_{10} |H_3(-f_1, f_2, f_2)| \approx 20 \log_{10} |K_3| - 20 \log_{10} |f_1| - 40 \log_{10} |f_2|. \tag{4-47}$$

Hence, because f_1 and f_2 are approximately equal over the range of higher frequencies, $|H_3(-f_1, f_2, f_2)|$ falls off at a rate of 60 dB/decade. The measured and predicted values are seen to be in good agreement over the frequency span extending from 2 to 60 MHz.

Two-Stage Tuned Amplifier: When strong frequency-dependent interactions exist between the nonlinearities and linear network elements of a circuit, the nonlinear transfer functions are likely to be considerably more complicated than those associated with the common-emitter amplifier. Nevertheless, certain features can still be explained in terms of the linear frequency response. We illustrate these remarks by examining the nonlinear transfer functions of the two-stage tuned amplifier shown in Fig. 4.8. The first stage is a common-emitter amplifier driving a low-pass interstage network. The second stage is a common-base amplifier driving a tuned load.

Linear Transfer Function: A plot of the linear transfer function magnitude is presented in Fig. 4.9. For source and load impedances of 50 Ω the amplifier is tuned to 19.75 MHz with a midband gain of 28 dB. The 3-dB bandwidth is about 2 MHz. It is seen that prediction and measurement are in excellent agreement over the frequency range from 6 to 70 MHz, especially in view of the fact that the gain varies in excess of 63 dB over this range. There is some divergence between measured and predicted performance below 6 MHz.

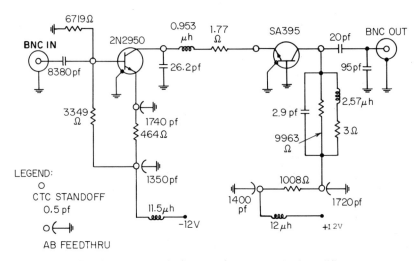

Fig. 4.8.   Schematic diagram of two-stage tuned amplifier.

Second-Order Nonlinear Transfer Function: The second-order nonlinear transfer function of the two-stage amplifier is investigated by using a two-tone test to generate an intermodulation frequency that remains constant at the tuned frequency of 19.75 MHz while the input frequencies, f_1 and f_2, are varied. Therefore, for f_1 below the tuned frequency f_1 and f_2 are constrained by the relation

$$f_1 + f_2 = 19.75 \text{ MHz} \tag{4-48}$$

and, for f_1 above the tuned frequency, the constraint is given by

$$f_1 - f_2 = 19.75 \text{ MHz}. \tag{4-49}$$

Computed and experimentally determined values of $|H_2(f_1, \pm f_2)|$ are plotted in Fig. 4.10 as a function of f_1 for the frequency range extending from 750 kHz to 80 MHz. The transfer function varies by some 35 dB over this range, peaking at 10, 30, and 55 MHz, and having minima at about 20 and 42 MHz. The symmetry about the peak at 10 MHz is readily explained. Note that 19.75 MHz/2 = 9.875 MHz $\approx$ 10 MHz. Since f_1 and f_2 are constrained by (4-48) in this range, the values tabulated in Table 4.3 apply. Because it is immaterial as to which input signal is assigned the frequency f_1, $H_2(f_1, f_2)$ is "symmetrical" about 9.875 MHz at the frequency pairs indicated in Table 4.3. The "symmetry" is upset by the null occurring at 19.75 MHz. When f_1 = 19.75 MHz, f_2 must equal zero. However, the circuit of Fig. 4.8 has a transmission zero at dc. Therefore, $H_2(f_1, f_2) = 0$ when f_1 = 19.75 MHz. The maxima at 30 and 55 MHz are difficult to explain. Nor is it obvious why $|H_2(f_1, f_2)|$ rises with a slope of 20 dB/

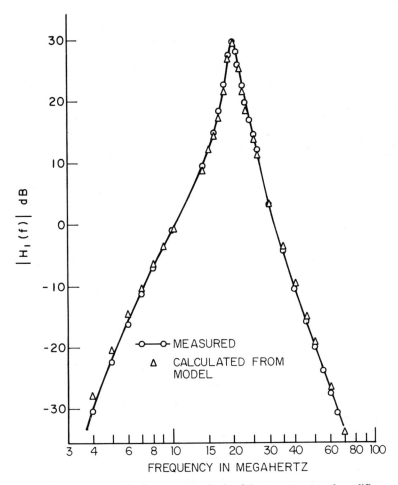

Fig. 4.9.   Linear transfer function magnitude of the two-stage tuned amplifier.

decade at low frequencies. We speculate that the null at 42 MHz is due to a transmission zero caused by a parallel resonance associated with the inductor in the interstage network. The agreement between predicted and measured values is good over the entire frequency range, although larger discrepancies exist at the lower frequencies. This is due to the greater inaccuracy in modeling the linear frequency response at low frequencies.

Third-Order Nonlinear Transfer Function: A two-tone test is also used to investigate the third-order nonlinear transfer function of the tuned amplifier. Once again, the input frequencies are varied such that a third-order intermodu-

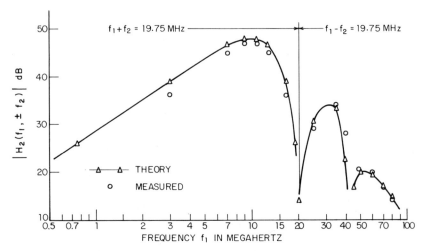

Fig. 4.10.   Second-order nonlinear transfer function magnitude of the two-stage tuned amplifier.

lation component is generated at the tuned frequency. For f_2 below 9.875 MHz, the frequency constraint is given by

$$2f_2 + f_1 = 19.75 \text{ MHz} \tag{4-50}$$

while, for f_2 above 9.875 MHz, f_1 is chosen such that

$$2f_2 - f_1 = 19.75 \text{ MHz.} \tag{4-51}$$

The corresponding nonlinear transfer function is $H_3(\pm f_1, f_2, f_2)$. Measured and predicted values of $|H_3(\pm f_1, f_2, f_2)|$ are plotted in Fig. 4.11 as a function of f_2 for frequencies between 5 and 50 MHz. Good agreement between predicted and measured values is obtained throughout this frequency interval in which the magnitude varies over a range of 50 dB. The transfer function has minima at 9.875 and 28 MHz and maxima at 7, 19.75, and 40 MHz. $f_1 = 0$ when $f_2 =$ 9.875 MHz. Hence the null at 9.875 MHz is due to the dc transmission zero of the linear portion of the circuit. When $f_2 = 19.75$ MHz, f_1 also equals 19.75 MHz. Therefore, the transfer function peaks at 19.75 MHz because both input tones are near the resonance of the linear frequency response for f_2 close to the

TABLE 4.3.   Values for f_1 and f_2, Centered About 9.875
MHz, when $f_1 + f_2 = 19.75$ MHz.

f_1(MHz)	8	9	9.875	10.75	11.75
f_2(MHz)	11.75	10.75	9.875	9	8

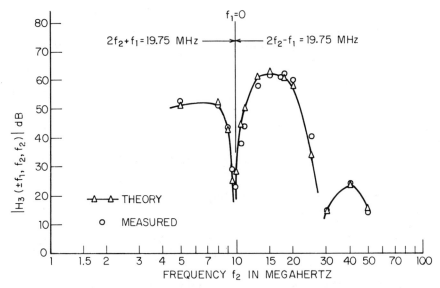

Fig. 4.11. Third-order nonlinear transfer function magnitude of the two-stage tuned amplifier.

tuned frequency. The remaining features of $\left|H_3(\pm f_1, f_2, f_2)\right|$ are not readily explained. Nevertheless, this transfer function does successfully predict the rather complicated frequency dependence of the third-order intermodulation component. Additional insight into nonlinear transfer functions is provided in succeeding chapters.

4.4 CASCADE RELATIONS FOR TWO NONINTERACTING NONLINEAR CIRCUITS

Many useful results that apply to linear transfer functions have their counterparts in the nonlinear transfer function approach. We illustrate this remark by considering the cascade connection of two noninteracting nonlinear systems, G and H, as shown in Fig. 4.12. The overall system is denoted by K.

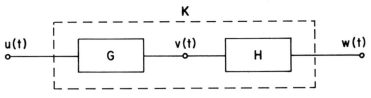

Fig. 4.12. Cascade connection of two noninteracting nonlinear systems.

Example 4.8

As a special case, we first assume the two stages to consist of zero-memory nonlinearities. Hence the input/output relationships for systems G and H, respectively, are given by

$$v(t) = \sum_{i=1}^{\infty} g_i u^i(t) \tag{4-52}$$

and

$$w(t) = \sum_{j=1}^{\infty} h_j v^j(t). \tag{4-53}$$

The objective is to obtain the input/output relationship for system K. This is accomplished by substituting (4-52) into (4-53) which results in

$$w(t) = \sum_{j=1}^{\infty} h_j \left\{ \sum_{i=1}^{\infty} g_i u^i(t) \right\}^j. \tag{4-54}$$

Carrying out the operations indicated in (4-54) and collecting terms up to fourth degree, we obtain

$$w(t) = g_1 h_1 u(t) + (g_2 h_1 + g_1^2 h_2) u^2(t) + (g_3 h_1 + 2 g_1 g_2 h_2 + g_1^3 h_3) u^3(t)$$
$$+ [g_4 h_1 + (2 g_1 g_3 + g_2^2) h_2 + 3 g_1^2 g_2 h_3 + g_1^4 h_4] u^4(t) + \cdots. \tag{4-55}$$

We see that the cascade connection of two zero-memory nonlinearities is, again, a zero-memory nonlinearity. It follows that the input/output relationship for system K is

$$w(t) = \sum_{l=1}^{\infty} k_l u^l(t) \tag{4-56}$$

where the first four coefficients are given by

$$k_1 = g_1 h_1$$
$$k_2 = g_2 h_1 + g_1^2 h_2$$
$$k_3 = g_3 h_1 + 2 g_1 g_2 h_2 + g_1^3 h_3$$
$$k_4 = g_4 h_1 + (2 g_1 g_3 + g_2^2) h_2 + 3 g_1^2 g_2 h_3 + g_1^4 h_4. \tag{4-57}$$

Higher order coefficients are obtained in like manner.

Now let G and H be weakly nonlinear systems that are characterized by their nth-order nonlinear transfer functions $G_n(f_1, \ldots, f_n)$ and $H_n(f_1, \ldots, f_n)$. The two systems are assumed to be noninteracting in the sense that the transfer functions of G remain unchanged before and after the cascade connection, as do those of H. In general, this is not the case. Because of loading by each system on the other, cascading two systems usually causes their transfer functions to change from those that existed before the connection. The cascade connection of two weakly nonlinear systems results in a weakly nonlinear system. Let the nth-order nonlinear transfer function of system K be denoted by $K_n(f_1, \ldots, f_n)$. The problem is to express the nonlinear transfer functions of K in terms of those for G and H.

Example 4.9

We first determine the linear transfer function, $K_1(f_1)$. For this purpose, we choose the input signal to system G to be the single complex exponential

$$u(t) = U_1 \exp[j2\pi f_1 t]. \tag{4-58}$$

$u(t)$, of course, is not a real signal because the complex conjugate term is not included as was done in (4-5). Nevertheless, it is a convenient mathematical signal to use in the analysis. Observe that (4-5) reduces to (4-58) by selecting $Q = 1$, $E_1 = 2U_1$, and by omitting negative frequency terms. $K_1(f_1)$ relates the output component of $w(t)$ at f_1 to the input that is given by (4-58). The output frequency at f_1 occurs only in the linear portion of $w(t)$ which, from (4-7), is

$$w_1(t) = U_1 K_1(f_1) \exp[j2\pi f_1 t]. \tag{4-59}$$

Working our way backwards through the cascade connection, $w_1(t)$ is the linear response of H to the component of $v(t)$ at f_1. If the complex amplitude of this input component is denoted by V_{f_1}, it follows that

$$w_1(t) = V_{f_1} H_1(f_1) \exp[j2\pi f_1 t]. \tag{4-60}$$

However, the component of $v(t)$ at f_1 is the linear response of G to the input $u(t)$. Therefore,

$$V_{f_1} = U_1 G_1(f_1). \tag{4-61}$$

Substituting (4-61) into (4-60), we obtain

$$w_1(t) = U_1 G_1(f_1) H_1(f_1) \exp[j2\pi f_1 t]. \tag{4-62}$$

Finally, by comparison of (4-59) with (4-62), we conclude

$$K_1(f_1) = G_1(f_1) H_1(f_1). \tag{4-63}$$

Equation (4-63) is recognized as the conventional cascade relationship for linear systems.

Example 4.10

We now determine the second-order transfer function $K_2(f_1, f_2)$. Here, it is convenient to choose the input signal to be

$$u(t) = U_1 \exp[j2\pi f_1 t] + U_2 \exp[j2\pi f_2 t] \qquad (4\text{-}64)$$

where f_1 and f_2 are assumed to be positive incommensurable frequencies. Equation (4-64) is a special case of (4-5) in which $Q = 2$, $E_1 = 2U_1$, $E_2 = 2U_2$, and negative frequency terms are omitted. $K_2(f_1, f_2)$ relates the output component of $w(t)$ at $f_1 + f_2$ to the two input components of $u(t)$ at f_1 and f_2. Because of our assumption of positive incommensurable frequencies, the output frequency at $f_1 + f_2$ occurs only in the second-order portion of $w(t)$. From (4-7)

$$w_2(t) = \sum_{q_1=1}^{2} \sum_{q_2=1}^{2} U_{q_1} U_{q_2} K_2(f_{q_1}, f_{q_2}) \exp[j2\pi(f_{q_1} + f_{q_2})t]. \qquad (4\text{-}65)$$

Let $w_2(t; f_1 + f_2)$ denote the sum of the components in $w_2(t)$ at $f_1 + f_2$. Utilizing the symmetry of $K_2(f_1, f_2)$, it follows that

$$w_2(t; f_1 + f_2) = 2U_1 U_2 K_2(f_1, f_2) \exp[j2\pi(f_1 + f_2)t]. \qquad (4\text{-}66)$$

As before, we work our way backwards through the system. The output at $f_1 + f_2$ is due to: (1) the linear response of H to the component of $v(t)$ at $f_1 + f_2$ and (2) the quadratic response of H to the components of $v(t)$ at f_1 and f_2. Let V_{f_k} denote the complex amplitude of the f_k component of $v(t)$. Then $w_2(t; f_1 + f_2)$ can be expressed as

$$w_2(t; f_1 + f_2) = [V_{f_1+f_2} H_1(f_1 + f_2) + 2V_{f_1} V_{f_2} H_2(f_1, f_2)] \exp[j2\pi(f_1 + f_2)t]. \qquad (4\text{-}67)$$

However, in terms of system G,

$$V_{f_1} = U_1 G_1(f_1)$$
$$V_{f_2} = U_2 G_1(f_2)$$
$$V_{f_1+f_2} = 2U_1 U_2 G_2(f_1, f_2). \qquad (4\text{-}68)$$

Substituting (4-68) into (4-67), we obtain

$$w_2(t; f_1 + f_2) = 2U_1 U_2 [G_2(f_1, f_2)H_1(f_1 + f_2)$$
$$+ G_1(f_1)G_1(f_2)H_2(f_1, f_2)] \exp[j2\pi(f_1 + f_2)t]. \qquad (4\text{-}69)$$

Comparing (4-66) with (4-69), it follows that

$$K_2(f_1, f_2) = G_2(f_1, f_2)H_1(f_1 + f_2) + G_1(f_1)G_1(f_2)H_2(f_1, f_2). \qquad (4\text{-}70)$$

Example 4.11

The third-order transfer function $K_3(f_1, f_2, f_3)$ is obtained in a similar manner. The input signal to system G is now assumed to be

$$u(t) = U_1 \exp[j2\pi f_1 t] + U_2 \exp[j2\pi f_2 t] + U_3 \exp[j2\pi f_3 t] \quad (4\text{-}71)$$

where f_1, f_2, and f_3 are positive and incommensurate. Note that (4-5) becomes (4-71) by choosing $Q = 3$, $E_1 = 2U_1$, $E_2 = 2U_2$, $E_3 = 2U_3$, and by omitting negative frequency terms. $K_3(f_1, f_2, f_3)$ is the nonlinear transfer function relating the $f_1 + f_2 + f_3$ component of $w(t)$ to the components of $u(t)$ at f_1, f_2, and f_3. Because the input frequencies are positive and incommensurate, the output frequency at $f_1 + f_2 + f_3$ occurs only in the third-order portion of $w(t)$. From (4-7)

$$w_3(t) = \sum_{q_1=1}^{3} \sum_{q_2=1}^{3} \sum_{q_3=1}^{3} U_{q_1} U_{q_2} U_{q_3} K_3(f_{q_1}, f_{q_2}, f_{q_3}) \exp[j2\pi(f_{q_1} + f_{q_2} + f_{q_3})t].$$

$$(4\text{-}72)$$

Let $w_3(t; f_1 + f_2 + f_3)$ denote the sum of the components in $w_3(t)$ at $f_1 + f_2 + f_3$. Utilizing the symmetry of the third-order nonlinear transfer function, as illustrated in (4-8), $w_3(t; f_1 + f_2 + f_3)$ is given by

$$w_3(t; f_1 + f_2 + f_3) = 6U_1 U_2 U_3 K_3(f_1, f_2, f_3) \exp[j2\pi(f_1 + f_2 + f_3)t].$$

$$(4\text{-}73)$$

With respect to system H, the output at $f_1 + f_2 + f_3$ is due to: (1) the linear response of H to the component of $v(t)$ at $f_1 + f_2 + f_3$, (2) all quadratic responses of H involving components of $v(t)$ at $f_j + f_k$ and f_l where $j \neq k \neq l$, and (3) the cubic response of H to the components of $v(t)$ at f_1, f_2, and f_3. In particular, if V_{f_k} denotes the complex amplitude of the $v(t)$ component at f_k, then

$$w_3(t; f_1 + f_2 + f_3) = [V_{f_1+f_2+f_3} H_1(f_1 + f_2 + f_3) + 2V_{f_1+f_2} V_{f_3} H_2(f_1 + f_2, f_3)$$
$$+ 2V_{f_1+f_3} V_{f_2} H_2(f_1 + f_3, f_2) + 2V_{f_2+f_3} V_{f_1} H_2(f_2 + f_3, f_1)$$
$$+ 6V_{f_1} V_{f_2} V_{f_3} H_3(f_1, f_2, f_3)] \exp[j2\pi(f_1 + f_2 + f_3)t].$$

$$(4\text{-}74)$$

Focusing attention on system G, the complex amplitudes in (4-74) are given by

$$V_{f_1} = U_1 G_1(f_1)$$
$$V_{f_2} = U_2 G_1(f_2)$$
$$V_{f_3} = U_3 G_1(f_3)$$
$$V_{f_1+f_2} = 2U_1 U_2 G_2(f_1, f_2)$$

$$V_{f_1+f_3} = 2U_1U_3G_2(f_1,f_3)$$

$$V_{f_2+f_3} = 2U_2U_3G_2(f_2,f_3)$$

$$V_{f_1+f_2+f_3} = 6U_1U_2U_3G_3(f_1,f_2,f_3). \qquad (4\text{-}75)$$

With the aid of (4-75), (4-74) becomes

$$\begin{aligned} w_3(t;f_1+f_2+f_3) = 6U_1U_2U_3 \{ &G_3(f_1,f_2,f_3)H_1(f_1+f_2+f_3) \\ &+ \tfrac{2}{3}\,[G_1(f_3)G_2(f_1,f_2)H_2(f_1+f_2,f_3) \\ &+ G_1(f_2)G_2(f_1,f_3)H_2(f_1+f_3,f_2) \\ &+ G_1(f_1)G_2(f_2,f_3)H_2(f_2+f_3,f_1)] \\ &+ G_1(f_1)G_1(f_2)G_1(f_3)H_3(f_1,f_2,f_3)\} \\ &\cdot \exp\,[j2\pi(f_1+f_2+f_3)\,t]. \qquad (4\text{-}76) \end{aligned}$$

Equation (4-76) is a rather long and involved expression. It can be simplified by first recognizing that symmetry of $G_2(f_j,f_k)$ enables terms involving the quadratic response of H to be rewritten as

$$\begin{aligned} \tfrac{1}{3}\,[&G_1(f_3)G_2(f_1,f_2)H_2(f_1+f_2,f_3) + G_1(f_3)G_2(f_2,f_1)H_2(f_2+f_1,f_3) \\ &+ G_1(f_2)G_2(f_1,f_3)H_2(f_1+f_3,f_2) + G_1(f_2)G_2(f_3,f_1)H_2(f_3+f_1,f_2) \\ &+ G_1(f_1)G_2(f_2,f_3)H_2(f_2+f_3,f_1) + G_1(f_1)G_2(f_3,f_2)H_2(f_3+f_2,f_1)]. \end{aligned}$$

Given $G_1(f_1)G_2(f_2,f_3)H_2(f_2+f_3,f_1)$, the above sum can be obtained by forming all possible permutations of the frequencies f_1, f_2, and f_3 and adding the results. In general, given a function of n different variables, the n variables can be permuted in $n!$ different ways. Let an overbar denote the arithmetic average of the $n!$ terms generated by all possible permutations of the n variables. For example,

$$\overline{A(f_1,f_2)} = \tfrac{1}{2}\,[A(f_1,f_2) + A(f_2,f_1)]$$

$$\begin{aligned} \overline{A(f_1,f_2,f_3)} = \tfrac{1}{6}\,[&A(f_1,f_2,f_3) + A(f_1,f_3,f_2) + A(f_2,f_1,f_3) \\ &+ A(f_2,f_3,f_1) + A(f_3,f_1,f_2) + A(f_3,f_2,f_1)]. \qquad (4\text{-}77) \end{aligned}$$

In terms of this notation,

$$\begin{aligned} w_3(t;f_1+f_2+f_3) = 6U_1U_2U_3 \{ &G_3(f_1,f_2,f_3)H_1(f_1+f_2+f_3) \\ &+ 2\,\overline{G_1(f_1)G_2(f_2,f_3)H_2(f_2+f_3,f_1)} \\ &+ G_1(f_1)G_1(f_2)G_1(f_3)H_3(f_1,f_2,f_3)\} \\ &\cdot \exp\,[j2\pi(f_1+f_2+f_3)\,t]. \qquad (4\text{-}78) \end{aligned}$$

We conclude from (4-73) and (4-78) that

$$K_3(f_1,f_2,f_3) = G_3(f_1,f_2,f_3)H_1(f_1 + f_2 + f_3)$$
$$+ 2 \overline{G_1(f_1)G_2(f_2,f_3)H_2(f_2 + f_3,f_1)}$$
$$+ G_1(f_1)G_1(f_2)G_1(f_3)H_3(f_1,f_2,f_3). \qquad (4\text{-}79)$$

The same procedure may be used to obtain higher order cascade relationships.

Example 4-12

The reader will find it instructive to verify that

$$K_4(f_1,f_2,f_3,f_4) = G_4(f_1,f_2,f_3,f_4)H_1(f_1 + f_2 + f_3 + f_4)$$
$$+ 2 \overline{G_1(f_1)G_3(f_2,f_3,f_4)H_2(f_1,f_2 + f_3 + f_4)}$$
$$+ \overline{G_2(f_1,f_2)G_2(f_3,f_4)H_2(f_1 + f_2,f_3 + f_4)}$$
$$+ 3 \overline{G_1(f_1)G_1(f_2)G_2(f_3,f_4)H_3(f_1,f_2,f_3 + f_4)}$$
$$+ G_1(f_1)G_1(f_2)G_1(f_3)G_1(f_4)H_4(f_1,f_2,f_3,f_4).$$
$$(4\text{-}80)$$

The general expression for $K_n(f_1,f_2,\ldots,f_n)$ is difficult to obtain. However, it should be noted that there is a close analogy between the zero-memory nonlinear system and the general nonlinear system with memory. By comparing (4-63), (4-70), (4-79), and (4-80) with (4-57), it is apparent that

$$G_n(f_1,f_2,\ldots,f_n) = g_n$$
$$H_n(f_1,f_2,\ldots,f_n) = h_n \qquad (4\text{-}81)$$

for zero-memory nonlinear systems. This analogy can be used to advantage in deriving and checking higher order relationships. Two special cases are generalized rather easily. When G is a linear system, its transfer functions above order one are identically zero and

$$K_n(f_1,f_2,\ldots,f_n) = G_1(f_1)G_1(f_2) \cdots G_1(f_n)H_n(f_1,f_2,\ldots,f_n).$$
$$(4\text{-}82)$$

Similarly, when H is a linear system, its transfer functions above order one are identically zero and

$$K_n(f_1,f_2,\ldots,f_n) = G_n(f_1,f_2,\ldots,f_n)H_1(f_1 + f_2 + \cdots + f_n).$$
$$(4\text{-}83)$$

Equation (4-82) indicates that the input components are linearly processed by system G and then combined nonlinearly in system H to produce an output component at $f_1 + f_2 + \cdots + f_n$. On the other hand, (4-83) indicates that the input components combine nonlinearly in system G to generate a component at $f_1 + f_2 + \cdots + f_n$ that is then linearly processed by system H.

Several relationships involving nonlinear transfer functions have been presented in this chapter. Given the nonlinear transfer functions, these relationships are readily utilized. However, we have yet to explain how nonlinear transfer functions may be obtained from a circuit description of a weakly nonlinear system. This matter is discussed next in Chapter 5.

5

Network Formulation
of Nonlinear
Transfer Functions

Nonlinearities inherent with electronic devices are potential sources of nonlinear behavior in electronic circuits. The nonlinear analysis of an electronic circuit requires that each electronic device be replaced by an equivalent nonlinear circuit model. This topic is discussed in detail in Chapter 7. Use of the nonlinear device models enables an electronic system to be modeled by an equivalent nonlinear circuit. In this chapter, assuming such a circuit has been generated, a procedure is presented for determining the nonlinear transfer functions of weakly nonlinear systems.

The chapter begins by discussing terminal relationships for nonlinear resistors, capacitors, inductors, and controlled sources. A simple weakly nonlinear circuit is solved to illustrate the various steps involved in evaluating nonlinear transfer functions. The procedure used, referred to as the harmonic input method, is then generalized to: (1) nodal analysis where the unknowns are node-to datum voltages, (2) loop analysis where the unknowns are mesh currents, and (3) mixed-variable analysis where the unknowns are an assortment of voltages and currents. In the harmonic input method nonlinearities manifest themselves as sources driving a linearized circuit. After developing a recursion relationship for the various-order sources, we conclude the chapter by formulating the harmonic input method in terms of solutions to equivalent linear networks where the desired nonlinear transfer functions appear as the unknown circuit variables.

5.1 NONLINEAR CIRCUIT ELEMENTS

Electronic devices and components in weakly nonlinear systems are typically operated over a localized region of their characteristics. Their performance is then adequately characterized using power series expansions about their quiescent operating points. In this section we discuss power series representations for nonlinear resistors, capacitors, inductors, and controlled sources. Various combinations of these nonlinear circuit elements suffice to model most of the nonlinearities encountered in electronic systems.

Nonlinear Resistor: A nonlinear resistor is a two-terminal element described by its characteristic curve in the current-voltage plane. When the voltage is a single-valued function of the current, as shown in Fig. 5.1(a), the resistor is said to be current controlled. On the other hand, when the characteristic curve is single valued in the voltage, as shown in Fig. 5.1(b), the resistor is said to be voltage controlled. A resistor is both current controlled and voltage controlled when its current-voltage relationship is monotonic as shown in Fig. 5.1(c).

The voltage of a current-controlled resistor may be expressed as

$$e_R(t) = r[i_R(t)] \tag{5-1}$$

where $r(\cdot)$ is a single-valued zero-memory functional. Let the quiescent operating point of the resistor be given by $e_R(t) = e_0$ and $i_R(t) = i_0$. Expanding (5-1) in a power series about the quiescent operating point, the terminal relationship becomes

$$e_R(t) - e_0 = \sum_{k=1}^{\infty} r_k [i_R(t) - i_0]^k \tag{5-2}$$

where r_k is the kth power series coefficient. To simplify notation, let the incremental voltage and current be denoted by

$$e_r(t) = e_R(t) - e_0 \qquad \text{and} \qquad i_r(t) = i_R(t) - i_0. \tag{5-3}$$

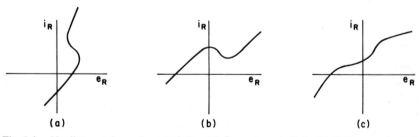

Fig. 5.1.   Nonlinear resistor characteristics. (a) Current controlled. (b) Voltage controlled. (c) Current controlled and voltage controlled.

Equation (5-2) then reduces to

$$e_r(t) = \sum_{k=1}^{\infty} r_k i_r^k(t).\tag{5-4}$$

For small enough inputs, the resistor behaves linearly and r_1 is the resistance of the linear resistor that would be included in a linear incremental equivalent circuit. The inverse of (5-1) is not considered since it is multivalued and difficult to handle.

In a similar manner, the current through a voltage-controlled resistor is given by

$$i_R(t) = g[e_R(t)]\tag{5-5}$$

where $g(\cdot)$ is a single-valued zero-memory functional. Expressing (5-5) in a power series about the quiescent operating point, we have

$$i_r(t) = \sum_{k=1}^{\infty} g_k e_r^k(t)\tag{5-6}$$

where $e_r(t)$ and $i_r(t)$ are the incremental variables defined in (5-3). The linear resistor that would be included in a linear incremental equivalent circuit is now characterized by the conductance g_1. The inverse of (5-5) is also ignored because it too is multivalued.

When a resistor is both current controlled and voltage controlled, either of the representations given by (5-1) and (5-5) may be used. Note that each representation is the single-valued inverse of the other. If there is an explicit function for the characteristic curve, then an explicit expression for the inverse may also exist.

Example 5.1

If

$$e_R(t) = \exp\left[i_R(t)\right] - 3,\tag{5-7}$$

then the inverse is

$$i_R(t) = \ln\left[e_R(t) + 3\right].\tag{5-8}$$

Given explicit expressions for the zero-memory functions $r(\cdot)$ and $g(\cdot)$, the power series expansions in (5-4) and (5-6) are obtained in a straightforward manner.

In some instances an explicit expression cannot be found for the characteristic curve. The coefficients in the power series expansions are then determined numerically. When the characteristic curve is monotonic, the coefficients in (5-4) and (5-6) are directly related as can be seen from a series reversion of (5-4). Specifically, assume the coefficients in (5-4) have been determined such that

$$e_r(t) = r_1 i_r(t) + r_2 i_r^2(t) + r_3 i_r^3(t) + r_4 i_r^4(t) + r_5 i_r^5(t) + \cdots. \tag{5-9}$$

Also, assume $i_r(t)$ is represented by the series

$$i_r(t) = g_1 e_r(t) + g_2 e_r^2(t) + g_3 e_r^3(t) + g_4 e_r^4(t) + g_5 e_r^5(t) + \cdots. \tag{5-10}$$

We wish to express the coefficients in (5-10) in terms of those in (5-9). To accomplish this, (5-10) is substituted into (5-9). This results in

$$\begin{aligned}
e_r(t) = {} & r_1 [g_1 e_r(t) + g_2 e_r^2(t) + g_3 e_r^3(t) + g_4 e_r^4(t) + g_5 e_r^5(t) + \cdots] \\
& + r_2 [g_1^2 e_r^2(t) + 2g_1 g_2 e_r^3(t) + (g_2^2 + 2g_1 g_3) e_r^4(t) \\
& + (2g_1 g_4 + 2g_2 g_3) e_r^5(t) + \cdots] \\
& + r_3 [g_1^3 e_r^3(t) + 3g_1^2 g_2 e_r^4(t) + (3g_1^2 g_3 + 3g_1 g_2^2) e_r^5(t) + \cdots] \\
& + r_4 [g_1^4 e_r^4(t) + 4g_1^3 g_2 e_r^5(t) + \cdots] \\
& + r_5 [g_1^5 e_r^5(t) + \cdots] + \cdots
\end{aligned} \tag{5-11}$$

where only terms up to fifth degree have been written out in detail. Equating the coefficients of $e_r(t)$, it is seen that

$$r_1 g_1 = 1. \tag{5-12}$$

Hence the first coefficient is

$$g_1 = \frac{1}{r_1}. \tag{5-13}$$

As expected, the linear conductance is the reciprocal of the linear resistance. g_2 is obtained by equating the coefficients of $e_r^2(t)$. Therefore,

$$r_1 g_2 + r_2 g_1^2 = 0. \tag{5-14}$$

It follows that

$$g_2 = -\frac{r_2}{r_1} g_1^2 = -\frac{r_2}{r_1^3}. \tag{5-15}$$

Equating the coefficients of $e_r^3(t)$ results in

$$r_1 g_3 + 2r_2 g_1 g_2 + r_3 g_1^3 = 0. \tag{5-16}$$

Solving for g_3 and making use of (5-13) and (5-15), we obtain

$$g_3 = -\frac{2r_2}{r_1} g_1 g_2 - \frac{r_3}{r_1} g_1^3 = \frac{2r_2^2 - r_1 r_3}{r_1^5}. \tag{5-17}$$

Higher order coefficients are obtained in like manner. In particular, it can be shown that

$$g_4 = \frac{5r_1r_2r_3 - r_1^2r_4 - 5r_2^3}{r_1^7} \tag{5-18}$$

and

$$g_5 = \frac{6r_1^2r_2r_4 + 3r_1^2r_3^2 + 14r_2^4 - r_1^3r_5 - 21r_1r_2^2r_3}{r_1^9}. \tag{5-19}$$

Note that the inverse to (5-9) does not exist when $r_1 = 0$. Even when a characteristic curve is multivalued, as in Fig. 5.1(a) and (b), the resistor may be both voltage and current controlled provided its characteristic curve is exercised over a monotonic region of the current-voltage plane. In our work we will assume this to be the case unless specified otherwise.

Nonlinear Capacitor: A nonlinear capacitor is a two-terminal element characterized by a curve in the charge versus voltage plane. It is said to be voltage controlled, charge controlled, or both voltage and charge controlled depending upon its characteristic curve as illustrated in Fig. 5.2. In weakly nonlinear systems it is likely that excitations are small enough such that nonlinear capacitors are operated over monotonic portions of their characteristics. Consequently, unless stated to the contrary, we assume capacitors to be both voltage and charge controlled.

The current through a capacitor equals the time rate of change of the stored charge. Making use of the chain rule for derivatives, we have

$$i_C(t) = \frac{dq_C(t)}{dt} = \frac{dq_C(t)}{de_C(t)} \frac{de_C(t)}{dt}. \tag{5-20}$$

The incremental capacitance is defined to be

$$c[e_C(t)] = \frac{dq_C(t)}{de_C(t)}. \tag{5-21}$$

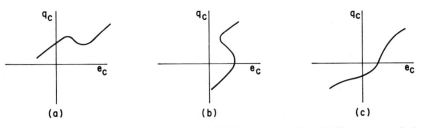

Fig. 5.2. Nonlinear capacitor characteristics. (a) Voltage controlled. (b) Charge controlled. (c) Voltage and charge controlled.

Hence at a given instant of time, the incremental capacitance is the slope of the charge-voltage characteristic at that instant. Inserting (5-21) into (5-20), the terminal relationship for a nonlinear capacitor is expressed as

$$i_C(t) = c[e_C(t)] \frac{de_C(t)}{dt}. \tag{5-22}$$

Let the quiescent operating point of the capacitor be given by $e_C(t) = e_0$, $q_C(t) = q_0$, and $i_C(t) = i_0$ and define the incremental variables

$$e_c(t) = e_C(t) - e_0, \quad q_c(t) = q_C(t) - q_0, \quad \text{and} \quad i_c(t) = i_C(t) - i_0. \tag{5-23}$$

Also, denote the incremental capacitance at the quiescent operating point by c_0. Expanding the incremental capacitance in a power series about the quiescent operating point, we obtain

$$c[e_C(t)] - c_0 = \sum_{k=1}^{\infty} c_k [e_C(t) - e_0]^k \tag{5-24}$$

where c_k is the kth power series coefficient. Introducing the incremental voltage into (5-24) and placing c_0 on the right-hand side of the equation results in

$$c[e_C(t)] = \sum_{k=0}^{\infty} c_k e_c^k(t). \tag{5-25}$$

Substitution of (5-25) into (5-22) then yields the current-voltage relationship

$$i_C(t) = \sum_{k=0}^{\infty} c_k e_c^k(t) \frac{de_C(t)}{dt}. \tag{5-26}$$

Equation (5-26) employs a mixture of total and incremental variables. To eliminate the mixture, use is made of (5-23) to rewrite (5-26) as

$$i_c(t) + i_0 = \sum_{k=0}^{\infty} c_k e_c^k(t) \frac{d[e_c(t) + e_0]}{dt}. \tag{5-27}$$

However, $i_0 = 0$ because a capacitor does not pass dc current. In addition, $de_0/dt = 0$ because e_0 is a constant. Therefore, the terminal relationship in terms of incremental variables becomes

$$i_c(t) = \sum_{k=0}^{\infty} c_k e_c^k(t) \frac{de_c(t)}{dt}. \tag{5-28}$$

Observe that the capacitance of a linear capacitor that would be inserted into a linear incremental equivalent circuit is given by c_0.

A second way to characterize the behavior of a nonlinear capacitor is to make a power series expansion of the charge-voltage characteristic. Let the charge be expressed as

$$q_C(t) = q[e_C(t)] \tag{5-29}$$

where $q(\cdot)$ is a single-valued zero-memory functional. Using incremental variables, the power series expansion about the quiescent operating point is

$$q_c(t) = \sum_{h=1}^{\infty} q_h e_c^h(t) \tag{5-30}$$

where q_h is the hth power series coefficient. The incremental capacitor current is then given by

$$i_c(t) = \frac{dq_c(t)}{dt} = \frac{d}{dt} \left\{ \sum_{h=1}^{\infty} q_h e_c^h(t) \right\} = \sum_{h=1}^{\infty} h q_h e_c^{h-1}(t) \frac{de_c(t)}{dt}. \tag{5-31}$$

Introducing the index

$$k = h - 1, \tag{5-32}$$

(5-31) is transformed into the equation

$$i_c(t) = \sum_{k=0}^{\infty} (k+1) q_{k+1} e_c^k(t) \frac{de_c(t)}{dt}. \tag{5-33}$$

Finally, comparison of (5-33) with (5-28) reveals that

$$c_k = (k+1) q_{k+1}; \qquad k = 0, 1, 2, \ldots. \tag{5-34}$$

In particular,

$$c_0 = q_1. \tag{5-35}$$

For a monotonic curve, it is just as reasonable to write

$$e_C(t) = s[q_C(t)] \tag{5-36}$$

where $s(\cdot)$ is a single-valued zero-memory functional. The corresponding power series expansion about the quiescent operating point is

$$e_c(t) = \sum_{h=1}^{\infty} s_h q_c^h(t) \tag{5-37}$$

where $e_c(t)$ and $q_c(t)$ are incremental variables and s_h is the hth power series coefficient. Since the incremental current is the time derivative of the incremental charge, as indicated in (5-31), it follows that

$$q_c(t) = \int_{-\infty}^{t} i_c(z)\, dz. \tag{5-38}$$

Substituting (5-38) into (5-37) the terminal relationship for the capacitor in terms of current and voltage becomes

$$e_c(t) = \sum_{h=1}^{\infty} s_h \left[\int_{-\infty}^{t} i_c(z)\, dz \right]^h. \tag{5-39}$$

The power series coefficients in (5-39) are related to those in (5-30).

Example 5.2

Making a series reversion of (5-30), the first three coefficients in (5-39) become

$$s_1 = \frac{1}{q_1} = \frac{1}{c_0}$$

$$s_2 = -\frac{q_2}{q_1^3}$$

$$s_3 = \frac{2q_2^2 - q_1 q_3}{q_1^5}. \tag{5-40}$$

Note that s_1 is the elastance of the linear capacitor that would be used in a linear incremental equivalent circuit. As expected, it is the reciprocal of the linear capacitance.

Nonlinear Inductor: A nonlinear inductor is a two-terminal element described by its characteristic curve in the flux linkage versus current plane as indicated in Fig. 5.3. We see that the nature of the curve determines whether the inductor is flux-linkage controlled, current controlled, or both. Unless specified otherwise, we will assume the operating regions to be monotonic so that inductors are both flux linkage and current controlled.

The terminal relationship for an inductor is dual to that of a capacitor. Thus the voltage across an inductor equals the time rate of change of the flux linkage linking the inductor. Defining the incremental inductance to be

$$l[i_L(t)] = \frac{d\lambda_L(t)}{di_L(t)}, \tag{5-41}$$

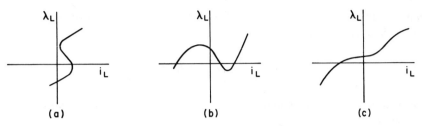

Fig. 5.3. Nonlinear inductor characteristics. (a) Flux-linkage controlled. (b) Current controlled. (c) Flux linkage and current controlled.

the voltage across the inductor can be expressed as

$$e_L(t) = \frac{d\lambda_L(t)}{dt} = \frac{d\lambda_L(t)}{di_L(t)} \frac{di_L(t)}{dt} = l[i_L(t)] \frac{di_L(t)}{dt}. \tag{5-42}$$

Let the quiescent operating point be given by $i_L(t) = i_0$, $\lambda_L(t) = \lambda_0$, and $e_L(t) = e_0$. The corresponding incremental variables are

$$i_l(t) = i_L(t) - i_0, \quad \lambda_l(t) = \lambda_L(t) - \lambda_0, \quad \text{and} \quad e_l(t) = e_L(t) - e_0. \tag{5-43}$$

Also, denote the incremental inductance at the quiescent operating point by l_0. Following a procedure dual to that leading from (5-24) to (5-28), the incremental inductance becomes

$$l[i_L(t)] = \sum_{k=0}^{\infty} l_k i_l^k(t) \tag{5-44}$$

and the voltage-current relationship for the inductor is given by

$$e_l(t) = \sum_{k=0}^{\infty} l_k i_l^k(t) \frac{di_l(t)}{dt} \tag{5-45}$$

where l_0 is the inductance of the linear inductor needed for a linear incremental equivalent circuit.

Dual to (5-29), the flux linkage may be expressed in terms of the inductor current as

$$\lambda_L(t) = \lambda[i_L(t)] \tag{5-46}$$

where $\lambda(\cdot)$ is a single-valued zero-memory functional. The corresponding power series expansion about the quiescent operating point is

$$\lambda_l(t) = \sum_{h=1}^{\infty} \lambda_h i_l^h(t) \tag{5-47}$$

where $\lambda_l(t)$ and $i_l(t)$ are incremental variables and λ_h is the hth power series coefficient. It follows that

$$e_l(t) = \frac{d\lambda_l(t)}{dt} = \sum_{h=1}^{\infty} h\lambda_h i_l^{h-1}(t)\frac{di_l(t)}{dt}. \tag{5-48}$$

With the change of index indicated by (5-32), the incremental inductor voltage becomes

$$e_l(t) = \sum_{k=0}^{\infty} (k+1)\lambda_{k+1}i_l^k(t)\frac{di_l(t)}{dt}. \tag{5-49}$$

From (5-45) and (5-49) we conclude that

$$l_k = (k+1)\lambda_{k+1}; \qquad k = 0, 1, 2, \ldots. \tag{5-50}$$

Observe that

$$l_0 = \lambda_1. \tag{5-51}$$

The inverse to (5-46) is

$$i_L(t) = \Gamma[\lambda_L(t)] \tag{5-52}$$

where $\Gamma(\cdot)$ is a single-valued zero-memory functional. In terms of incremental variables, the power series expansion about the quiescent operating point is

$$i_l(t) = \sum_{h=1}^{\infty} \Gamma_h \lambda_l^h(t) \tag{5-53}$$

where Γ_h is the hth power series coefficient. Since

$$\lambda_l(t) = \int_{-\infty}^{t} e_l(z)\, dz, \tag{5-54}$$

we conclude that

$$i_l(t) = \sum_{h=1}^{\infty} \Gamma_h \left[\int_{-\infty}^{t} e_l(z)\, dz\right]^h \tag{5-55}$$

where the first three coefficients in (5-55) are related to those in (5-47) according to the identities

$$\Gamma_1 = \frac{1}{\lambda_1} = \frac{1}{l_0}$$

$$\Gamma_2 = -\frac{\lambda_2}{\lambda_1^3}$$

$$\Gamma_3 = \frac{2\lambda_2^2 - \lambda_1\lambda_3}{\lambda_1^5}. \tag{5-56}$$

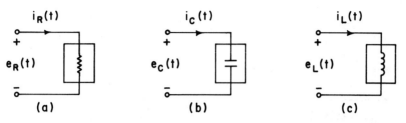

Fig. 5.4. Circuit symbols. (a) Nonlinear resistor. (b) Nonlinear capacitor. (c) Nonlinear inductor.

We see that Γ_1 is the reciprocal inductance of the linear inductor that would be placed in a linear incremental equivalent circuit.

The circuit symbols used in this text for nonlinear resistors, capacitors, and inductors are shown in Fig. 5.4. They are identical to the conventional linear circuit symbols except for the rectangular boxes placed around the elements. Section 5.1 is now concluded by discussing the class of circuit elements known as nonlinear controlled (or dependent) sources.

Nonlinear Controlled Sources: A nonlinear controlled (or dependent) source is a two-terminal element whose terminal voltage or current at any instant of time is a nonlinear function of the voltage e_X or the current i_X in another part of the circuit. e_X is called a control voltage while i_X is referred to as a control current. If i_X or e_X pertains to a particular element X in a circuit, this element is called the controlling element. The control equation is the defining equation for the controlled source in terms of its control variable. Figure 5.5 shows the circuit symbols for the four types of nonlinear controlled sources along with their control equations. Note that $\mu(\cdot)$, $r(\cdot)$, $g(\cdot)$, and $\alpha(\cdot)$ are each assumed to be single-valued zero-memory functionals.

Let the incremental variables associated with e_X, i_X, e_{CS}, and i_{CS} be denoted, respectively, by e_x, i_x, e_{cs}, and i_{cs}. Then power series expansions about the quiescent operating points of the controlled sources result in the following control equations.

(a) Voltage-controlled voltage source

$$e_{cs}(t) = \sum_{k=1}^{\infty} \mu_k e_x^k(t). \tag{5-57}$$

(b) Current-controlled voltage source

$$e_{cs}(t) = \sum_{k=1}^{\infty} r_{ks} i_x^k(t). \tag{5-58}$$

(c) Voltage-controlled current source

$$i_{cs}(t) = \sum_{k=1}^{\infty} g_{ks} e_x^k(t). \tag{5-59}$$

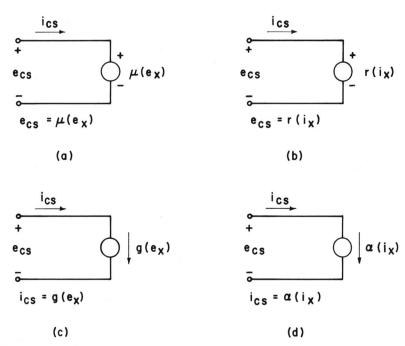

Fig. 5.5. Controlled source circuit symbols. (a) Voltage-controlled voltage source. (b) Current-controlled voltage source. (c) Voltage-controlled current source. (d) Current-controlled current source.

(d) Current-controlled current source

$$i_{cs}(t) = \sum_{k=1}^{\infty} \alpha_k i_x^k(t). \tag{5-60}$$

The linear terms in each expansion define the linear controlled sources to be used in linear incremental equivalent circuits. The linear coefficients μ_1, r_{1s}, g_{1s}, and α_1 are referred to as the voltage amplification factor, the mutual resistance, the mutual conductance, and the current amplification factor, respectively.

Having introduced the nonlinear circuit elements necessary for modeling electronic circuits, we now discuss a technique for determining the nonlinear transfer functions of weakly nonlinear systems.

5.2 ANALYSIS OF A SIMPLE WEAKLY NONLINEAR CIRCUIT

Before discussing the general procedure for obtaining nonlinear transfer functions, the key elements involved are illustrated by working out the details to a relatively simple problem. Consider the nonlinear incremental circuit of Fig. 5.6

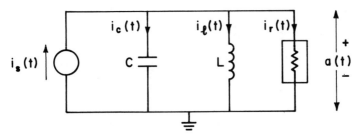

Fig. 5.6.   Weakly nonlinear incremental circuit involving a single node-to-datum voltage.

consisting of the parallel combination of a linear capacitor, a linear inductor, a nonlinear resistor, and an independent current source. All currents and voltages are incremental quantities with the incremental node-to-datum voltage denoted by $a(t)$. The current-voltage relationship for the nonlinear resistor is assumed to be

$$i_r(t) = g_1 a(t) + g_2 a^2(t). \qquad (5\text{-}61)$$

The problem is to determine the nonlinear transfer functions associated with $a(t)$. Once these are known, the nonlinear transfer functions for $i_c(t), i_l(t)$, and $i_r(t)$ are readily evaluated.

Circuit Equation and Equivalent Circuit: Application of Kirchhoff's current law to the top node in the figure yields the equation

$$i_s(t) = i_c(t) + i_l(t) + i_r(t)$$

$$= C \frac{da(t)}{dt} + \Gamma \int_{-\infty}^{t} a(z)\, dz + g_1 a(t) + g_2 a^2(t) \qquad (5\text{-}62)$$

where $\Gamma = 1/L$ is the reciprocal inductance. Rearrangement of (5-62) such that only terms linear in $a(t)$ appear on one side of the equation results in

$$C \frac{da(t)}{dt} + \Gamma \int_{-\infty}^{t} a(z)\, dz + g_1 a(t) = i_s(t) - g_2 a^2(t). \qquad (5\text{-}63)$$

This equation can be interpreted in terms of the network shown in Fig. 5.7. Observe that the nonlinear resistor has been replaced by a linear resistor, having conductance g_1, in parallel with a voltage-controlled current source. In this manner, the nonlinearity manifests itself as a controlled source exciting the linearized circuit. At this point, it is convenient to introduce operator notation. In particular, let p be the differential operator d/dt such that

$$pa(t) = \frac{da(t)}{dt} \quad \text{and} \quad \frac{1}{p} a(t) = \int_{-\infty}^{t} a(z)\, dz. \qquad (5\text{-}64)$$

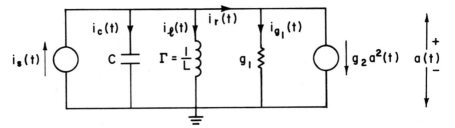

Fig. 5.7.   Equivalent circuit of that shown in Fig. 5.6.

In terms of operator notation, (5-63) simplifies to

$$\left[pC + \frac{\Gamma}{p} + g_1 \right] a(t) = i_s(t) - g_2 a^2(t) \tag{5-65}$$

or, equivalently,

$$Y(p)\, a(t) = i_s(t) - g_2 a^2(t) \tag{5-66}$$

where $Y(p)$ is the linear integro-differential operator

$$Y(p) = pC + \frac{\Gamma}{p} + g_1. \tag{5-67}$$

Harmonic Input Method: The method for determining the nonlinear transfer function is recursive in nature. The linearized circuit is first "probed" by the single complex exponential $\exp[j2\pi f_1 t]$. This permits determination of the linear transfer function. Then the sum of two complex exponentials, given by $\exp[j2\pi f_1 t] + \exp[j2\pi f_2 t]$, is applied. This yields the second-order transfer function in terms of the linear transfer function. The procedure continues with one additional complex exponential being added to the input at each step. Therefore, at step n, the input consists of the sum $\exp[j2\pi f_1 t] + \cdots + \exp[j2\pi f_n t]$. The nonlinear transfer function of order n is then found to be constructed from all of the previously determined lower order nonlinear transfer functions.

Example 5.3

The procedure for determining the linear transfer function of the circuit in Fig. 5.7 begins by selecting the excitation to be

$$i_s(t) = \exp[j2\pi f_1 t]. \tag{5-68}$$

Observe that (5-68) is a special case of (4-5) where $Q = 1, E_1 = 2$, and negative frequency terms are omitted. If the nth-order transfer function of $a(t)$ is denoted by $A_n(f_1, f_2, \ldots, f_n)$, it follows from (4-7) that the nth-order component of

$a(t)$ is expressed as

$$a_n(t) = A_n(f_1, \ldots, f_1) \exp [j2\pi(nf_1) t]. \tag{5-69}$$

Hence, the total node-to-datum voltage is given by

$$a(t) = \sum_{n=1}^{N} a_n(t) = \sum_{n=1}^{N} A_n(f_1, \ldots, f_1) \exp [j2\pi(nf_1) t] \tag{5-70}$$

where terms above Nth-order are assumed to be negligible. We see that the output frequencies consist of the original input frequency and its harmonics.

Substitution of (5-68) and (5-70) into (5-66) results in

$$Y(p) \left\{ \sum_{n=1}^{N} A_n(f_1, \ldots, f_1) \exp [j2\pi(nf_1) t] \right\} = \exp [j2\pi f_1 t]$$

$$- g_2 \left\{ \sum_{n=1}^{N} A_n(f_1, \ldots, f_1) \exp [j2\pi(nf_1) t] \right\}^2. \tag{5-71}$$

Note that application of the operators p and $1/p$ to $\exp [j2\pi(nf_1) t]$ yields

$$p\{\exp [j2\pi(nf_1) t]\} = j2\pi(nf_1) \exp [j2\pi(nf_1) t]$$

$$\frac{1}{p} \{\exp [j2\pi(nf_1) t]\} = \frac{1}{j2\pi(nf_1)} \exp [j2\pi(nf_1) t]. \tag{5-72}$$

Consequently, when the integro-differential operator $Y(p)$ is applied to $\exp [j2\pi(nf_1) t]$, the effect is to replace $Y(p)$ by

$$Y(j2\pi nf_1) = \left[j2\pi(nf_1) C + \frac{\Gamma}{j2\pi(nf_1)} + g_1 \right] \tag{5-73}$$

where $Y(j2\pi nf_1)$ is recognized as the admittance of the linearized network evaluated at the nth harmonic of the input. In addition, carrying out the square on the right-hand side of the equation, (5-71) becomes

$$\sum_{n=1}^{N} Y(j2\pi nf_1) A_n(f_1, \ldots, f_1) \exp [j2\pi(nf_1) t] = \exp [j2\pi f_1 t]$$

$$- g_2 [A_1(f_1)]^2 \exp [j2\pi(2f_1) t] - 2g_2 A_1(f_1) A_2(f_1, f_1)$$

$$\cdot \exp [j2\pi(3f_1) t] - \cdots - g_2 [A_N(f_1, \ldots, f_1)]^2 \exp [j2\pi(2Nf_1) t]. \tag{5-74}$$

However, exponential functions have the important property that they are linearly independent. Specifically, if $f_1 \neq f_2 \neq \cdots \neq f_m$, then there exists no set of constants $C_1, C_2, \ldots, C_m$ (at least one of which is not zero) such that

$$C_1 \exp [j2\pi f_1 t] + C_2 \exp [j2\pi f_2 t] + \cdots + C_m \exp [j2\pi f_m t] = 0 \quad (5\text{-}75)$$

for all values of t. It follows that each distinct frequency component in (5-74) must be individually satisfied by the equation. Equating the coefficients of $\exp [j2\pi f_1 t]$ results in

$$Y(j2\pi f_1) A_1(f_1) = 1. \qquad (5\text{-}76)$$

Observe that only first-order terms are involved in the solution for the linear transfer function. Thus $A_1(f_1)$ is given by

$$A_1(f_1) = \frac{1}{Y(j2\pi f_1)} = Z(j2\pi f_1) \qquad (5\text{-}77)$$

where $Z(j2\pi f_1)$ is the impedance of the linearized network. Equation (5-77) is a reasonable result because it states that the first-order component of the voltage response is related to the current excitation by the impedance of the linearized network.

Example 5.4

To obtain the second-order transfer function of the circuit in Fig. 5.7, the excitation is chosen to be

$$i_s(t) = \exp [j2\pi f_1 t] + \exp [j2\pi f_2 t] \qquad (5\text{-}78)$$

where f_1 and f_2 are assumed to be both positive and incommensurable frequencies. Equation (5-78) is obtained from (4-5) by selecting $Q = 2$, $E_1 = E_2 = 2$, and by omitting negative frequency terms. Using (4-7) the nth-order component of $a(t)$ is given by

$$a_n(t) = \sum_{q_1=1}^{2} \cdots \sum_{q_n=1}^{2} A_n(f_{q_1}, \ldots, f_{q_n}) \exp [j2\pi (f_{q_1} + \cdots + f_{q_n}) t].$$
$$(5\text{-}79)$$

Assuming components above Nth-order are insignificant, the total node-to-datum voltage $a(t)$ is simply the sum of the various-order components up to order N. This sum, along with (5-78), is then substituted into (5-66).

Having made these substitutions, the second-order transfer function $A_2(f_1, f_2)$ is determined by equating the coefficient of $\exp [j2\pi (f_1 + f_2) t]$ on both sides of the equation. Because the input frequencies f_1 and f_2 are positive and incommensurable, intermodulation components at $f_1 + f_2$ can be generated only in terms of second order. Therefore, as far as the evaluation of $A_2(f_1, f_2)$ is concerned, terms that are not of second order may be ignored. Rewriting (5-66) such that only second-order terms are shown explicitly, we have

$$Y(p)[a_2(t) + \cdots] = -g_2 a_1^2(t) + \cdots. \qquad (5\text{-}80)$$

The independent source term $i_s(t)$ is ignored since its components are at f_1 and f_2 as opposed to $f_1 + f_2$. Making use of (5-79), note that

$$a_2(t) = \sum_{q_1=1}^{2} \sum_{q_2=1}^{2} A_2(f_{q_1}, f_{q_2}) \exp\left[j2\pi(f_{q_1} + f_{q_2})t\right],$$

$$a_1^2(t) = \left\{\sum_{q_1=1}^{2} A_1(f_{q_1}) \exp\left[j2\pi f_{q_1}t\right]\right\}^2$$

$$= \sum_{q_1=1}^{2} \sum_{q_2=1}^{2} A_1(f_{q_1}) A_1(f_{q_2}) \exp\left[j2\pi(f_{q_1} + f_{q_2})t\right]. \qquad (5\text{-}81)$$

Substituting (5-81) into (5-80) and carrying out the operations implied by the integro-differential operator $Y(p)$, we obtain

$$\sum_{q_1=1}^{2} \sum_{q_2=1}^{2} Y[j2\pi(f_{q_1} + f_{q_2})] A_2(f_{q_1}, f_{q_2}) \exp\left[j2\pi(f_{q_1} + f_{q_2})t\right] + \cdots$$

$$= -g_2 \sum_{q_1=1}^{2} \sum_{q_2=1}^{2} A_1(f_{q_1}) A_1(f_{q_2}) \exp\left[j2\pi(f_{q_1} + f_{q_2})t\right] + \cdots. \qquad (5\text{-}82)$$

In each double summation there are two identical terms involving $\exp[j2\pi(f_1 + f_2)t]$ that occur for $q_1 = 1$, $q_2 = 2$, and $q_1 = 2$, $q_2 = 1$. Equating the coefficients of these terms results in the equation

$$2Y[j2\pi(f_1 + f_2)] A_2(f_1, f_2) = -2g_2 A_1(f_1) A_1(f_2). \qquad (5\text{-}83)$$

Solving for $A_2(f_1, f_2)$, we have

$$A_2(f_1, f_2) = -\frac{g_2 A_1(f_1) A_1(f_2)}{Y[j2\pi(f_1 + f_2)]}$$

$$= -g_2 A_1(f_1) A_1(f_2) A_1(f_1 + f_2) \qquad (5\text{-}84)$$

where use has been made of (5-77) in order to eliminate $Y[j2\pi(f_1 + f_2)]$. We see that the second-order transfer function may be determined from the linear transfer function.

Example 5.5

The third-order transfer function of the circuit in Fig. 5.7 is determined next. Choose the input to be

$$i_s(t) = \sum_{q_1=1}^{3} \exp\left[j2\pi f_{q_1}t\right] \qquad (5\text{-}85)$$

where, as in Example 5.4, the input frequencies are assumed to be positive and incommensurable. Recognizing that (5-85) stems from (4-5) with $Q = 3$, $E_1 = E_2 = E_3 = 2$, and negative frequency terms omitted, it follows from (4-7) that the nth-order component of $a(t)$ is

$$a_n(t) = \sum_{q_1=1}^{3} \cdots \sum_{q_n=1}^{3} A_n(f_{q_1}, \ldots, f_{q_n}) \exp \left[j2\pi(f_{q_1} + \cdots + f_{q_n}) t \right].$$

$$(5-86)$$

Once again, $a(t)$ is the sum of its various-order components. This sum, along with (5-85), is then substituted into (5-66).

To determine the third-order transfer function $A_3(f_1, f_2, f_3)$, we equate the coefficients of $\exp\left[j2\pi(f_1 + f_2 + f_3)t\right]$ on both sides of the equation. However, now the positive and incommensurable nature of the input frequencies requires that intermodulation components at $f_1 + f_2 + f_3$ be generated only in third-order terms. Rewriting (5-66) such that only third-order terms are shown explicitly, we have

$$Y(p)[a_3(t) + \cdots] = -2g_2 a_1(t) a_2(t) + \cdots. \qquad (5-87)$$

As with the second-order case, the independent source term is ignored because it does not contain components at the frequency of interest. From (5-86), it follows that

$$a_3(t) = \sum_{q_1=1}^{3} \sum_{q_2=1}^{3} \sum_{q_3=1}^{3} A_3(f_{q_1}, f_{q_2}, f_{q_3}) \exp\left[j2\pi(f_{q_1} + f_{q_2} + f_{q_3})t\right],$$

$$a_1(t) a_2(t) = \left\{ \sum_{q_1=1}^{3} A_1(f_{q_1}) \exp\left[j2\pi f_{q_1} t\right] \right\}$$

$$\cdot \left\{ \sum_{q_1=1}^{3} \sum_{q_2=1}^{3} A_2(f_{q_1}, f_{q_2}) \exp\left[j2\pi(f_{q_1} + f_{q_2})t\right] \right\}$$

$$= \sum_{q_1=1}^{3} \sum_{q_2=1}^{3} \sum_{q_3=1}^{3} A_1(f_{q_1}) A_2(f_{q_2}, f_{q_3})$$

$$\cdot \exp\left[j2\pi(f_{q_1} + f_{q_2} + f_{q_3})t\right]. \qquad (5-88)$$

Substituting (5-88) into (5-87) and carrying out the operations implied by the integro-differential operator $Y(p)$, (5-87) becomes

$$\sum_{q_1=1}^{3} \sum_{q_2=1}^{3} \sum_{q_3=1}^{3} Y[j2\pi(f_{q_1} + f_{q_2} + f_{q_3})] A_3(f_{q_1}, f_{q_2}, f_{q_3})$$

$$\cdot \exp\left[j2\pi(f_{q_1} + f_{q_2} + f_{q_3})t\right] + \cdots$$

$$= -2g_2 \sum_{q_1=1}^{3} \sum_{q_2=1}^{3} \sum_{q_3=1}^{3} A_1(f_{q_1}) A_2(f_{q_2}, f_{q_3})$$

$$\cdot \exp\left[j2\pi(f_{q_1} + f_{q_2} + f_{q_3})t\right] + \cdots. \tag{5-89}$$

In each triple summation there are six terms involving $\exp\left[j2\pi(f_1 + f_2 + f_3)t\right]$. These arise from the $3! = 6$ permutations of the indices $q_1 = 1$, $q_2 = 2$, and $q_3 = 3$. Equating the coefficients of these terms produces the equation

$$6Y[j2\pi(f_1 + f_2 + f_3)] A_3(f_1, f_2, f_3)$$

$$= -2g_2[A_1(f_1) A_2(f_2, f_3) + A_1(f_1) A_2(f_3, f_2) + A_1(f_2) A_2(f_1, f_3)$$

$$+ A_1(f_2) A_2(f_3, f_1) + A_1(f_3) A_2(f_1, f_2) + A_1(f_3) A_2(f_2, f_1)].$$

$$\tag{5-90}$$

The right-hand side of (5-90) contains all permutations of $A_1(f_1) A_2(f_2, f_3)$ generated by permuting the frequencies f_1, f_2, and f_3. As in (4-77), let an overbar denote the arithmetic average of the permuted terms. Thus

$$\overline{A_1(f_1) A_2(f_2, f_3)} = \tfrac{1}{6}[A_1(f_1) A_2(f_2, f_3) + A_1(f_1) A_2(f_3, f_2)$$

$$+ A_1(f_2) A_2(f_1, f_3) + A_1(f_2) A_2(f_3, f_1)$$

$$+ A_1(f_3) A_2(f_1, f_2) + A_1(f_3) A_2(f_2, f_1)]. \tag{5-91}$$

With this notation, (5-90) becomes

$$6Y[j2\pi(f_1 + f_2 + f_3)] A_3(f_1, f_2, f_3) = -12g_2 \overline{A_1(f_1) A_2(f_2, f_3)}. \tag{5-92}$$

Using (5-77) to express $Y[j2\pi(f_1 + f_2 + f_3)]$ in terms of $A_1(f_1 + f_2 + f_3)$ and solving for $A_3(f_1, f_2, f_3)$, the third-order nonlinear transfer function is given by

$$A_3(f_1, f_2, f_3) = -2g_2 A_1(f_1 + f_2 + f_3) \overline{A_1(f_1) A_2(f_2, f_3)}. \tag{5-93}$$

Therefore, it is possible to determine $A_3(f_1, f_2, f_3)$ in terms of the previously determined first- and second-order transfer functions.

Higher order nonlinear transfer functions of the circuit in Fig. 5.7 are found in a similar manner. To determine $A_n(f_1, \ldots, f_n)$, the excitation consists of

$$i_s(t) = \sum_{q_1=1}^{n} \exp\left[j2\pi f_{q_1} t\right] \tag{5-94}$$

where the input frequencies are positive and incommensurable. The nth-order component of $a(t)$ is then given by

$$a_n(t) = \sum_{q_1=1}^{n} \cdots \sum_{q_n=1}^{n} A_n(f_{q_1}, \ldots, f_{q_n}) \exp [j2\pi(f_{q_1} + \cdots + f_{q_n}) t]. \quad (5\text{-}95)$$

An equation for $A_n(f_1, \ldots, f_n)$ is obtained by equating the coefficients in (5-66) of $\exp [j2\pi(f_1 + \cdots + f_n) t]$. These arise only in terms of nth order which are in the form of n-fold summations. The terms involving $\exp [j2\pi(f_1 + \cdots + f_n) t]$ are generated by the $n!$ permutations of the n indices $q_1 = 1, q_2 = 2, \ldots, q_n = n$. Finally, solution for $A_n(f_1, \ldots, f_n)$ results in an expression for the nth-order nonlinear transfer function as a function of the lower order transfer functions.

The requirement that the input frequencies be positive and incommensurable was introduced to ensure that distinct frequency mixes produce distinct intermodulation frequencies. Although this simplifies the analysis, it is not necessary. It can be shown that the analytical expressions obtained for the nonlinear transfer functions are valid even when the requirement is not satisfied.

Example 5.6

$A_2(f_1, -f_2)$ and $A_3(f_2, f_2, f_2)$ may be evaluated from (5-84) and (5-93), respectively, to yield

$$A_2(f_1, -f_2) = -g_2 A_1(f_1) A_1(-f_2) A_1(f_1 - f_2)$$

$$A_3(f_2, f_2, f_2) = -2g_2 A_1(3f_2) \overline{A_1(f_2) A_2(f_2, f_2)}$$

$$= -2g_2 A_1(3f_2) A_1(f_2) A_2(f_2, f_2)$$

$$= 2g_2^2 A_1^3(f_2) A_1(2f_2) A_1(3f_2). \quad (5\text{-}96)$$

Therefore, even though $A_n(f_1, \ldots, f_n)$ is determined on the basis of positive and incommensurable input frequencies, it is allowable to evaluate the resulting analytical expressions using any set of values for $f_1, \ldots, f_n$.

To complete the discussion of the circuit in Fig. 5.7, we show how the nonlinear transfer functions of $i_c(t), i_l(t),$ and $i_{g_1}(t)$ are readily determined from those of $a(t)$.

Example 5.7

Let the nth-order components of the currents be denoted by $i_{cn}(t), i_{ln}(t),$ and $i_{g_1 n}(t)$, respectively. Because $C, L,$ and g_1 refer to linear circuit elements, the nth-order component of each current depends only upon the nth-order component of each element voltage. With reference to (5-95), we have

$$i_{cn}(t) = C\frac{da_n(t)}{dt} = \sum_{q_1=1}^{n} \cdots \sum_{q_n=1}^{n} j2\pi(f_{q_1} + \cdots + f_{q_n})\,CA_n(f_{q_1}, \ldots, f_{q_n})$$

$$\cdot \exp\left[j2\pi(f_{q_1} + \cdots + f_{q_n})\,t\right],$$

$$i_{ln}(t) = \frac{1}{L}\int_{-\infty}^{t} a_n(z)\,dz = \sum_{q_1=1}^{n} \cdots \sum_{q_n=1}^{n} \frac{A_n(f_{q_1}, \ldots, f_{q_n})}{j2\pi(f_{q_1} + \cdots + f_{q_n})\,L}$$

$$\cdot \exp\left[j2\pi(f_{q_1} + \cdots + f_{q_n})\,t\right],$$

$$i_{g_1 n}(t) = g_1 a_n(t) = \sum_{q_1=1}^{n} \cdots \sum_{q_n=1}^{n} g_1 A_n(f_{q_1}, \ldots, f_{q_n})$$

$$\cdot \exp\left[j2\pi(f_{q_1} + \cdots + f_{q_n})\,t\right]. \tag{5-97}$$

However, if the nth-order transfer functions of $i_c(t)$, $i_l(t)$, and $i_{g_1}(t)$ are denoted by $I_{cn}(f_1, \ldots, f_n)$, $I_{ln}(f_1, \ldots, f_n)$, and $I_{g_1 n}(f_1, \ldots, f_n)$, respectively, it is also true that

$$i_{cn}(t) = \sum_{q_1=1}^{n} \cdots \sum_{q_n=1}^{n} I_{cn}(f_{q_1}, \ldots, f_{q_n}) \exp\left[j2\pi(f_{q_1} + \cdots + f_{q_n})\,t\right]$$

$$i_{ln}(t) = \sum_{q_1=1}^{n} \cdots \sum_{q_n=1}^{n} I_{ln}(f_{q_1}, \ldots, f_{q_n}) \exp\left[j2\pi(f_{q_1} + \cdots + f_{q_n})\,t\right]$$

$$i_{g_1 n}(t) = \sum_{q_1=1}^{n} \cdots \sum_{q_n=1}^{n} I_{g_1 n}(f_{q_1}, \ldots, f_{q_n}) \exp\left[j2\pi(f_{q_1} + \cdots + f_{q_n})\,t\right].$$

$$\tag{5-98}$$

By comparison of (5-97) with (5-98), we conclude that

$$I_{cn}(f_1, \ldots, f_n) = j2\pi(f_1 + \cdots + f_n)\,CA_n(f_1, \ldots, f_n)$$

$$I_{ln}(f_1, \ldots, f_n) = \frac{A_n(f_1, \ldots, f_n)}{j2\pi(f_1 + \cdots + f_n)\,L}$$

$$I_{g_1 n}(f_1, \ldots, f_n) = g_1 A_n(f_1, \ldots, f_n). \tag{5-99}$$

These are recognized to be simple generalizations of the usual frequency-domain current-voltage relationships from linear circuit theory.

An interesting feature of the circuit in Fig. 5.6 is the existence of third-order responses and above even though the nonlinear resistor is of second degree. Had the circuit been modeled in the form of the classical approach of Chapter 3, responses above second order would have been identically zero. This illustrates, once again, the more general capability of the nonlinear transfer function approach.

5.3   NODAL ANALYSIS OF WEAKLY NONLINEAR CIRCUITS

When analyzing complicated networks, it is desirable to have systematic methods for formulating network equations. In general, these are based upon incidence, loop, and cut-set matrices that are derived from topological considerations. Because we choose not to interrupt the flow of our discussion by going deeply into the topic of network topology, we extend the harmonic input method, introduced in Section 5.2, to circuit analysis techniques that do not require an extensive topological background. This means that our discussion on the network formulation of nonlinear transfer functions is somewhat restricted. Nevertheless, the basic ideas presented should be adequate for most applications.

In this section the harmonic input method is applied to nodal analysis. Since electronic circuits tend to contain many more loops than nodes, nodal analysis is a frequently used technique. In general, we assume the network to consist of $K + 1$ nodes. One of the nodes, usually the ground connection, is assigned to be the reference or datum node. Voltage symbols are then assigned to each of the remaining K nodes. To preclude the use of an excessive amount of subscripts, we designate: (1) the nodes by $a, b, \ldots, k$, (2) the corresponding node-to-datum voltages by $a(t), b(t), \ldots, k(t)$, and (3) the associated nth-order transfer functions by $A_n(f_1, \ldots, f_n), B_n(f_1, \ldots, f_n), \ldots, K_n(f_1, \ldots, f_n)$. This is illustrated for the four-node circuit of Fig. 5.8 where the ground is chosen as the datum node. Since we are interested in the nonlinear incremental analysis of electronic circuits, all currents and voltages are assumed to be incremental variables.

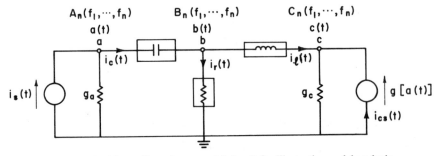

Fig. 5.8.   Weakly nonlinear incremental circuit for illustrating nodal analysis.

In order to perform nodal analysis, it is necessary that each nonlinear resistor, capacitor, and inductor have an admittance representation in which the element current is expressed as a function of the element voltage. Should an admittance representation not exist, as is the case for a current-controlled resistor, nodal analysis is not possible. As far as controlled sources are concerned, only voltage-controlled current sources have an admittance representation. The harmonic input method, in conjunction with nodal analysis, is illustrated in Example 5.8. Following the example, techniques are discussed for including controlled sources that do not have an admittance representation.

Example 5.8

Network Equations and Equivalent Circuit: In nodal analysis the network equations are obtained by writing Kirchhoff's current law at each of the K nodes other than the reference node. For the network of Fig. 5.8, Kirchhoff's current law is applied at nodes a, b, and c, respectively, resulting in the three equations

$$g_a a(t) + i_c(t) = i_s(t)$$

$$-i_c(t) + i_r(t) + i_l(t) = 0$$

$$-i_l(t) + g_c c(t) = i_{cs}(t). \tag{5-100}$$

The next step is to expand in a power series each current through a nonlinear circuit element where the series variables are node-to-datum voltages. With reference to (5-6), (5-28), (5-55), and (5-59), the desired expansions for the nonlinear elements in the circuit of Fig. 5.8 are

$$i_r(t) = \sum_{k=1}^{\infty} g_k b^k(t)$$

$$i_c(t) = \sum_{k=0}^{\infty} c_k [a(t) - b(t)]^k \, \frac{d[a(t) - b(t)]}{dt}$$

$$i_l(t) = \sum_{k=1}^{\infty} \Gamma_k \left\{ \int_{-\infty}^{t} [b(z) - c(z)] \, dz \right\}^k$$

$$i_{cs}(t) = \sum_{k=1}^{\infty} g_{ks} a^k(t). \tag{5-101}$$

Separating each power series into a linear term plus second-order terms and higher suggests that each nonlinear circuit element may be replaced by a linear circuit element in parallel with a nonlinear voltage-controlled current source. This is shown in Fig. 5.9 for the circuit of Fig. 5.8. With this interpretation, the

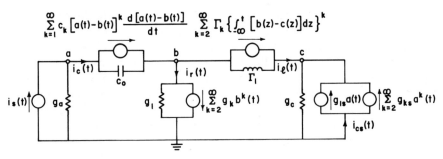

Fig. 5.9. Equivalent circuit of nonlinear network shown in Fig. 5.8.

nonlinearities manifest themselves as controlled sources driving an augmented linear circuit.

After substituting the power series expansions into the K Kirchhoff current law equations, the equations are rearranged such that only terms linear in the node-to-datum voltages appear on the left-hand side. The differential operator $p = d/dt$ is also introduced. Carrying out these steps for the equations in (5-101) and (5-100), we obtain

$$[g_a + pc_0]\,a(t) - pc_0 b(t) = i_s(t) - \sum_{k=1}^{\infty} c_k [a(t) - b(t)]^k p[a(t) - b(t)]$$

$$- pc_0 a(t) + \left[g_1 + pc_0 + \frac{\Gamma_1}{p}\right] b(t) - \frac{\Gamma_1}{p} c(t)$$

$$= - \sum_{k=2}^{\infty} g_k b^k(t) - \sum_{k=2}^{\infty} \Gamma_k \left\{\frac{1}{p} [b(t) - c(t)]\right\}^k$$

$$+ \sum_{k=1}^{\infty} c_k [a(t) - b(t)]^k p[a(t) - b(t)]$$

$$- g_{1s} a(t) - \frac{\Gamma_1}{p} b(t) + \left[g_c + \frac{\Gamma_1}{p}\right] c(t) = \sum_{k=2}^{\infty} \Gamma_k \left\{\frac{1}{p} [b(t) - c(t)]\right\}^k + \sum_{k=2}^{\infty} g_{ks} a^k(t).$$

$$(5\text{-}102)$$

The equations are further simplified through the introduction of matrix notation. Define

$[Y(\cdot)]$ $K \times K$ node-admittance matrix of the augmented linear circuit;

$e(t)$ $K \times 1$ voltage vector consisting of the K node-to-datum voltages;

$i_s(t)$ $K \times 1$ independent current vector consisting of all independent current excitations;

$j(t)$ $K \times 1$ nonlinear current vector consisting of all second-order terms and higher arising from the power series expansion of the nonlinearities.

The K Kirchhoff current law equations then assume the form

$$[Y(p)] \, e(t) = i_s(t) + j(t). \tag{5-103}$$

With regard to the equations in (5-102),

$$[Y(p)] = \begin{bmatrix} (g_a + pc_0) & -pc_0 & 0 \\ -pc_0 & g_1 + pc_0 + \dfrac{\Gamma_1}{p} & -\dfrac{\Gamma_1}{p} \\ -g_{1s} & -\dfrac{\Gamma_1}{p} & \left(g_c + \dfrac{\Gamma_1}{p}\right) \end{bmatrix},$$

$$e(t) = \begin{bmatrix} a(t) \\ b(t) \\ c(t) \end{bmatrix}, \quad i_s(t) = \begin{bmatrix} i_s(t) \\ 0 \\ 0 \end{bmatrix},$$

$$j(t) = \begin{bmatrix} -\displaystyle\sum_{k=1}^{\infty} c_k [a(t) - b(t)]^k p [a(t) - b(t)] \\[2ex] -\displaystyle\sum_{k=2}^{\infty} g_k b^k(t) - \sum_{k=2}^{\infty} \Gamma_k \left\{ \dfrac{1}{p} [b(t) - c(t)] \right\}^k \\[1ex] \qquad + \displaystyle\sum_{k=1}^{\infty} c_k [a(t) - b(t)]^k p [a(t) - b(t)] \\[2ex] \displaystyle\sum_{k=2}^{\infty} \Gamma_k \left\{ \dfrac{1}{p} [b(t) - c(t)] \right\}^k + \sum_{k=2}^{\infty} g_{ks} a^k(t) \end{bmatrix}. \tag{5-104}$$

In general, nonlinear transfer functions relate the output of a weakly nonlinear system to its input. For simplicity in nodal analysis, we assume the input to be a single independent current source applied at node a. It follows that every row of the $K \times 1$ independent current vector $i_s(t)$ is identically zero except for the first row.

Linear Transfer Function: As indicated in Section 5.2, the harmonic input method is recursive in nature. The linear transfer functions are first determined by selecting the independent current vector to be

$$i_s(t) = \begin{bmatrix} \exp(j 2\pi f_1 t) \\ 0 \\ \vdots \\ 0 \end{bmatrix}. \tag{5-105}$$

The K node-to-datum voltages are then given by

$$a(t) = \sum_{n=1}^{N} a_n(t)$$

$$\vdots$$

$$k(t) = \sum_{n=1}^{N} k_n(t) \tag{5-106}$$

where the nth-order components are

$$a_n(t) = A_n(f_1, \ldots, f_1) \exp [j2\pi(nf_1)t]$$

$$\vdots$$

$$k_n(t) = K_n(f_1, \ldots, f_1) \exp [j2\pi(nf_1)t]. \tag{5-107}$$

After substituting equations (5-105) through (5-107) into (5-103), carrying out the operations implied by the matrix integro-differential operator $[Y(p)]$, and equating coefficients of $\exp [j2\pi f_1 t]$, the linear transfer functions are seen to satisfy the matrix equation

$$[Y(j2\pi f_1)] \begin{bmatrix} A_1(f_1) \\ B_1(f_1) \\ \vdots \\ K_1(f_1) \end{bmatrix} = \begin{bmatrix} 1 \\ 0 \\ \vdots \\ 0 \end{bmatrix}. \tag{5-108}$$

Since components at f_1 appear only in first-order terms, the nonlinear current vector $j(t)$ does not contribute to (5-108) and may be ignored as far as the equation for the linear transfer functions is concerned. Inverting the node-admittance matrix of the linearized circuit, the linear transfer functions are given by

$$\begin{bmatrix} A_1(f_1) \\ B_1(f_1) \\ \vdots \\ K_1(f_1) \end{bmatrix} = [Y(j2\pi f_1)]^{-1} \begin{bmatrix} 1 \\ 0 \\ \vdots \\ 0 \end{bmatrix}. \tag{5-109}$$

These are identical to the transfer functions that would be obtained by a conventional linear analysis of the linearized circuit.

   Higher Order Nonlinear Transfer Functions: The second-order nonlinear transfer functions are determined next and the procedure is continued recursively for each of the higher order transfer functions. In general, to determine the nth-order transfer functions, the independent current vector is selected to be

$$i_s(t) = \begin{bmatrix} \sum_{q_1=1}^{n} \exp(j2\pi f_{q_1} t) \\ 0 \\ \vdots \\ 0 \end{bmatrix} \qquad (5\text{-}110)$$

where the input frequencies are assumed to be positive and incommensurable. The node-to-datum voltages are again given by (5-106). However, now the nth-order components are expressed as

$$a_n(t) = \sum_{q_1=1}^{n} \cdots \sum_{q_n=1}^{n} A_n(f_{q_1}, \ldots, f_{q_n}) \exp[j2\pi(f_{q_1} + \cdots + f_{q_n})t]$$

$$\vdots$$

$$k_n(t) = \sum_{q_1=1}^{n} \cdots \sum_{q_n=1}^{n} K_n(f_{q_1}, \ldots, f_{q_n}) \exp[j2\pi(f_{q_1} + \cdots + f_{q_n})t].$$

$$(5\text{-}111)$$

Following the same approach as that used in evaluating the linear transfer functions, (5-110), (5-106), and (5-111) are substituted into (5-103). The operations implied by the matrix integro-differential operator $[Y(p)]$ are carried out and the coefficients of $\exp[j2\pi(f_1 + \cdots + f_n)t]$ are equated. Because of the assumption of positive and incommensurable input frequencies, components at $f_1 + \cdots + f_n$ occur only in nth-order terms. Therefore, the independent current vector may be ignored as far as the equation for the nonlinear transfer functions is concerned. The nth-order terms are in the form of n-fold summations and the components involving $\exp[j2\pi(f_1 + \cdots + f_n)t]$ are generated from the $n!$ permutations of the indices $q_1 = 1, q_2 = 2, \ldots, q_n = n$. With respect to the voltage vector $e(t)$, all $n!$ terms involving $\exp[j2\pi(f_1 + \cdots + f_n)t]$ are identical. Focusing on the coefficients of $\exp[j2\pi(f_1 + \cdots + f_n)t]$, it follows that $e(t)$ gives rise to the coefficient vector

$$E_n(f_1, \ldots, f_n) = \begin{bmatrix} n! A_n(f_1, \ldots, f_n) \\ n! B_n(f_1, \ldots, f_n) \\ \vdots \\ n! K_n(f_1, \ldots, f_n) \end{bmatrix} = n! \begin{bmatrix} A_n(f_1, \ldots, f_n) \\ B_n(f_1, \ldots, f_n) \\ \vdots \\ K_n(f_1, \ldots, f_n) \end{bmatrix}.$$

$$(5\text{-}112)$$

In general, the $n!$ terms contained in each row of the nonlinear current vector $j(t)$ are not identical. Since their evaluation becomes rather complicated for values of

n greater than or equal to three, we simply denote, at this point of our presentation, the coefficient vector arising from $\mathbf{j}(t)$ by

$$
\mathbf{J}_n(f_1,\ldots,f_n) =
\begin{bmatrix}
J_{an}(f_1,\ldots,f_n) \\
J_{bn}(f_1,\ldots,f_n) \\
\vdots \\
J_{kn}(f_1,\ldots,f_n)
\end{bmatrix}
\tag{5-113}
$$

where $J_{hn}(f_1,\ldots,f_n)$ represents the sum of the coefficients of $\exp[j2\pi(f_1 + \cdots + f_n)t]$ arising from the nonlinear circuit elements connected to node h; $h = a, b, \ldots, k$. A general algorithm for evaluating $J_{hn}(f_1,\ldots,f_n)$ is derived in Section 5.6. In terms of the coefficient vectors $\mathbf{E}_n(f_1,\ldots,f_n)$ and $\mathbf{J}_n(f_1, \ldots,f_n)$, the matrix equation for the nth-order transfer functions is given by

$$
[Y(j2\pi(f_1 + \cdots + f_n))]\,\mathbf{E}_n(f_1,\ldots,f_n) = \mathbf{J}_n(f_1,\ldots,f_n). \tag{5-114}
$$

Note that the matrix integro-differential operator $[Y(p)]$ has been replaced by the node-admittance matrix of the linearized circuit, evaluated at $f_1 + \cdots + f_n$. Substituting (5-112) and (5-113) into (5-114) and solving for the nonlinear transfer functions, we obtain

$$
\begin{bmatrix}
A_n(f_1,\ldots,f_n) \\
B_n(f_1,\ldots,f_n) \\
\vdots \\
K_n(f_1,\ldots,f_n)
\end{bmatrix}
= [Y(j2\pi(f_1 + \cdots + f_n))]^{-1}
\begin{bmatrix}
J_{an}(f_1,\ldots,f_n)/n! \\
J_{bn}(f_1,\ldots,f_n)/n! \\
\vdots \\
J_{kn}(f_1,\ldots,f_n)/n!
\end{bmatrix}.
$$

$$\tag{5-115}$$

The node-admittance matrix of the linearized circuit is seen to play a central role in determining the nonlinear transfer functions. The solution procedure is straightforward with the major difficulty being associated with the evaluation of the coefficient vector $\mathbf{J}_n(f_1,\ldots,f_n)$. In Section 5.6 the entries in $\mathbf{J}_n(f_1, \ldots, f_n)$ are shown to depend entirely on previously determined lower order transfer functions. The solution given by (5-115) can also be shown to be valid even when the input frequencies $f_1,\ldots,f_n$ are not positive and incommensurable.

As pointed out earlier, nodal analysis requires an admittance representation for each circuit element. Consequently, at the outset of our discussion, we allowed only voltage-controlled current sources. Other types of controlled sources may be included provided they can be converted to equivalent voltage-controlled current sources. This conversion is first demonstrated for current-controlled current sources.

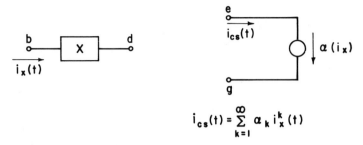

$$i_{cs}(t) = \sum_{k=1}^{\infty} \alpha_k\, i_x^k(t)$$

Fig. 5.10. Current-controlled current source with controlling element X.

Current-Controlled Current Source: Consider a circuit in which there is a current-controlled current source connected between nodes e and g, as shown in Fig. 5.10. Let the controlling element X be connected between nodes b and d. Assuming X is either a linear resistor with conductance G_x, a linear capacitor with capacitance C_x, or a linear inductor with reciprocal inductance Γ_x, the control current $i_x(t)$ may be expressed in terms of the voltage across the control element according to the relations

$$i_x(t) = G_x\,[b(t) - d(t)],$$

$$i_x(t) = C_x\,\frac{d[b(t) - d(t)]}{dt}\ ,$$

or

$$i_x(t) = \Gamma_x \int_{-\infty}^{t} [b(z) - d(z)]\, dz. \tag{5-116}$$

When these are substituted into the control equation, we obtain

$$i_{cs}(t) = \sum_{k=1}^{\infty} \alpha_k\, \{G_x\,[b(t) - d(t)]\}^k,$$

$$i_{cs}(t) = \sum_{k=1}^{\infty} \alpha_k\, \left\{C_x\,\frac{d[b(t) - d(t)]}{dt}\right\}^k,$$

or

$$i_{cs}(t) = \sum_{k=1}^{\infty} \alpha_k\, \left\{\Gamma_x \int_{-\infty}^{t} [b(z) - d(z)]\, dz\right\}^k. \tag{5-117}$$

Therefore, current-controlled current sources may be converted into equivalent voltage-controlled current sources by using the appropriate control equation from (5-117).

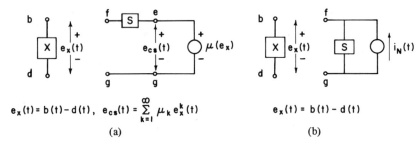

$$e_x(t) = b(t) - d(t), \quad e_{cs}(t) = \sum_{k=1}^{\infty} \mu_k e_x^k(t)$$

(a)

$$e_x(t) = b(t) - d(t)$$

(b)

Fig. 5.11. (a) Voltage-controlled voltage source with controlling element X and series element S. (b) Norton equivalent.

Voltage-Controlled Voltage Source: A voltage-controlled voltage source is now assumed to be connected between nodes e and g, as shown in Fig. 5.11(a). Typically, there will be a circuit element S in series with each voltage source. The voltage-controlled voltage source is then converted to a voltage-controlled current source by applying Norton's theorem as shown in Fig. 5.11(b). Assuming the series circuit element to be either a linear resistor, capacitor, or inductor with element values G_s, C_s, and Γ_s, respectively, the Norton equivalent current source is given by either

$$i_N(t) = G_s e_{cs}(t) = G_s \sum_{k=1}^{\infty} \mu_k e_x^k(t),$$

$$i_N(t) = C_s \frac{de_{cs}(t)}{dt} = C_s \sum_{k=1}^{\infty} k\mu_k e_x^{k-1}(t) \frac{de_x(t)}{dt},$$

or

$$i_N(t) = \Gamma_s \int_{-\infty}^{t} e_{cs}(z)\, dz = \Gamma_s \sum_{k=1}^{\infty} \mu_k \int_{-\infty}^{t} e_x^k(z)\, dz. \qquad (5\text{-}118)$$

Equations (5-118) are recognized to be the control equations for the equivalent voltage-controlled current sources. Finally, by substituting

$$e_x(t) = b(t) - d(t) \qquad (5\text{-}119)$$

into (5-118), we obtain the desired control equations in terms of node-to-datum voltages.

Current-Controlled Voltage Source: As illustrated in Fig. 5.12, current-controlled voltage sources may also be converted by means of Norton's theorem. However, depending upon whether the series element is a resistor, capacitor, or

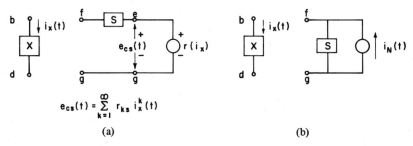

$$e_{cs}(t) = \sum_{k=1}^{\infty} r_{ks} i_x^k (t)$$

(a)   (b)

Fig. 5.12.   (a) Current-controlled voltage source with controlling element X and series element S. (b) Norton equivalent.

inductor, the Norton equivalent current source is now given by either

$$i_N(t) = G_s e_{cs}(t) = G_s \sum_{k=1}^{\infty} r_{ks} i_x^k(t),$$

$$i_N(t) = C_s \frac{de_{cs}(t)}{dt} = C_s \sum_{k=1}^{\infty} k r_{ks} i_x^{k-1}(t) \frac{di_x(t)}{dt},$$

or

$$i_N(t) = \Gamma_s \int_{-\infty}^{t} e_{cs}(z)\,dz = \Gamma_s \sum_{k=1}^{\infty} r_{ks} \int_{-\infty}^{t} i_x^k(z)\,dz. \qquad (5\text{-}120)$$

Assuming the controlling element X to be linear, the desired control equations in terms of node-to-datum voltages are obtained by substituting the appropriate equation from (5-116) into (5-120).

We see, therefore, that all types of controlled sources may be included in nodal analysis provided we first carry out the necessary preliminary manipulations.

5.4   LOOP ANALYSIS OF WEAKLY NONLINEAR CIRCUITS

Whereas nodal analysis makes use of Kirchhoff's current law, loop analysis employs Kirchhoff's voltage law. As a result, each circuit element is required to have an impedance representation in which the element voltage is expressed as a function of the element current. Should an impedance representation not exist, as is the case for a voltage-controlled resistor, loop analysis is not possible. As far as controlled sources are concerned, only current-controlled voltage sources have an impedance representation. Loop analysis is useful when a circuit contains elements having impedance, but not admittance, representations. In addition, loop analysis is preferable to nodal analysis when a network contains many more nodes than loops.

Although many different sets of loops may be chosen in loop analysis, for simplicity, we restrict ourselves to the set of loops known as meshes. By definition, a mesh is a closed loop that does not enclose any branches within its interior. In this section we consider planar-connected networks for which there are, in general, K meshes. In analogy with the notation used for nodal analysis, we designate: (1) the meshes by $a, b, \ldots, k$, (2) the corresponding mesh currents by $a(t), b(t), \ldots, k(t)$, and (3) the associated nth-order transfer functions by $A_n(f_1, \ldots, f_n), B_n(f_1, \ldots, f_n), \ldots, K_n(f_1, \ldots, f_n)$. This notation is illustrated for the circuit shown in Fig. 5.13 where, as before, all currents and voltages are assumed to be incremental variables.

Loop analysis is, in fact, the dual analytical procedure to nodal analysis. Therefore, a great deal of similarity exists between the two approaches. To further emphasize this similarity, the network in Fig. 5.13 is chosen to be the dual of that shown in Fig. 5.8. The use of loop analysis with the harmonic input method is illustrated in Example 5.9. Because of the resulting close correspondence to the equations of Example 5.8, detailed derivations are not provided. Instead, we simply give needed definitions and the main results. The reader is referred to Section 5.3 for the missing details.

Example 5.9

Network Equations and Equivalent Circuit: In loop analysis, where attention is focused on the meshes of a circuit, the network equations are obtained by writing Kirchhoff's voltage law around each of the K meshes. For the network of Fig. 5.13, we obtain

$$r_a a(t) + e_l(t) = e_s(t)$$

$$-e_l(t) + e_r(t) + e_c(t) = 0$$

$$-e_c(t) + r_c c(t) = e_{cs}(t). \tag{5-121}$$

The next step is to expand in a power series each voltage across a nonlinear circuit element where the series variables are mesh currents. An equivalent circuit is then generated in which each nonlinear circuit element is replaced by

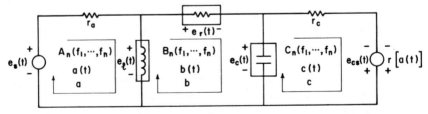

Fig. 5.13. Weakly nonlinear incremental circuit for illustrating loop analysis.

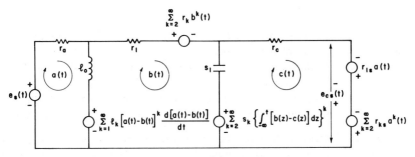

Fig. 5.14.   Equivalent circuit of nonlinear network shown in Fig. 5.13.

a linear circuit element in series with a current-controlled voltage source. By referring to the power series expansions in (5-4), (5-45), (5-39), and (5-58), the equivalent circuit for the network of Fig. 5.13 becomes that shown in Fig. 5.14. As with nodal analysis, the nonlinearities manifest themselves as controlled sources driving the augmented linear circuit.

The equivalent circuit is then used to write the K Kirchhoff voltage law equations such that only terms linear in the mesh currents appear on the left-hand side. The differential operator $p = d/dt$ is also introduced. For the circuit of Fig. 5.14, application of these steps gives rise to the equations

$$[r_a + pl_0]\, a(t) - pl_0 b(t) = e_s(t) - \sum_{k=1}^{\infty} l_k [a(t) - b(t)]^k p\, [a(t) - b(t)]$$

$$- pl_0 a(t) + \left[r_1 + pl_0 + \frac{s_1}{p} \right] b(t) - \frac{s_1}{p} c(t)$$

$$= - \sum_{k=2}^{\infty} r_k b^k(t) - \sum_{k=2}^{\infty} s_k \left\{ \frac{1}{p}\, [b(t) - c(t)] \right\}^k$$

$$+ \sum_{k=1}^{\infty} l_k [a(t) - b(t)]^k p\, [a(t) - b(t)]$$

$$- r_{1s} a(t) - \frac{s_1}{p} b(t) + \left[r_c + \frac{s_1}{p} \right] c(t) = \sum_{k=2}^{\infty} s_k \left\{ \frac{1}{p}\, [b(t) - c(t)] \right\}^k + \sum_{k=2}^{\infty} r_{ks} a^k(t).$$

$$(5\text{-}122)$$

In terms of matrix notation, the K Kirchhoff voltage law equations can, in general, be written as

$$[Z(p)]\, \mathbf{i}(t) = \mathbf{e}_s(t) + \mathbf{v}(t) \qquad (5\text{-}123)$$

where

$[Z(\cdot)]$ $K \times K$ loop-impedance matrix of the augmented linear circuit;

$i(t)$ $K \times 1$ current vector consisting of the K mesh currents;

$e_s(t)$ $K \times 1$ independent voltage vector consisting of all independent voltage excitations;

$v(t)$ $K \times 1$ nonlinear voltage vector consisting of all second-order terms and higher arising from the power series expansions of the nonlinearities.

Nonlinear Transfer Functions: Having expressed the network equations in the form of (5-123), the harmonic input method is then used to solve for the nonlinear transfer functions. In particular, assuming the independent voltage source appears only in loop a, the linear transfer functions are given by

$$
\begin{bmatrix} A_1(f_1) \\ B_1(f_1) \\ \vdots \\ K_1(f_1) \end{bmatrix} = [Z(j2\pi f_1)]^{-1} \begin{bmatrix} 1 \\ 0 \\ \vdots \\ 0 \end{bmatrix}. \tag{5-124}
$$

As in nodal analysis, these are identical to the transfer functions that would be obtained by a conventional linear analysis of the linearized circuit. Higher order transfer functions are found recursively. In analogy with (5-113), the coefficient vector arising from the nonlinear voltage vector $v(t)$ is defined to be

$$
\mathbf{V}_n(f_1, \ldots, f_n) = \begin{bmatrix} V_{an}(f_1, \ldots, f_n) \\ V_{bn}(f_1, \ldots, f_n) \\ \vdots \\ V_{kn}(f_1, \ldots, f_n) \end{bmatrix}. \tag{5-125}
$$

However, now $V_{hn}(f_1, \ldots, f_n)$ represents the sum of the coefficients of $\exp[j2\pi(f_1 + \cdots + f_n)t]$ due to the nonlinear circuit elements appearing in loop h; $h = a, b, \ldots, k$. A general algorithm for evaluating $V_{hn}(f_1, \ldots, f_n)$ is derived in Section 5.6. There it is shown that $V_{hn}(f_1, \ldots, f_n)$ depends only on nonlinear transfer functions up to order $(n - 1)$. In general, the solution for the nth-order transfer functions is expressed as

$$
\begin{bmatrix} A_n(f_1, \ldots, f_n) \\ B_n(f_1, \ldots, f_n) \\ \vdots \\ K_n(f_1, \ldots, f_n) \end{bmatrix} = [Z(j2\pi(f_1 + \cdots + f_n))]^{-1} \begin{bmatrix} V_{an}(f_1, \ldots, f_n)/n! \\ V_{bn}(f_1, \ldots, f_n)/n! \\ \vdots \\ V_{kn}(f_1, \ldots, f_n)/n! \end{bmatrix}.
$$

$$\tag{5-126}$$

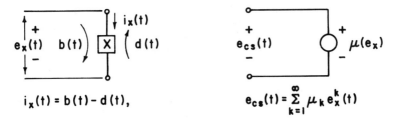

$$i_x(t) = b(t) - d(t),$$

$$e_{cs}(t) = \sum_{k=1}^{\infty} \mu_k e_x^k(t)$$

Fig. 5.15.   Voltage-controlled voltage source with controlling element X.

For loop analysis, we see that the loop-impedance matrix of the linearized circuit plays a central role in determining the nonlinear transfer functions.

Because loop analysis requires an impedance representation for each circuit element, all controlled sources were initially required to be current-controlled voltage sources. This requirement may be relaxed provided the other types of controlled sources can be appropriately converted.

Voltage-Controlled Voltage Source: The easiest source to convert is the voltage-controlled voltage source shown in Fig. 5.15. Assuming the controlling element X is either a linear resistor with resistance R_x, a linear inductor with inductance L_x, or a linear capacitor with elastance S_x, the control equation may be written as either

$$e_{cs}(t) = \sum_{k=1}^{\infty} \mu_k \{R_x [b(t) - d(t)]\}^k,$$

$$e_{cs}(t) = \sum_{k=1}^{\infty} \mu_k \left\{ L_x \frac{d[b(t) - d(t)]}{dt} \right\}^k,$$

or

$$e_{cs}(t) = \sum_{k=1}^{\infty} \mu_k \left\{ S_x \int_{-\infty}^{t} [b(z) - d(z)] \, dz \right\}^k. \tag{5-127}$$

Using the appropriate control equation, we see that a voltage-controlled voltage source may be converted to the desired current-controlled voltage source.

Current-Controlled Current Source: Now consider the current-controlled current source illustrated in Fig. 5.16(a). Typically, a circuit element P will be connected in parallel with each current source. By applying Thévenin's theorem, it is possible to convert the current-controlled current source to an equivalent current-controlled voltage source, as indicated in Fig. 5.16(b). Assuming the parallel circuit element to be either a linear resistor, inductor, or capacitor with

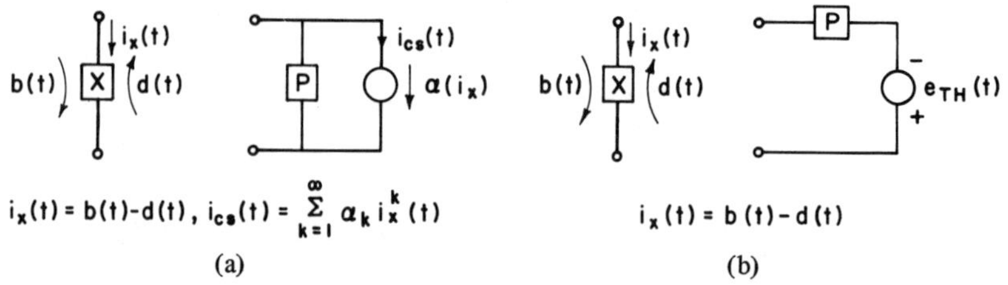

$$i_x(t) = b(t) - d(t), \quad i_{cs}(t) = \sum_{k=1}^{\infty} \alpha_k i_x^k(t)$$

(a)

$$i_x(t) = b(t) - d(t)$$

(b)

Fig. 5.16. (a) Current-controlled current source with controlling element X and parallel element P. (b) Thévenin equivalent.

element values R_p, L_p, and S_p, respectively, the Thévenin equivalent voltage source is given by either

$$e_{TH}(t) = R_p i_{cs}(t) = R_p \sum_{k=1}^{\infty} \alpha_k i_x^k(t),$$

$$e_{TH}(t) = L_p \frac{di_{cs}(t)}{dt} = L_p \sum_{k=1}^{\infty} k\alpha_k i_x^{k-1}(t) \frac{di_x(t)}{dt},$$

or

$$e_{TH}(t) = S_p \int_{-\infty}^{t} i_{cs}(z)\, dz = S_p \sum_{k=1}^{\infty} \alpha_k \int_{-\infty}^{t} i_x^k(z)\, dz. \qquad (5\text{-}128)$$

The desired control equations in terms of mesh currents are obtained by substituting

$$i_x(t) = b(t) - d(t) \qquad (5\text{-}129)$$

into (5-128).

Voltage-Controlled Current Source: Thévenin's theorem may also be used to convert a voltage-controlled current source, as shown in Fig. 5.17. However, now the corresponding expression for the Thévenin equivalent voltage source is either

$$e_{TH}(t) = R_p i_{cs}(t) = R_p \sum_{k=1}^{\infty} g_{ks} e_x^k(t),$$

$$e_{TH}(t) = L_p \frac{di_{cs}(t)}{dt} = L_p \sum_{k=1}^{\infty} k g_{ks} e_x^{k-1}(t) \frac{de_x(t)}{dt},$$

or

$$e_{TH}(t) = S_p \int_{-\infty}^{t} i_{cs}(z)\, dz = S_p \sum_{k=1}^{\infty} g_{ks} \int_{-\infty}^{t} e_x^k(z)\, dz. \qquad (5\text{-}130)$$

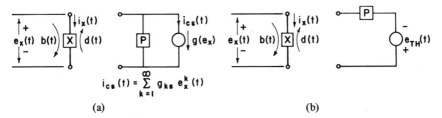

$$i_{cs}(t) = \sum_{k=1}^{\infty} g_{ks}\, e_x^k(t)$$

(a)   (b)

Fig. 5.17. (a) Voltage-controlled current source with controlling element X and parallel element P. (b) Thévenin equivalent.

For a linear controlling element X, the control equation in (5-130) is expressed in terms of mesh currents by replacing $e_x(t)$ with either

$$e_x(t) = R_x[b(t) - d(t)],$$

$$e_x(t) = L_x \frac{d[b(t) - d(t)]}{dt},$$

or

$$e_x(t) = S_x \int_{-\infty}^{t} [b(z) - d(z)]\, dz. \tag{5-131}$$

In this way, an impedance representation is generated for voltage-controlled current sources.

This completes our discussion of loop analysis. In the next section we consider mixed-variable analysis where both voltages and currents appear as unknowns in the network equations.

5.5 MIXED-VARIABLE ANALYSIS OF WEAKLY NONLINEAR CIRCUITS

As we have seen, the success of writing node or loop equations, respectively, depends on the existence of admittance and impedance representations for each circuit element. Elements that operate in a monotonic region of their characteristic curves possess both admittance and impedance representations. Although series reversion may be used to convert between the two representations, one of the series expansions may be preferable because of its convergence properties (i.e., one series may require fewer terms than the other for an acceptable characterization). We have also shown that certain preliminary manipulations may be used to convert controlled sources from one representation to another. In some instances, however, such manipulations may be undesirable. In addition, some elements have only admittance representations, as with voltage-controlled

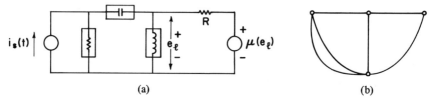

Fig. 5.18.   (a) Network. (b) Network graph.

capacitors, while other elements have only impedance representations, as with current-controlled inductors. It is desirable, therefore, to have a technique for writing network equations that utilizes circuit element relationships in their natural form. Such a scheme is presented in this section.

Trees and Cotrees: The first step in the procedure is to construct a network graph. This is accomplished by replacing each circuit element with a line having a small circle at each end. The circles are called nodes, or terminals, and the line segments between nodes are called branches.

Example 5.10

The network graph for the circuit of Fig. 5.18(a) is shown in Fig. 5.18(b). We use n_b to denote the number of branches and n_t the number of nodes in the network graph. For the network graph of Fig. 5.18(b), we have $n_b = 6$ and $n_t = 4$.

To assist us in writing the network equations, we next separate the network graph into two parts: (1) a tree and (2) a cotree. A tree is any set of branches that connects all the nodes of a network without forming any closed paths. It follows that all trees contain $n_t - 1$ branches. The set of branches not in the tree is called the complement of the tree or the cotree. Therefore, all cotrees consist of $n_b - (n_t - 1) = n_b - n_t + 1$ branches. The branches in a tree are called tree branches or twigs; those branches that are not in a tree are called links. For a given network graph it is possible, in general, to draw several different trees.

Example 5.11

A number of trees for the network graph of Fig. 5.18(b) are presented in Fig. 5.19. Note that each tree (solid lines) defines a unique cotree (broken lines).

Selection of a Proper Tree: Tree branch voltages and link currents are particularly significant in that all branch voltages can be expressed in terms of any set of tree branch voltages while all branch currents can be expressed in terms of

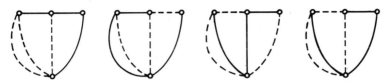

Fig. 5.19. Possible trees (solid lines) and cotrees (broken lines) for the network graph of Fig. 5.18(b).

any set of link currents. With this in mind, we propose to write a system of network equations in which the tree branch voltages and corresponding link currents appear as the unknowns. For the process to be successful, the tree and cotree must be selected so that circuit elements are appropriately located. Specifically, the tree must contain all branches corresponding to: (1) independent voltage sources, (2) controlled current sources, (3) voltage-controlled resistors, (4) voltage-controlled capacitors, (5) flux-linkage-controlled inductors, and (6) elements whose voltages appear as control voltages elsewhere in the circuit.

On the other hand, the cotree must contain all branches corresponding to: (1) independent current sources, (2) controlled voltage sources, (3) current-controlled resistors, (4) current-controlled inductors, (5) charge-controlled capacitors, and (6) elements whose currents appear as control currents elsewhere in the circuit.

Branches corresponding to elements that have both admittance and impedance representations may be placed either in the tree or cotree. In most practical networks the requirements listed above for the selection of a proper tree are usually met. Should they not be satisfied, the difficulties encountered can usually be overcome by placing small parasitic elements at strategic points in the network to simulate either lead resistances or shunt conductances of physical elements. To illustrate the selection of a proper tree, we now consider two cases relative to the network of Fig. 5.18(a). In one the selection procedure proceeds routinely whereas a difficulty does arise in the other.

Example 5.12

First, let the characteristic curves of the nonlinear elements in the network of Fig. 5.18(a) be monotonic in nature. Then both admittance and impedance representations exist for each of the nonlinear elements as well as for the linear resistor R. Following the rules for selecting a proper tree, the branch corresponding to the inductor must be in the tree because its voltage is used as a control voltage. On the other hand, the branches corresponding to the independent current source and voltage-controlled voltage source must be in the cotree. One acceptable tree, therefore, consists of the branches corresponding to the nonlinear resistor, nonlinear inductor, and linear resistor. Another is the tree constructed from the branches corresponding to the nonlinear capacitor, nonlinear inductor, and linear resistor.

Example 5.13

For the second case, the inductor in the network of Fig. 5.18(a) is assumed to be current controlled. This results in a problem because there are now two conflicting requirements as to where the branch corresponding to the inductor should be placed. The current-controlled aspect of the inductor requires that its branch appear in a cotree. Simultaneously, use of the inductor voltage as a control voltage requires that the branch be placed in a tree. This conflict is resolved by placing a small conductance in shunt with the inductor, as shown in Fig. 5.20. The parasitic conductance G_p is assumed to be sufficiently small so as to have a negligible effect on the distribution of currents and voltages in the network. Nevertheless, the control voltage across the inductor is equal to the voltage across G_p. Therefore, G_p can be considered as the control element. This allows G_p to be placed in the tree instead of the inductor. Consequently, a proper tree may be constructed from the branches corresponding to the nonlinear resistor, parasitic conductance, and linear resistor.

Formulation of the Mixed-Variable Equations: Having chosen a proper tree, the network equations are obtained by carrying out the following steps:

(1) Assign current and voltage symbols to each branch of the network graph.
(2) For each tree branch that does not correspond to an independent voltage source, use Kirchhoff's current law to express the tree branch current as a sum of link currents.
(3) For each link that does not correspond to an independent current source, use Kirchhoff's voltage law to express the link voltage as a sum of tree branch voltages.
(4) Eliminate all tree branch currents and all link voltages from the equations obtained in steps 2 and 3 by substituting the appropriate source, control,

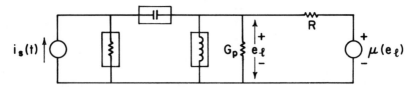

Fig. 5.20. Circuit of Fig. 5.18(a) with parasitic conductance placed in shunt with nonlinear inductor.

and element equations. The variables in the resulting equations should be the link currents and tree branch voltages.

In general, if there are n_v independent voltage sources and n_c independent current sources, the number of equations obtained by the above procedure is given by

$K =$ (number of tree branch voltages) $-$ (number of independent voltage sources)

  $+$ (number of links) $-$ (number of independent current sources)

$$= (n_t - 1) - (n_v) + (n_b - n_t + 1) - (n_c) = n_b - n_v - n_c. \qquad (5\text{-}132)$$

Therefore, the number of equations equals the number of branches reduced by the number of independent sources. This is a major drawback since the number of equations for the mixed-variable approach is typically much greater than the number of equations needed in either nodal or loop analysis. Of the $n_b - n_v - n_c$ variables appearing in the mixed-variable equations, $n_t - n_v - 1$ are tree branch voltages and $n_b - n_t - n_c + 1$ are link currents. Application of mixed-variable analysis to the harmonic input method is now illustrated in Example 5.14.

Example 5.14

Network Equations and Equivalent Circuit: Once again, consider the network of Fig. 5.18(a). Assume both admittance and impedance representations to exist for each of the nonlinear elements. A network graph with assigned current and voltage symbols is shown in Fig. 5.21. Solid lines indicate a proper tree while broken lines are used for the corresponding cotree. As in our discussion of nodal and loop analysis, numerical subscripts are reserved to designate the order of a current or voltage component. Therefore, the link currents are denoted by $a(t)$, $b(t)$, and $f(t)$ and the tree branch voltages are denoted by $c(t)$, $d(t)$, and $e(t)$.

Performing steps 2 and 3 of the procedure, we obtain the following set of

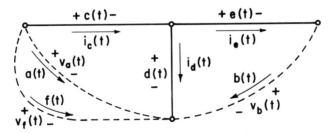

Fig. 5.21. Network graph, tree (solid lines), cotree (broken lines), current and voltage symbols for circuit of Fig. 5.18(a).

equations:

$$i_c(t) = -a(t) - f(t)$$

$$i_d(t) = -a(t) - b(t) - f(t)$$

$$i_e(t) = b(t)$$

$$v_a(t) = c(t) + d(t)$$

$$v_b(t) = d(t) - e(t). \tag{5-133}$$

Note that an equation is not written for branch f since it corresponds to an independent current source. The appropriate source, control, and element equations needed for step 4 are

$$f(t) = -i_s(t)$$

$$v_b(t) = \mu[d(t)] = \sum_{k=1}^{\infty} \mu_k d^k(t)$$

$$i_c(t) = \sum_{k=0}^{\infty} c_k c^k(t) \frac{dc(t)}{dt}$$

$$i_d(t) = \sum_{k=1}^{\infty} \Gamma_k \left[\int_{-\infty}^{t} d(z)\, dz \right]^k$$

$$i_e(t) = \frac{1}{R} e(t)$$

$$v_a(t) = \sum_{k=1}^{\infty} r_k a^k(t) \tag{5-134}$$

where use has been made of (5-57), (5-28), (5-55), and (5-4), respectively, to get power series expansions for the nonlinear elements. Substituting (5-134) into (5-133), the network equations become

$$\sum_{k=0}^{\infty} c_k c^k(t) \frac{dc(t)}{dt} = -a(t) + i_s(t)$$

$$\sum_{k=1}^{\infty} \Gamma_k \left[\int_{-\infty}^{t} d(z)\, dz \right]^k = -a(t) - b(t) + i_s(t)$$

$$\frac{1}{R} e(t) = b(t)$$

$$\sum_{k=1}^{\infty} r_k a^k(t) = c(t) + d(t)$$

$$\sum_{k=1}^{\infty} \mu_k d^k(t) = d(t) - e(t). \qquad (5\text{-}135)$$

Finally, introducing the differential operator $p = d/dt$ and rearranging the equations so that only terms linear in $a(t)$, $b(t)$, $c(t)$, $d(t)$, and $e(t)$ appear on the left-hand side, we have

$$a(t) + pc_0 c(t) = i_s(t) - \sum_{k=1}^{\infty} c_k c^k(t) \, pc(t)$$

$$a(t) + b(t) + \frac{\Gamma_1}{p} d(t) = i_s(t) - \sum_{k=2}^{\infty} \Gamma_k \left[\frac{1}{p} d(t) \right]^k$$

$$-b(t) + \frac{1}{R} e(t) = 0$$

$$-r_1 a(t) + c(t) + d(t) = \sum_{k=2}^{\infty} r_k a^k(t)$$

$$[\mu_1 - 1] \, d(t) + e(t) = - \sum_{k=2}^{\infty} \mu_k d^k(t). \qquad (5\text{-}136)$$

The equations in (5-136) suggest the nonlinear incremental equivalent circuit shown in Fig. 5.22. Just as happened in nodal and loop analysis, the nonlinearities manifest themselves as controlled sources driving the linearized circuit. Although there are six branches in the network of Fig. 5.18(a), there are only five equations in (5-136) since one of the branches is an independent current

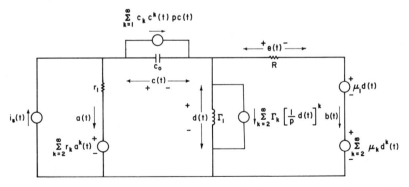

Fig. 5.22. Nonlinear incremental equivalent circuit of network shown in Fig. 5.18(a).

source. Once the unknown tree branch voltages and link currents have been determined, currents and voltages can be evaluated anywhere in the circuit.

In general, the $K = n_b - n_v - n_c$ network equations can be written, using matrix notation, in the form

$$[M(p)]\, \mathbf{x}(t) = \mathbf{g}_s(t) + \mathbf{u}(t) \tag{5-137}$$

where

$[M(\cdot)]$ $K \times K$ mixed-variable immittance matrix of the augmented linear circuit;

$\mathbf{x}(t)$ $K \times 1$ mixed-variable vector consisting of the K unknown link currents and tree branch voltages;

$\mathbf{g}_s(t)$ $K \times 1$ mixed-variable generator vector consisting of all independent current and voltage excitations;

$\mathbf{u}(t)$ $K \times 1$ nonlinear mixed-variable vector consisting of all second-order terms and higher arising from the power series expansions of the nonlinearities.

For the equations in (5-136),

$$\mathbf{x}(t) = \begin{bmatrix} a(t) \\ b(t) \\ c(t) \\ d(t) \\ e(t) \end{bmatrix}, \qquad \mathbf{g}_s(t) = \begin{bmatrix} i_s(t) \\ i_s(t) \\ 0 \\ 0 \\ 0 \end{bmatrix},$$

$$\mathbf{u}(t) = \begin{bmatrix} -\displaystyle\sum_{k=1}^{\infty} c_k c^k(t)\, pc(t) \\[2ex] -\displaystyle\sum_{k=2}^{\infty} \Gamma_k \left[\dfrac{1}{p} d(t) \right]^k \\[2ex] 0 \\[2ex] \displaystyle\sum_{k=2}^{\infty} r_k a^k(t) \\[2ex] -\displaystyle\sum_{k=2}^{\infty} \mu_k d^k(t) \end{bmatrix},$$

$$[M(p)] = \begin{bmatrix} 1 & 0 & pc_0 & 0 & 0 \\ 1 & 1 & 0 & \Gamma_1/p & 0 \\ 0 & -1 & 0 & 0 & 1/R \\ -r_1 & 0 & 1 & 1 & 0 \\ 0 & 0 & 0 & (\mu_1 - 1) & 1 \end{bmatrix}. \tag{5-138}$$

Nonlinear Transfer Functions: Having expressed the network equations in the form of (5-137), the next step is to apply the harmonic input method to solve for the nonlinear transfer functions. As in nodal and loop analysis, the linear transfer functions are identical to those that would be obtained by a conventional linear mixed-variable analysis of the linearized circuit. Higher order transfer functions are found recursively. In analogy with (5-113) and (5-125), the coefficient vector arising from the nonlinear mixed-variable vector $u(t)$ is denoted by

$$\mathbf{U}_n(f_1, \ldots, f_n) = \begin{bmatrix} U_{an}(f_1, \ldots, f_n) \\ U_{bn}(f_1, \ldots, f_n) \\ \vdots \\ U_{kn}(f_1, \ldots, f_n) \end{bmatrix}. \tag{5-139}$$

A general algorithm for generating the entries in $\mathbf{U}_n(f_1, \ldots, f_n)$ is presented in Section 5.6. In terms of this vector, the nth-order nonlinear transfer functions are given by

$$\begin{bmatrix} A_n(f_1, \ldots, f_n) \\ B_n(f_1, \ldots, f_n) \\ \vdots \\ K_n(f_1, \ldots, f_n) \end{bmatrix} = [M(j2\pi(f_1 + \cdots + f_n))]^{-1} \begin{bmatrix} U_{an}(f_1, \ldots, f_n)/n! \\ U_{bn}(f_1, \ldots, f_n)/n! \\ \vdots \\ U_{kn}(f_1, \ldots, f_n)/n! \end{bmatrix}.$$

$$\tag{5-140}$$

Once again, the form of the solution emphasizes the need for a good linearized model of the nonlinear system if the higher order transfer functions are to be accurately evaluated.

5.6 ALGORITHM FOR EVALUATING $\mathbf{J}_n(f_1, \ldots, f_n)$, $\mathbf{V}_n(f_1, \ldots, f_n)$, AND $\mathbf{U}_n(f_1, \ldots, f_n)$

In our discussions of nodal, loop, and mixed-variable analysis, the nonlinear transfer functions were shown to depend upon the nonlinear vectors $\mathbf{J}_n(f_1, \ldots, f_n)$, $\mathbf{V}_n(f_1, \ldots, f_n)$, and $\mathbf{U}_n(f_1, \ldots, f_n)$, respectively. The elements of these vectors are coefficients of $\exp[j2\pi(f_1 + \cdots + f_n)t]$ and are obtained from the power series expansions associated with the nonlinear circuit elements. Because of the assumption of positive and incommensurable input frequencies, intermodulation components at $(f_1 + \cdots + f_n)$ are generated only in nth-order terms. Since evaluation of the coefficients of $\exp[j2\pi(f_1 + \cdots + f_n)t]$ becomes increasingly tedious for large values of n, a recursive relationship is developed in this section to simplify the procedure.

Zero-Memory Nonlinearity: As has been shown, the power series expansions associated with the nonlinear circuit elements involve various quantities raised to powers. We first consider zero-memory expansions of the form

$$z(t) = \sum_{r=2}^{\infty} a_r w^r(t) \tag{5-141}$$

where the linear term has been omitted because it contributes to the augmented linear circuit. In the nonlinear transfer function approach, the nth-order component of $w(t)$ is denoted by $w_n(t)$ and $w(t)$ is represented as

$$w(t) = \sum_{n=1}^{N} w_n(t). \tag{5-142}$$

Therefore, substitution of (5-142) into (5-141) results in

$$z(t) = \sum_{r=2}^{\infty} a_r \left[\sum_{n=1}^{N} w_n(t) \right]^r \tag{5-143}$$

and the rth-degree term of the series contains components of order r, given by $a_r w_1^r(t)$, up to components of order Nr, given by $a_r w_N^r(t)$. If the nth-order component of $z(t)$ is denoted by $z_n(t)$, it follows that (5-143) may be expressed as

$$z(t) = \sum_{n=2}^{\infty} z_n(t) \tag{5-144}$$

where

$$z_2(t) = a_2 w_1^2(t)$$
$$z_3(t) = a_3 w_1^3(t) + 2a_2 w_1(t) w_2(t)$$
$$z_4(t) = a_4 w_1^4(t) + 3a_3 w_1^2(t) w_2(t) + a_2 [w_2^2(t) + 2w_1(t) w_3(t)]. \tag{5-145}$$

In general, the nth-order component of $z(t)$ is generated from terms in (5-141) having degrees 2 through n. This is emphasized by expressing $z_n(t)$ as

$$z_n(t) = \sum_{r=2}^{n} a_r w_{n,r}(t) \tag{5-146}$$

where $w_{n,r}(t)$ denotes a component of order n that is generated by the rth-degree term in (5-141). Note that

$$w_{2,2}(t) = w_1^2(t),$$
$$w_{3,2}(t) = 2w_1(t) w_2(t),$$
$$w_{3,3}(t) = w_1^3(t),$$

$$w_{4,2}(t) = w_2^2(t) + 2w_1(t) w_3(t),$$

$$w_{4,3}(t) = 3w_1^2(t) w_2(t),$$

$$w_{4,4}(t) = w_1^4(t). \tag{5-147}$$

Recall that the entries in $\mathbf{J}_n(f_1, \ldots, f_n)$, $\mathbf{V}_n(f_1, \ldots, f_n)$, and $\mathbf{U}_n(f_1, \ldots, f_n)$ arise only from nth-order terms. The form of (5-146) is especially significant because it reveals the structure of $z_n(t)$. To ease the task of evaluating $z_n(t)$, we now develop a recursive relationship for $w_{n,r}(t)$. In particular, we show that

$$w_{n,r}(t) = \sum_{i=1}^{n-r+1} w_i(t) w_{n-i,r-1}(t) \tag{5-148}$$

where

$$1 \leqslant i \leqslant N \tag{5-149}$$

and

$$r \leqslant n \leqslant Nr. \tag{5-150}$$

The inequality on i exists because the components of $w(t)$ above order N are assumed to be negligible, as indicated by (5-142). The summation in (5-148) terminates with $i = N$ when $n - r + 1 > N$. The inequality on n exists because only components from order r up to order Nr are capable of being generated by the rth-degree term of (5-141).

Derivation of Recursive Relationship for $w_{n,r}(t)$: The derivation of the recursion relation begins by defining

$$A(\alpha) = \sum_{j=1}^{N} \alpha^j w_j(t) \tag{5-151}$$

and

$$B(\alpha) = \left[\sum_{j=1}^{N} \alpha^j w_j(t) \right]^{r-1}. \tag{5-152}$$

The dummy variable α serves to keep track of various-order terms. Note that

$$A(\alpha) B(\alpha) = \left[\sum_{j=1}^{N} \alpha^j w_j(t) \right]^r. \tag{5-153}$$

Because the summation in (5-153) is raised to the rth power, $A(\alpha) B(\alpha)$ contains components from order r up to order Nr. In addition, a component of order k has α^k as a multiplier. This is reflected by rewriting (5-153) as

$$A(\alpha) B(\alpha) = \sum_{k=r}^{Nr} \alpha^k w_{k,r}(t). \tag{5-154}$$

In general, $w_{k,r}(t)$ is of kth order and is composed of terms containing r factors that are functions of time, as illustrated in (5-147). Note that

$$w_{k,r}(t) = 0 \quad \text{for} \quad r > k \tag{5-155}$$

since an r-fold product of factors, each of which has order greater than or equal to unity, cannot yield a term having order less than r.

Our objective is to find a recursion relation for $w_{n,r}(t)$. Differentiating (5-154) n times with respect to α, where n satisfies the inequality in (5-150), there results

$$\frac{d^n}{d\alpha^n} [A(\alpha) B(\alpha)] = \sum_{k=n}^{Nr} k(k-1) \cdots (k-n+1) \alpha^{k-n} w_{k,r}(t). \tag{5-156}$$

Observe that the lower limit in (5-156) is $k = n$ since the nth derivative of α^k with respect to α is identically zero for $k < n$. For $\alpha = 0$ in (5-156) the only term that is nonzero arises from $k = n$. It follows that

$$\frac{d^n}{d\alpha^n} [A(\alpha) B(\alpha)] \bigg|_{\alpha=0} = n! \, w_{n,r}(t). \tag{5-157}$$

On the other hand, Leibnitz's theorem for n-fold differentiation of the product of two functions states that

$$\frac{d^n}{d\alpha^n} [A(\alpha) B(\alpha)] = \sum_{i=0}^{n} \binom{n}{i} \frac{d^i A(\alpha)}{d\alpha^i} \frac{d^{n-i} B(\alpha)}{d\alpha^{n-i}}. \tag{5-158}$$

With $\alpha = 0$, (5-158) becomes

$$\frac{d^n}{d\alpha^n} [A(\alpha) B(\alpha)] \bigg|_{\alpha=0} = \sum_{i=0}^{n} \binom{n}{i} \frac{d^i A(\alpha)}{d\alpha^i} \bigg|_{\alpha=0} \frac{d^{n-i} B(\alpha)}{d\alpha^{n-i}} \bigg|_{\alpha=0}. \tag{5-159}$$

Reasoning as in (5-156), the ith derivative of $A(\alpha)$, where i satisfies the inequality in (5-149), is given by

$$\frac{d^i A(\alpha)}{d\alpha^i} = \sum_{j=i}^{N} (j) (j-1) \cdots (j-i+1) \alpha^{j-i} w_j(t). \tag{5-160}$$

When $\alpha = 0$, only the term corresponding to $j = i$ is nonzero. Therefore,

$$\frac{d^i A(\alpha)}{d\alpha^i} \bigg|_{\alpha=0} = i! \, w_i(t). \tag{5-161}$$

Equation (5-159) also requires the $(n-i)$th derivative of $B(\alpha)$. From the definition of $B(\alpha)$ in (5-152), it follows that $B(\alpha)$ contains components from order

$(r - 1)$ up to order $N(r - 1)$. In analogy with (5-154), we express $B(\alpha)$ as

$$B(\alpha) = \sum_{k=r-1}^{N(r-1)} \alpha^k w_{k,r-1}(t). \tag{5-162}$$

Provided $(n - i)$ satisfies the inequality

$$r - 1 \leqslant n - i \leqslant N(r - 1), \tag{5-163}$$

the $(n - i)$th derivative of $B(\alpha)$ with respect to α is

$$\frac{d^{n-i}B(\alpha)}{d\alpha^{n-i}} = \sum_{k=n-i}^{N(r-1)} k(k - 1) \cdots (k - n + i + 1) \alpha^{k-n+i} w_{k,r-1}(t). \tag{5-164}$$

Letting $\alpha = 0$ in (5-164), the only nonzero term in the sum is the term for which $k = n - i$. It follows that

$$\left. \frac{d^{n-i}B(\alpha)}{d\alpha^{n-i}} \right|_{\alpha=0} = (n - i)!\, w_{n-i,r-1}(t). \tag{5-165}$$

The remaining term to be evaluated in (5-159) is the binomial coefficient which, by definition, is

$$\binom{n}{i} = \frac{n!}{i!(n - i)!}. \tag{5-166}$$

Consequently, making use of (5-161), (5-165), and (5-166), (5-159) reduces to

$$\left. \frac{d^n}{d\alpha^n} [A(\alpha)\, B(\alpha)] \right|_{\alpha=0} = \sum_{i=0}^{n} n!\, w_i(t)\, w_{n-i,r-1}(t). \tag{5-167}$$

Equating (5-157) with (5-167) and cancelling $n!$ on each side of the equation, there results

$$w_{n,r}(t) = \sum_{i=0}^{n} w_i(t)\, w_{n-i,r-1}(t). \tag{5-168}$$

However, $w_0(t)$ is identically zero, as evidenced by the representation for $w(t)$ in (5-142). Therefore, the summation in (5-168) can begin with i equal to unity. In addition, the inequality in (5-163) requires $i \leqslant n - r + 1$. Hence the summation in (5-168) terminates with i equal to $n - r + 1$. The desired recursive relationship becomes

$$w_{n,r}(t) = \sum_{i=1}^{n-r+1} w_i(t)\, w_{n-i,r-1}(t) \tag{5-169}$$

as was stated in (5-148). In using the recursive relationship, it is helpful to know that

$$w_{n,1}(t) = w_n(t). \tag{5-170}$$

This is verified by comparing (5-153) with (5-154) for $r = 1$.

As a check on the recursive relationship, observe that the second-, third-, and fourth-order terms in (5-147) are generated by application of (5-169) in conjunction with (5-170). We further illustrate (5-169) by using it to generate the fifth-order terms in (5-141) arising from $r = 2, 3, 4, 5$.

Example 5.15

With reference to (5-146),

$$z_5(t) = \sum_{r=2}^{5} a_r w_{5,r}(t) \tag{5-171}$$

where

$$w_{5,2}(t) = \sum_{i=1}^{4} w_i(t) \, w_{5-i,1}(t) = w_1(t) \, w_{4,1}(t) + w_2(t) \, w_{3,1}(t)$$

$$+ w_3(t) \, w_{2,1}(t) + w_4(t) \, w_{1,1}(t) = w_1(t) \, w_4(t)$$

$$+ w_2(t) \, w_3(t) + w_3(t) \, w_2(t) + w_4(t) \, w_1(t)$$

$$= 2[w_1(t) \, w_4(t) + w_2(t) \, w_3(t)],$$

$$w_{5,3}(t) = \sum_{i=1}^{3} w_i(t) \, w_{5-i,2}(t) = w_1(t) \, w_{4,2}(t) + w_2(t) \, w_{3,2}(t)$$

$$+ w_3(t) \, w_{2,2}(t) = w_1(t)[w_2^2(t) + 2w_1(t) \, w_3(t)]$$

$$+ w_2(t)[2w_1(t) \, w_2(t)] + w_3(t)[w_1^2(t)]$$

$$= 3[w_1^2(t) \, w_3(t) + w_1(t) \, w_2^2(t)],$$

$$w_{5,4}(t) = \sum_{i=1}^{2} w_i(t) \, w_{5-i,3}(t) = w_1(t) \, w_{4,3}(t) + w_2(t) \, w_{3,3}(t)$$

$$= w_1(t)[3w_1^2(t) \, w_2(t)] + w_2(t)[w_1^3(t)]$$

$$= 4w_1^3(t) \, w_2(t),$$

$$w_{5,5}(t) = \sum_{i=1}^{1} w_i(t) \, w_{5-i,4}(t) = w_1(t) \, w_{4,4}(t) = w_1(t)[w_1^4(t)]$$

$$= w_1^5(t). \tag{5-172}$$

Evaluation of Entries in the Nonlinear Vectors: Having determined an algorithm for generating the nth-order components in zero-memory expansions of the form of (5-141), we return to the problem of evaluating entries in the nonlinear vectors $\mathbf{J}_n(f_1, \ldots, f_n)$, $\mathbf{V}_n(f_1, \ldots, f_n)$, and $\mathbf{U}_n(f_1, \ldots, f_n)$. In the harmonic input method the nth-order transfer functions are obtained by assuming the input to be a sum of n complex exponential time functions where the input frequencies are assumed to be positive and incommensurable. As a result, the components of the nonlinear vectors are obtained from the coefficients of $\exp\left[j2\pi(f_1 + \cdots + f_n)t\right]$. These are found only in the nth-order components. They arise in (5-141) from terms of degree 2 through n, as indicated by (5-146).

The basic building block in the nth-order component of $z(t)$ is $w_{n,r}(t)$. Each term in $w_{n,r}(t)$ is of nth order and is composed of the product of r functions of time. In particular, a typical term in $w_{n,r}(t)$ may be expressed in the form $Cw_{n_1}(t) w_{n_2}(t) \cdots w_{n_r}(t)$ where C is a constant and the subscripts are positive integers for which

$$n_1 + n_2 + \cdots + n_r = n. \tag{5-173}$$

In terms of the nonlinear transfer function approach the kth factor of this product is given by

$$w_{n_k}(t) = \sum_{q_1=1}^{n} \cdots \sum_{q_{n_k}=1}^{n} W_{n_k}(f_{q_1}, \ldots, f_{q_{n_k}}) \exp\left[j2\pi(f_{q_1} + \cdots + f_{q_{n_k}})t\right]$$

$$\tag{5-174}$$

where $W_n(f_1, \ldots, f_n)$ is the nth-order transfer function associated with $w(t)$. It follows that the typical term, $Cw_{n_1}(t) w_{n_2}(t) \ldots w_{n_r}(t)$, is an n-fold sum. Specifically, we have

$$Cw_{n_1}(t)w_{n_2}(t) \ldots w_{n_r}(t) = C \sum_{q_1=1}^{n} \cdots \sum_{q_n=1}^{n} W_{n_1}(f_{q_1}, \ldots, f_{q_{n_1}})$$
$$\cdot W_{n_2}(f_{q_{n_1+1}}, \ldots, f_{q_{n_1+n_2}})$$
$$\cdots W_{n_r}(f_{q_{n-n_r+1}}, \ldots, f_{q_n})$$
$$\cdot \exp\left[j2\pi(f_{q_1} + \cdots + f_{q_n})t\right]. \tag{5-175}$$

The only terms of interest in (5-175) are those whose frequency is $(f_1 + f_2 + \cdots + f_n)$. Obviously, selecting the summation indices to be

$$q_1 = 1, \qquad q_2 = 2, \ldots, q_n = n \tag{5-176}$$

yields this particular intermodulation frequency, as do all $n!$ permutations of the indices. Summing the corresponding $n!$ terms in (5-175) and using the overbar, as in (4-77), to denote the arithmetic average of the permuted terms, their sum is given by

$$n!\, C\overline{W_{n_1}(f_1,\ldots,f_{n_1})\, W_{n_2}(f_{n_1+1},\ldots,f_{n_1+n_2})\cdots W_{n_r}(f_{n-n_r+1},\ldots,f_n)}$$
$$\cdot \exp\left[j2\pi(f_1+f_2+\cdots+f_n)t\right]. \quad (5\text{-}177)$$

Focusing on the coefficient of $\exp\left[j2\pi(f_1+\cdots+f_n)t\right]$, the contribution from the typical term in $w_{n,r}(t)$ to the nonlinear vectors is of the form

$$n!\, C\overline{W_{n_1}(f_1,\ldots,f_{n_1})\, W_{n_2}(f_{n_1+1},\ldots,f_{n_1+n_2})\cdots W_{n_r}(f_{n-n_r+1},\ldots,f_n)}$$

$$(5\text{-}178)$$

where the constraint in (5-173) is satisfied. Note that the factor of $n!$ is needed to compensate for the division by $n!$ which occurs in the arithmetic average indicated by the overbar.

Given a zero-memory nonlinearity, as shown in (5-141), we now summarize the procedure for evaluating entries in the nonlinear vectors. First, the nth-order component is expressed in the form of (5-146). $w_{n,r}(t)$ is then evaluated for $r = 2, 3, \ldots, n$ using the recursive relationship in (5-169). A typical term in $w_{n,r}(t)$ is of the form $Cw_{n_1}(t)\,w_{n_2}(t)\ldots w_{n_r}(t)$. In general, the factor $w_{n_1}(t)$ is replaced by $W_{n_1}(f_1,\ldots,f_{n_1})$, the factor $w_{n_2}(t)$ is replaced by $W_{n_2}(f_{n_1+1}, \ldots, f_{n_1+n_2})$, $\ldots$, and the factor $w_{n_r}(t)$ is replaced by $W_{n_r}(f_{n-n_r+1},\ldots,f_n)$. Finally, the entry that is inserted into the nonlinear vector is obtained by taking an arithmetic average over all $n!$ permutations and multiplying by $n!$, as indicated in (5-178).

Example 5.16

The above procedure is now illustrated for the circuit of Fig. 5.18(a) that was introduced in our discussion of mixed-variable analysis. The network equations are given by (5-137) and (5-138). Assume that the first- and second- order transfer functions have been previously determined and we are now interested in solving for the third-order transfer functions. According to (5-140), the solution is

$$
\begin{bmatrix} A_3(f_1,f_2,f_3) \\ B_3(f_1,f_2,f_3) \\ C_3(f_1,f_2,f_3) \\ D_3(f_1,f_2,f_3) \\ E_3(f_1,f_2,f_3) \end{bmatrix} =
\begin{bmatrix}
1 & 0 & j2\pi(f_1+f_2+f_3)c_0 & 0 & 0 \\
1 & 1 & 0 & \dfrac{\Gamma_1}{j2\pi(f_1+f_2+f_3)} & 0 \\
0 & -1 & 0 & 0 & \dfrac{1}{R} \\
-r_1 & 0 & 1 & 1 & 0 \\
0 & 0 & 0 & (\mu_1-1) & 1
\end{bmatrix}^{-1}
\begin{bmatrix} U_{a3}(f_1,f_2,f_3)/3! \\ U_{b3}(f_1,f_2,f_3)/3! \\ U_{c3}(f_1,f_2,f_3)/3! \\ U_{d3}(f_1,f_2,f_3)/3! \\ U_{e3}(f_1,f_2,f_3)/3! \end{bmatrix}
$$

$$(5\text{-}179)$$

where $U_{a3}(f_1, f_2, f_3)$ through $U_{e3}(f_1, f_2, f_3)$ remain to be specified. These terms are generated by the expansions in the nonlinear mixed-variable vector $u(t)$ which is specified in (5-138). Observe that only the last two entries in $u(t)$ are zero-memory expansions. In particular, $U_{a3}(f_1, f_2, f_3)$ arises from $\sum_{k=2}^{\infty} r_k a^k(t)$ and $U_{e3}(f_1, f_2, f_3)$ arises from $-\sum_{k=2}^{\infty} \mu_k d^k(t)$.

We now develop expressions for $U_{a3}(f_1, f_2, f_3)$ and $U_{e3}(f_1, f_2, f_3)$. Let

$$z(t) = \sum_{k=2}^{\infty} r_k a^k(t). \tag{5-180}$$

From (5-146) and (5-147)

$$z_3(t) = r_2 a_{3,2}(t) + r_3 a_{3,3}(t)$$
$$= 2r_2 a_1(t) a_2(t) + r_3 a_1^3(t). \tag{5-181}$$

Hence, following the procedure outlined above, we conclude that

$$U_{a3}(f_1, f_2, f_3)/3! = 2r_2 \overline{A_1(f_1) A_2(f_2, f_3)} + r_3 \overline{A_1(f_1) A_1(f_2) A_1(f_3)}. \tag{5-182}$$

In an identical manner,

$$U_{e3}(f_1, f_2, f_3)/3! = -[2\mu_2 \overline{D_1(f_1) D_2(f_2, f_3)} + \mu_3 \overline{D_1(f_1) D_1(f_2) D_1(f_3)}]. \tag{5-183}$$

Capacitive Nonlinearity: The zero-memory expansion of (5-141) is appropriate only for resistive and controlled-source nonlinearities as described in (5-4), (5-6), and (5-57) through (5-60). As seen from (5-28) and (5-45), it is not applicable to capacitive and inductive nonlinearities because the nonlinear portions of both the current through a capacitor and the voltage across an inductor are expressed in the form

$$v(t) = \sum_{r=1}^{\infty} a_r w^r \frac{dw(t)}{dt}. \tag{5-184}$$

Once again, the linear term is omitted since it is accounted for in the augmented linear circuit. The series in (5-184) may be rewritten as

$$v(t) = \frac{d}{dt} \left\{ \sum_{r=1}^{\infty} \frac{a_r}{r+1} w^{r+1}(t) \right\}$$

$$= \frac{d}{dt} \left\{ \sum_{r=2}^{\infty} \frac{a_{r-1}}{r} w^r(t) \right\}. \tag{5-185}$$

Except for coefficients, the series within the brackets of (5-185) is identical to that in (5-141). Consequently, if attention is focused on intermodulation components at $f_1 + f_2 + \cdots + f_n$, the time derivative is applied to terms that are similar in appearance to (5-177). It follows that series of the type in (5-184) contribute entries to the nonlinear vectors $\mathbf{J}_n(f_1, \ldots, f_n)$, $\mathbf{V}_n(f_1, \ldots, f_n)$, and $\mathbf{U}_n(f_1, \ldots, f_n)$ having the form of

$$j2\pi(f_1 + f_2 + \cdots + f_n)$$
$$\cdot n! \; C\overline{W_{n_1}(f_1, \ldots, f_{n_1}) \; W_{n_2}(f_{n_1+1}, \ldots, f_{n_1+n_2}) \ldots W_{n_r}(f_{n-n_r+1}, \ldots, f_n)}.$$

$$(5\text{-}186)$$

Example 5.17

With regard to the circuit of Fig. 5.18(a), the first entry in the nonlinear mixed-variable vector $\mathbf{u}(t)$ corresponds to the current through the nonlinear capacitor. Specifically, as seen from (5-138), the first entry in $\mathbf{u}(t)$ is given by

$$v(t) = - \sum_{k=1}^{\infty} c_k c^k(t) \frac{dc(t)}{dt}$$

$$= - \frac{d}{dt} \left\{ \sum_{k=2}^{\infty} \frac{c_{k-1}}{k} c^k(t) \right\}. \qquad (5\text{-}187)$$

With reference to (5-182) and applying the procedure outlined above, we conclude that

$$U_{a3}(f_1, f_2, f_3)/3! = -j2\pi(f_1 + f_2 + f_3)\left[2\frac{c_1}{2} \overline{C_1(f_1) C_2(f_2, f_3)} \right.$$

$$\left. + \frac{c_2}{3} \overline{C_1(f_1) C_1(f_2) C_1(f_3)} \right]. \qquad (5\text{-}188)$$

It is interesting to note from (5-34) that $c_1/2$ and $c_2/3$ are equal, respectively, to the coefficients q_2 and q_3 that appear in the series expansion for the charge given by (5-30).

Inductive Nonlinearity: Finally, as shown in (5-39) and (5-55), the nonlinear portions of the voltage across a capacitor and the current through an inductor are expressed in the form

$$s(t) = \sum_{r=2}^{\infty} \left[\int_{-\infty}^{t} w(z) \, dz \right]^r \qquad (5\text{-}189)$$

A typical nth-order component resulting from a term of rth degree in (5-189) is now given by

$$C\left[\int_{-\infty}^{t} w_{n_1}(z)\,dz\right]\left[\int_{-\infty}^{t} w_{n_2}(z)\,dz\right]\cdots\left[\int_{-\infty}^{t} w_{n_r}(z)\,dz\right]$$

where C is a constant and the indices $n_1, n_2, \ldots, n_r$ satisfy (5-173). Making use of (5-174), note that

$$\int_{-\infty}^{t} w_{n_k}(z)\,dz = \sum_{q_1=1}^{n}\cdots\sum_{q_{n_k}=1}^{n}\frac{W_{n_k}(f_{q_1},\ldots,f_{q_{n_k}})}{j2\pi(f_{q_1}+\cdots+f_{q_{n_k}})}$$

$$\cdot\exp\left[j2\pi(f_{q_1}+\cdots+f_{q_{n_k}})t\right]. \qquad (5\text{-}190)$$

Consequently, we conclude that series of the type in (5-189) contribute entries to the nonlinear vectors having the form

$$n!\,C\,\frac{W_{n_1}(f_1,\cdots,f_{n_1})}{j2\pi(f_1+\cdots+f_{n_1})}\frac{W_{n_2}(f_{n_1+1},\cdots,f_{n_1+n_2})}{j2\pi(f_{n_1+1}+\cdots+f_{n_1+n_2})}\cdots\frac{W_{n_r}(f_{n-n_r+1},\ldots,f_n)}{j2\pi(f_{n-n_r+1}+\cdots+f_n)}.$$

$$(5\text{-}191)$$

Example 5.18

As before, consider the nonlinear circuit of Fig. 5.18(a). We now focus on the second entry of $\mathbf{u}(t)$ in (5-138) that is

$$s(t) = -\sum_{k=2}^{\infty}\Gamma_k\left[\int_{-\infty}^{t} d(z)\,dz\right]^k. \qquad (5\text{-}192)$$

Appealing to (5-182) and (5-191), it follows that

$$U_{b3}(f_1, f_2, f_3)/3! = -\left[2\Gamma_2\,\frac{\overline{D_1(f_1)}}{j2\pi f_1}\frac{D_2(f_2,f_3)}{j2\pi(f_2+f_3)} + \Gamma_3\,\frac{\overline{D_1(f_1)}}{j2\pi f_1}\frac{D_1(f_2)}{j2\pi f_2}\frac{D_1(f_3)}{j2\pi f_3}\right].$$

$$(5\text{-}193)$$

We see, therefore, that it is straightforward to determine the entries needed for the nonlinear vectors $\mathbf{J}_n(f_1, \ldots, f_n)$, $\mathbf{V}_n(f_1, \ldots, f_n)$, and $\mathbf{U}_n(f_1, \ldots, f_n)$. The key step in carrying out the procedures developed in this section is use of the recursive relationship in (5-169) for evaluating $w_{n,r}(t)$.

Algorithms for Direct Evaluation of Nonlinear Vectors: As is suggested by the results obtained in Examples 5.16, 5.17, and 5.18, the nonlinear vector entries may be obtained directly by means of suitable recursion relations. Since their development closely parallels the discussion just given, the algorithms are stated here without proof.

Zero-Memory Nonlinearity: For the zero-memory nonlinearity of (5-141) the entry in an nth-order nonlinear vector may be expressed as

$$n! \sum_{r=2}^{n} a_r \overline{W_{n,r}(f_1, \ldots, f_n)} \tag{5-194}$$

where, as usual, the overbar denotes the arithmetic average of the $n!$ permutations and $W_{n,r}(f_1, \ldots, f_n)$ may be generated by the recursive relation

$$W_{n,r}(f_1, \ldots, f_n) = \sum_{i=1}^{n-r+1} W_i(f_1, \ldots, f_i) W_{n-i,r-1}(f_{i+1}, \ldots, f_n). \tag{5-195}$$

For $r = 1$ and $r > n$, $W_{n,r}(f_1, \ldots, f_n)$ has the properties

$$W_{n,r}(f_1, \ldots, f_n) = \begin{cases} W_n(f_1, \ldots, f_n); & r = 1 \\ 0; & r > n. \end{cases} \tag{5-196}$$

Capacitive Nonlinearity: For the capacitive nonlinearity of (5-184) the entry in an nth-order nonlinear vector may be expressed as

$$n! \, j2\pi(f_1 + \cdots + f_n) \sum_{r=2}^{n} \frac{a_{r-1}}{r} \, \overline{W_{n,r}(f_1, \ldots, f_n)} \tag{5-197}$$

where $W_{n,r}(f_1, \ldots, f_n)$ may again be generated by the recursive relation of (5-195).

Inductive Nonlinearity: For the inductive nonlinearity of (5-189) the entry in an nth-order nonlinear vector may be expressed as

$$n! \sum_{r=2}^{n} a_r \overline{X_{n,r}(f_1, \ldots, f_n)} \tag{5-198}$$

where a recursive relation for $X_{n,r}(f_1, \ldots, f_n)$ is given by

$$X_{n,r}(f_1, \ldots, f_n) = \sum_{i=1}^{n-r+1} \frac{W_i(f_1, \ldots, f_i)}{j2\pi(f_1 + \cdots + f_i)} X_{n-i,r-1}(f_{i+1}, \ldots, f_n). \tag{5-199}$$

For $r = 1$ and $r > n$, $X_{n,r}(f_1, \ldots, f_n)$ has the properties

$$X_{n,r}(f_1, \ldots, f_n) = \begin{cases} \dfrac{W_n(f_1, \ldots, f_n)}{j2\pi(f_1 + \cdots + f_n)}; & r = 1 \\ 0; & r > n. \end{cases} \tag{5-200}$$

5.7   A SECOND LOOK AT THE HARMONIC INPUT METHOD

Depending upon whether nodal, loop, or mixed-variable analysis is used to formulate the network equations of a weakly nonlinear circuit, the nth-order nonlinear transfer functions are given either by (5-115), (5-126), or (5-140). In each case the equations can be interpreted in terms of a sinusoidal steady-state solution at frequency $f_1 + \cdots + f_n$ of a linearized circuit excited by sources contained in the nonlinear vectors $\mathbf{J}_n(f_1, \ldots, f_n)/n!$, $\mathbf{V}_n(f_1, \ldots, f_n)/n!$, and $\mathbf{U}_n(f_1, \ldots, f_n)/n!$, respectively, where the unknowns are the desired nonlinear transfer functions. The sources are readily determined using the procedure given in Section 5.6.

Example 5.19

We illustrate our remarks with the circuit shown in Fig. 5.23. For simplicity, the nonlinear capacitor and inductor are assumed to be characterized by

$$i_c(t) = \sum_{k=0}^{1} c_k a^k(t) \frac{da(t)}{dt}$$

$$i_l(t) = \sum_{k=1}^{2} \Gamma_k \left[\int_{-\infty}^{t} b(z) \, dz \right]^k . \tag{5-201}$$

The objective is to determine the nonlinear transfer functions associated with the node-to-datum voltages $a(t)$ and $b(t)$.

For this purpose, we need only consider the solutions to a set of equivalent sinusoidal steady-state linear circuit problems. The equivalent circuits required for the first-, second-, and third-order transfer functions are shown in Fig. 5.24. Note that the desired nonlinear transfer functions appear as unknown complex node-to-datum voltages. The independent current source is used as an excitation only in the equivalent circuit for the first-order transfer functions. It does not occur in the equivalent circuits for the higher order transfer functions. There the excitations are provided by the nonlinearities that manifest themselves as current sources driving the linearized network. Observe that, except for the sources, the same linearized circuit occurs in each case.

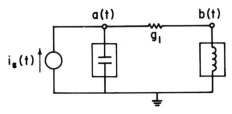

Fig. 5.23.   Weakly nonlinear circuit with nonlinear capacitor and nonlinear inductor.

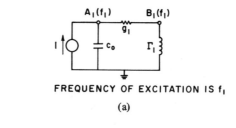

FREQUENCY OF EXCITATION IS f_1

(a)

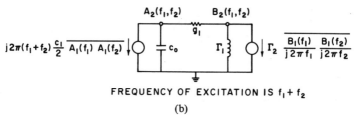

FREQUENCY OF EXCITATION IS $f_1 + f_2$

(b)

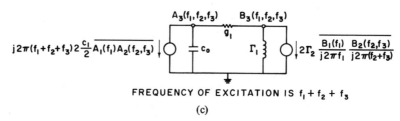

FREQUENCY OF EXCITATION IS $f_1 + f_2 + f_3$

(c)

Fig. 5.24. Equivalent circuits for determining (a) first-order, (b) second-order and (c) third-order transfer functions of the network in Fig. 5.23.

The first-order transfer functions are determined by assigning unit amplitude to the independent current source and applying nodal analysis at frequency f_1 in order to obtain the network equations. The second-order transfer functions are similarly determined. However, now the nodal analysis is carried out at frequency $f_1 + f_2$ and the excitations are given in terms of the linear transfer functions and nonlinear series coefficients. The same procedure is performed for the third-order transfer functions at frequency $f_1 + f_2 + f_3$ with the excitations specified in terms of the previously determined first- and second-order transfer functions. The process is readily continued for the higher order transfer functions.

Although the unknowns in Example 5.19 were node-to-datum voltages, any convenient set of independent voltages and currents may be chosen as the independent variables. Also, any convenient technique may be used to write the network equations. In each instance solution for the nonlinear transfer functions can be treated in terms of equivalent linear circuits driven by known excitations. This is a fortunate circumstance. Not only is the solution procedure straightforward, but intuition gained from linear circuit theory is also useful. We also see that modeling of the linearized circuit is extremely important if the nonlinear transfer functions are to be accurately determined.

6

Nonlinear Distortion and Interference Criteria

Nonlinearities inherent with electronic devices may cause severe degradation in the performance of communication and electronic circuits. The nonlinear effects experienced may be broadly classified into two categories. Those that produce waveform distortion of the desired signal, in the absence of interference, are referred to as nonlinear distortion effects while those that involve one or more undesired input signals are known as nonlinear interference effects. When unwanted nonlinear effects occur, it is desirable to be able to judge the severity of the resulting distortion and/or interference.

Criteria for this purpose are presented in this chapter. In particular, we discuss the nonlinear phenomena of harmonic generation, gain compression/expansion, intermodulation, desensitization, cross modulation, and spurious response. Generally speaking, harmonic generation and gain compression/expansion fall into the category of nonlinear distortion effects. On the other hand, desensitization, cross modulation, and spurious response are normally classified as nonlinear interference effects. Intermodulation may be classified either way depending upon whether desired signal components alone or interfering signal components are involved in the frequency mix. In this chapter the nonlinear transfer function approach is used to derive analytical expressions for criteria that are introduced to measure the extent of these various effects.

6.1   HARMONIC GENERATION

A convenient method for investigating linearity in an electronic circuit consists of applying a sinusoidal signal and measuring relative amplitudes of harmonic components that are generated. For an ideally linear circuit, the output would be a purely sinusoidal signal having the same frequency as the input. However, nonlinearities present in an actual circuit cause the output waveform to be distorted. In particular, the output is a periodic signal containing sinusoidal components at the fundamental input frequency and harmonics thereof.

Specifically, let the input to a weakly nonlinear system be

$$x(t) = |E_1| \cos (2\pi f_1 t + \theta_1). \tag{6-1}$$

Then the output is a periodic signal whose Fourier series expansion may be written as

$$y(t) = |A_0| + |A_1| \cos [2\pi f_1 t + \alpha_1] + |A_2| \cos [2\pi (2f_1) t + \alpha_2]$$
$$+ |A_3| \cos [2\pi (3f_1) t + \alpha_3] + \cdots. \tag{6-2}$$

The magnitudes of the various harmonics have relatively little meaning by themselves. Rather it is their fraction of the fundamental magnitude that serves as a measure of the nonlinear distortion. Hence the fractional harmonic distortion for the kth harmonic is defined by

$$k\text{hd} = \frac{|A_k|}{|A_1|}. \tag{6-3}$$

We recognize that $k$hd represents an awkward choice of notation. However, because so many different criteria are introduced in this chapter, we elect to use a notation, cumbersome as it is, that is suggestive of the nonlinear effect being characterized. In this particular instance, $k$hd is a mnemonic that stands for "kth harmonic distortion." The percentage harmonic distortion for the kth harmonic is obtained by multiplying (6-3) by 100 percent.

Another quantity of interest is total harmonic distortion, thd, which is also referred to as the distortion factor. This is defined as the ratio of the root-mean-square voltage of the output, with the fundamental and dc components removed, to the magnitude of the output fundamental voltage. Therefore, thd is given by

$$\text{thd} = \frac{[|A_2|^2 + |A_3|^2 + \cdots]^{1/2}}{|A_1|}$$
$$= [(2\text{hd})^2 + (3\text{hd})^2 + \cdots]^{1/2}. \tag{6-4}$$

Multiplication of (6-4) by 100 percent results in the total harmonic distortion being expressed in percent. Assuming $y(t)$ to be a voltage, the power output at the fundamental frequency is

$$p_1 = \tfrac{1}{2} |A_1|^2 G_L(f_1) \tag{6-5}$$

where $G_L(f)$ is the real part of the load admittance. Since the total ac output power, p_L, is the sum of powers in the fundamental and higher harmonics, it follows that

$$p_L = \tfrac{1}{2} [|A_1|^2 G_L(f_1) + |A_2|^2 G_L(2f_1) + |A_3|^2 G_L(3f_1) + \cdots]. \tag{6-6}$$

When the load is a pure resistor with conductance G_L, (6-6) reduces to

$$\begin{aligned} p_L &= \tfrac{1}{2} [|A_1|^2 + |A_2|^2 + |A_3|^2 + \cdots] G_L \\ &= [1 + (2\text{hd})^2 + (3\text{hd})^2 + \cdots] p_1 \\ &= [1 + (\text{thd})^2] p_1. \end{aligned} \tag{6-7}$$

Therefore, for a purely resistive load, the total ac output power may be expressed in terms of the total harmonic distortion and the power output at the fundamental frequency.

Example 6.1

For moderate amounts of distortion, little error is made in using only the fundamental power to calculate the total ac output power. As an example, if the total harmonic distortion is 10 percent, then

$$p_L = [1 + (0.1)^2] p_1 = 1.01 p_1 \approx p_1. \tag{6-8}$$

Although the total ac output power is relatively unchanged, the harmonic distortion may nevertheless still be troublesome.

The nonlinear transfer function approach is now used to obtain an analytical expression for the kth harmonic distortion. Once again, we assume that the input to the weakly nonlinear system is given by (6-1). The total nth-order portion of the response, $y_n(t)$, is obtained by summing all nth-order terms resulting from distinct frequency mixes, as indicated by (4-16). Because the excitation is a single sinusoid, the input frequencies are $-f_1$ and f_1. As a result, a frequency mix of order n is characterized by the frequency mix vector $\mathbf{m} = (m_{-1}, m_1)$ where

$$m_{-1} + m_1 = n. \tag{6-9}$$

With reference to (4-15), the nth-order portion of the response that corresponds to a frequency mix represented by $\mathbf{m}$ is expressed as

$$y_n(t;\mathbf{m}) = \frac{(n;\mathbf{m})}{2^n} (E_1^*)^{m-1} (E_1)^{m_1} H_n(\underbrace{-f_1, \ldots, -f_1}_{m_{-1}}, \underbrace{f_1, \ldots, f_1}_{m_1})$$

$$\cdot \exp\left[j2\pi(m_1 - m_{-1})f_1 t\right]. \tag{6-10}$$

Equation (6-10) clearly reveals that the output consists of either dc or harmonics of the input depending upon the values of the positive integers m_1 and m_{-1}.

Contributions to the kth harmonic result when

$$m_1 - m_{-1} = k. \tag{6-11}$$

Solution of the simultaneous equations given by (6-9) and (6-11) for m_{-1} and m_1 yields

$$m_{-1} = \frac{n - k}{2}$$

$$m_1 = \frac{n + k}{2}. \tag{6-12}$$

Since m_{-1} is constrained to be nonnegative, it follows that n must be greater than or equal to k. In addition, because m_{-1} and m_1 are integers, n must be odd when k is odd and even when k is even. Therefore, odd harmonics are generated by all odd-order terms and even harmonics are generated by all even-order terms for which $n \geqslant k$.

By combining (6-10) with its complex conjugate, as is done in (4-22), the sinusoidal response at $(m_1 - m_{-1})f_1$ becomes

$$\hat{y}_n(t;\mathbf{m}) = 2 \operatorname{Re}\{y_n(t;\mathbf{m})\}$$

$$= \operatorname{Re}\left\{\frac{(n;\mathbf{m})}{2^{n-1}} (E_1^*)^{m-1} (E_1)^{m_1} H_n(\underbrace{-f_1, \ldots, -f_1}_{m_{-1}}, \underbrace{f_1, \ldots, f_1}_{m_1})\right.$$

$$\left. \cdot \exp\left[j2\pi(m_1 - m_{-1})f_1 t\right]\right\} \tag{6-13}$$

where, as usual, $\operatorname{Re}\{\cdot\}$ denotes the operation of taking the real part of the expression inside the brackets. As pointed out above, the kth harmonic is obtained by summing the contributions from various-order responses. We express $\hat{y}_n(t;\mathbf{m})$ in the form of (6-13) because the addition of complex numbers is easier than the addition of trigonometric functions.

Example 6.2

Making use of (6-13), the leading terms of the first, second, and third harmonics are given by

$$y(t;f_1) = \hat{y}_1(t;0,1) + \hat{y}_3(t;1,2) + \hat{y}_5(t;2,3) + \cdots$$

$$= \mathrm{Re}\,\{(E_1) \exp\,[j2\pi(f_1)\,t]\,[H_1(f_1) + \tfrac{3}{4}|E_1|^2 H_3(-f_1,f_1,f_1)$$

$$+ \tfrac{5}{8}|E_1|^4 H_5(-f_1,-f_1,f_1,f_1,f_1) + \cdots]\},$$

$$y(t;2f_1) = \hat{y}_2(t;0,2) + \hat{y}_4(t;1,3) + \hat{y}_6(t;2,4) + \cdots$$

$$= \mathrm{Re}\,\{(E_1)^2 \exp\,[j2\pi(2f_1)\,t]\,[\tfrac{1}{2}H_2(f_1,f_1) + \tfrac{1}{2}|E_1|^2 H_4(-f_1,f_1,f_1,f_1)$$

$$+ \tfrac{15}{32}|E_1|^4 H_6(-f_1,-f_1,f_1,f_1,f_1,f_1) + \cdots]\},$$

$$y(t;3f_1) = \hat{y}_3(t;0,3) + \hat{y}_5(t;1,4) + \hat{y}_7(t;2,5) + \cdots$$

$$= \mathrm{Re}\,\{(E_1)^3 \exp\,[j2\pi(3f_1)\,t]\,[\tfrac{1}{4}H_3(f_1,f_1,f_1)$$

$$+ \tfrac{5}{16}|E_1|^2 H_5(-f_1,f_1,f_1,f_1,f_1)$$

$$+ \tfrac{21}{64}|E_1|^4 H_7(-f_1,-f_1,f_1,f_1,f_1,f_1,f_1) + \cdots]\}. \tag{6-14}$$

The results in Example 6.2 are readily extended. In fact, the leading terms in the expression for the kth harmonic are

$$y(t;kf_1) = \hat{y}_k(t;0,k) + \hat{y}_{k+2}(t;1,k+1) + \hat{y}_{k+4}(t;2,k+2) + \cdots$$

$$= \mathrm{Re}\,\bigg\{(E_1)^k \exp\,[j2\pi(kf_1)\,t]\,\bigg[\frac{1}{2^{k-1}}\,H_k(\underbrace{f_1,\ldots,f_1}_{k})$$

$$+ \frac{k+2}{2^{k+1}}\,|E_1|^2 H_{k+2}(\underbrace{-f_1,f_1,\ldots,f_1}_{k+1})$$

$$+ \frac{(k+4)(k+3)}{2^{k+4}}\,|E_1|^4 H_{k+4}(\underbrace{-f_1,-f_1,f_1,\ldots,f_1}_{k+2}) + \cdots\bigg]\bigg\}. \tag{6-15}$$

The summations inside the rectangular brackets should be carried out vectorially because the nonlinear transfer functions are, in general, complex quantities.

From (6-15) the magnitude of the kth harmonic is

$$|A_k| = |E_1|^k \left| \frac{1}{2^{k-1}} \underbrace{H_k(f_1, \ldots, f_1)}_{k} + \frac{k+2}{2^{k+1}} |E_1|^2 H_{k+2}(\underbrace{-f_1, f_1, \ldots, f_1)}_{k+1} \right.$$

$$\left. + \frac{(k+4)(k+3)}{2^{k+4}} |E_1|^4 H_{k+4}(-f_1, -f_1, \underbrace{f_1, \ldots, f_1)}_{k+2} + \cdots \right|. \qquad (6\text{-}16)$$

For small values of E_1 (i.e., $|E_1| \ll 1$) $|A_k|$ is dominated by its leading term. Then (6-16) simplifies to

$$|A_k| \approx \frac{|E_1|^k \left| H_k(\underbrace{f_1, \ldots, f_1)}_{k} \right|}{2^{k-1}}. \qquad (6\text{-}17)$$

We see that the magnitude of the kth harmonic is proportional to the kth power of the input magnitude for small signals. The fractional harmonic distortion for the kth harmonic is then approximated by

$$k\text{hd} \approx \frac{|E_1|^{k-1} \left| H_k(\underbrace{f_1, \ldots, f_1)}_{k} \right|}{2^{k-1} |H_1(f_1)|}. \qquad (6\text{-}18)$$

When the nonlinear transfer functions are known, (6-18) may be used in (6-4) to evaluate the total harmonic distortion. Alternatively, (6-18) may be used along with measurements of $|E_1|$, $|H_1(f_1)|$, and $k$hd to determine the magnitude of the kth-order nonlinear transfer function evaluated for the kth harmonic. Observe that knowledge of $|H_k(f_1, \ldots, f_1)|$ provides an incomplete characterization of a system's nonlinear behavior. The latter requires use of the more general transfer function $H_k(f_1, f_2, \ldots, f_k)$ in which the frequency variables are not constrained to be equal.

Before closing this section, a few words of caution are in order. Although total harmonic distortion is a measure of the extent to which a system is behaving nonlinearly, it is not necessarily indicative of bothersome nonlinear behavior. Recall that all musical instruments and voices produce sounds that are rich in harmonics. In fact, some sounds are made more enjoyable by enhancing their harmonics. Also, signals are usually far from sinusoidal and may not even be periodic. Although sinusoidal considerations can give results that are helpful in practice, they may also lead to false conclusions when these results are generalized to other waveforms.

6.2   GAIN COMPRESSION/EXPANSION

The gain compression/expansion ratio is another measure of a system's linearity. When driven hard enough, all electronic circuits exhibit saturation. To study this effect, let the excitation, once again, be the single sinusoid given by (6-1). The response of interest is the fundamental response which, from (6-2) and (6-14), is expressed as

$$
\begin{aligned}
y(t; f_1) &= |A_1| \cos[2\pi f_1 t + \alpha_1] \\
&= \mathrm{Re}\,\{(E_1)\exp[j2\pi(f_1)\,t]\,[H_1(f_1) + \tfrac{3}{4}|E_1|^2 H_3(-f_1, f_1, f_1) \\
&\quad + \tfrac{5}{8}|E_1|^4 H_5(-f_1, -f_1, f_1, f_1, f_1) + \cdots]\}.
\end{aligned} \tag{6-19}
$$

The voltage gain at the fundamental frequency is defined to be

$$
\begin{aligned}
\mathrm{vg} = \left|\frac{A_1}{E_1}\right| &= \left| H_1(f_1) + \frac{3}{4}|E_1|^2 H_3(-f_1, f_1, f_1) \right. \\
&\quad \left. + \frac{5}{8}|E_1|^4 H_5(-f_1, -f_1, f_1, f_1, f_1) + \cdots \right|.
\end{aligned} \tag{6-20}
$$

For a linear system all of the nonlinear transfer functions above order one are identically zero and the voltage gain is a constant equal to the magnitude of the linear transfer function evaluated at f_1. The output amplitude A_1 is then linearly proportional to the input amplitude E_1. However, the nonlinear transfer functions are not zero for a weakly nonlinear system. The voltage gain then depends nonlinearly upon the magnitude of the input signal.

Example 6.3

A typical response curve is shown in Fig. 6.1(a). The dashed straight line, having a slope equal to $|H_1(f_1)|$, corresponds to the response curve of an ideally

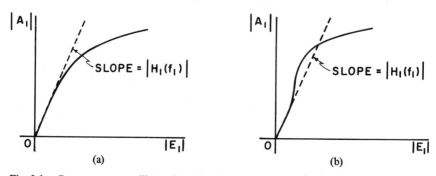

Fig. 6.1.   Response curves illustrating (a) gain compression and (b) both gain expansion and gain compression.

linear system. However, practical systems experience saturation which results in the gain falling below that of the linear system for large enough input amplitudes, as shown in the figure. This effect is referred to as gain compression. In some cases, as indicated in Fig. 6.1(b), the gain actually increases before saturation sets in. An increase in gain due to the nonlinear behavior of the system is called gain expansion. Both gain compression and gain expansion are accounted for by the analytical expression in (6-20). Which of the two nonlinear effects will occur depends upon the angles of the nonlinear transfer functions relative to the angle of $H_1(f_1)$. Also, because of the manner in which $|E_1|$ appears in (6-20), various terms become significant in increasing order as $|E_1|$ is increased in value. As an example, the voltage gain curve of Fig. 6.1(b) suggests that the angles of the nonlinear transfer functions are such that the term involving $H_3(-f_1, f_1, f_1)$ causes the gain to increase while the higher order terms collectively cause the gain to decrease. The term involving $H_3(-f_1, f_1, f_1)$ comes into play first. However, for larger values of $|E_1|$, its effect is negated by the higher order terms. Because the system gain is not constant but various as a function of the input amplitude, waveform distortion may result when either gain compression or expansion occur.

When a system saturates, (6-20) is not a convenient expression for the voltage gain because a large number of terms are typically required for an acceptable approximation. As a result, (6-20) is usually used in conjunction with suitably small inputs to predict the onset of nonlinear behavior. Effects above third order are then negligible and (6-20) reduces to

$$\text{vg} = \left| \frac{A_1}{E_1} \right| \approx \left| H_1(f_1) + \frac{3}{4} |E_1|^2 H_3(-f_1, f_1, f_1) \right|$$

$$= |H_1(f_1)| \left| 1 + \frac{3}{4} |E_1|^2 \frac{H_3(-f_1, f_1, f_1)}{H_1(f_1)} \right|. \tag{6-21}$$

The gain compression/expansion ratio, denoted by gcer, is defined to be

$$\text{gcer} = \left| \frac{A_1}{E_1 H_1(f_1)} \right| = \left| 1 + \frac{3}{4} |E_1|^2 \frac{H_3(-f_1, f_1, f_1)}{H_1(f_1)} \right|. \tag{6-22}$$

Hence the gain compression/expansion ratio is the ratio of the actual magnitude of the fundamental response to the magnitude that would have existed had the system been perfectly linear. This ratio, of course, is unity for an ideally linear system. It is less than unity when $\text{Re}\{H_3(-f_1, f_1, f_1)/H_1(f_1)\} < 0$ and we speak of gain compression. On the other hand, the ratio is greater than unity when $\text{Re}\{H_3(-f_1, f_1, f_1)/H_1(f_1)\} \geq 0$. The system is then said to experience gain expansion. The gain compression/expansion ratio may be interpreted as the magnitude of

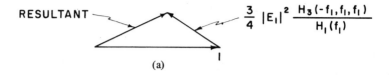

(a)

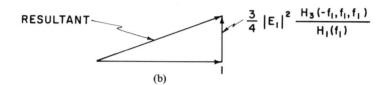

(b)

(c)

Fig. 6.2.   Interpretation of the gain compression/expansion ratio as the magnitude of a resultant vector. (a) $\text{Re}\{H_3(-f_1, f_1, f_1)/H_1(f_1)\} < 0$ and gcer < 1. (b) $\text{Re}\{H_3(-f_1, f_1, f_1)/H_1(f_1)\} = 0$ and gcer > 1. (c) $\text{Re}\{H_3(-f_1, f_1, f_1)/H_1(f_1)\} > 0$ and gcer > 1.

the resultant vector obtained by adding the vector $(\frac{3}{4})|E_1|^2 H_3(-f_1, f_1, f_1)/H_1(f_1)$ to the unit vector. This is illustrated in Fig. 6.2. Note that the angle of $H_3(-f_1, f_1, f_1)$ relative to the angle of $H_1(f_1)$ is important in determining whether compression or expansion results. For a given magnitude of the third-order transfer function, the greatest compression occurs when the two angles differ by $180°$ while the greatest expansion occurs when the two angles are equal.

When the input amplitude is small enough such that

$$\frac{3}{4}|E_1|^2 \left|\frac{H_3(-f_1, f_1, f_1)}{H_1(f_1)}\right| \ll 1, \qquad (6\text{-}23)$$

the expression in (6-22) is readily simplified. For this purpose, let B_r and B_i denote the real and imaginary parts, respectively, of $(\frac{3}{4})|E_1|^2 H_3(-f_1, f_1, f_1)/H_1(f_1)$. Hence

$$\frac{3}{4}|E_1|^2 \frac{H_3(-f_1, f_1, f_1)}{H_1(f_1)} = B_r + jB_i. \qquad (6\text{-}24)$$

Substitution of (6-24) into (6-22) yields

$$\text{gcer} = \left|1 + B_r + jB_i\right| = [(1 + B_r)^2 + (B_i)^2]^{1/2}. \qquad (6\text{-}25)$$

For the case in which n is not an integer, the binomial theorem states that

$$(a + b)^n = a^n + \sum_{k=1}^{\infty} \frac{n(n - 1) \cdots (n - k + 1)}{k!} a^{n-k} b^k. \tag{6-26}$$

If we let

$$n = \tfrac{1}{2}, \quad a = (1 + B_r)^2, \quad b = B_i^2, \tag{6-27}$$

then application of (6-26) to (6-25) results in

$$\text{gcer} = (1 + B_r) + \frac{1}{2} \frac{B_i^2}{(1 + B_r)} - \frac{1}{8} \frac{B_i^4}{(1 + B_r)^3} + \cdots. \tag{6-28}$$

However, because of the assumption in (6-23), B_r and B_i are each very much less than unity. It follows that the gain compression/expansion ratio may be approximated by

$$\text{gcer} \approx 1 + B_r = 1 + \frac{3}{4} |E_1|^2 \, \text{Re} \left\{ \frac{H_3(-f_1, f_1, f_1)}{H_1(f_1)} \right\}. \tag{6-29}$$

The approximate expression in (6-29) is convenient for predicting the onset of nonlinear behavior. Assuming suitably small input amplitudes, (6-29) is most inaccurate when the angle of $H_3(-f_1, f_1, f_1)$ differs from the angle of $H_1(f_1)$ by $\pm 90°$. Then, as shown in Fig. 6.2(b), a small amount of expansion arises from $B_i = (\tfrac{3}{4})|E_1|^2 \, \text{Im} \{H_3(-f_1, f_1, f_1)/H_1(f_1)\}$ which is neglected in (6-29). As a matter of fact, when $H_3(-f_1, f_1, f_1)$ and $H_1(f_1)$ are nearly in quadrature, the gain compression/expansion ratio is not a dependable indicator for the onset of nonlinear behavior, irrespective of the approximations leading to (6-29). This is true because, as can be seen from Fig. 6.2(b), the magnitude of the resultant vector is relatively insensitive to small changes in B_i when the component vectors are perpendicular.

In practice, the gain compression/expansion ratio is usually expressed in decibels. Because gcer involves the ratio of two voltages, the conversion is accomplished by means of (2-67). We use uppercase letters to denote quantities expressed in decibels. Therefore, assuming small input levels such that (6-29) is valid, the gain compression/expansion ratio in decibels is given by

$$\text{GCER} = 20 \log_{10} \text{gcer}$$

$$\approx 20 \log_{10} \left[1 + \frac{3}{4} |E_1|^2 \, \text{Re} \left\{ \frac{H_3(-f_1, f_1, f_1)}{H_1(f_1)} \right\} \right] \text{dB}. \tag{6-30}$$

The available power of the input signal is easily inserted into (6-30) by recalling from (2-89) that

$$p_{AS} = \frac{|E_1|^2}{8R_S} \tag{6-31}$$

where R_S is the real part of the source impedance. Hence

$$\text{GCER} \approx 20 \log_{10} \left[1 + 6R_S p_{AS} \, \text{Re} \left\{\frac{H_3(-f_1, f_1, f_1)}{H_1(f_1)}\right\}\right] \text{dB}. \qquad (6\text{-}32)$$

Because of the small-signal assumption, (6-30) and (6-32) should not be used to compute gain compression/expansion ratios in excess of 1 dB.

In some applications it is desirable to invert (6-32) in order to predict the available input power needed to produce a specified amount of compression or expansion. The first step in the inversion is to express (6-32) in terms of natural logarithms by making use of the relationship

$$\log_{10} x = [\log_{10} e] \ln x = 0.43 \ln x \qquad (6\text{-}33)$$

where $e = 2.72$ is the Naperian base. Therefore, (6-32) may also be expressed as

$$\text{GCER} \approx 8.6 \ln \left[1 + 6R_S p_{AS} \, \text{Re} \left\{\frac{H_3(-f_1, f_1, f_1)}{H_1(f_1)}\right\}\right] \text{dB}. \qquad (6\text{-}34)$$

However, for $x \ll 1$, it is known that

$$\ln (1 + x) \approx x. \qquad (6\text{-}35)$$

At the onset of nonlinear behavior, it is likely that

$$6R_S p_{AS} \, \text{Re} \left\{\frac{H_3(-f_1, f_1, f_1)}{H_1(f_1)}\right\} \ll 1. \qquad (6\text{-}36)$$

Utilizing (6-35) in (6-34), it follows that

$$\text{GCER} \approx 51.6 R_S p_{AS} \, \text{Re} \left\{\frac{H_3(-f_1, f_1, f_1)}{H_1(f_1)}\right\} \text{dB}. \qquad (6\text{-}37)$$

Solving for p_{AS}, we have

$$p_{AS} \approx \frac{\text{GCER}}{51.6 R_S \, \text{Re} \left\{\dfrac{H_3(-f_1, f_1, f_1)}{H_1(f_1)}\right\}}. \qquad (6\text{-}38)$$

Finally, expressing the available power in decibels referred to 1 mW, there results

$$P_{AS} = 10 \log_{10} \frac{p_{AS}}{10^{-3}} \approx 10 \log_{10} \left[\frac{19.4}{R_S} \frac{\text{GCER}}{\text{Re} \left\{\dfrac{H_3(-f_1, f_1, f_1)}{H_1(f_1)}\right\}}\right]$$

$$= -10 \log_{10} \left[\frac{1}{\text{GCER}} \, \text{Re} \left\{\frac{H_3(-f_1, f_1, f_1)}{H_1(f_1)}\right\}\right] + 10 \log_{10} \left[\frac{19.4}{R_S}\right] \text{dBm}.$$

$$(6\text{-}39)$$

At the upper extreme of 1-dB compression or expansion, the error in the calculation of P_{AS} from (6-39) typically does not exceed $\frac{1}{4}$ dB. When comparing the
linearity of various circuits, the power required to produce 1 dB of compression
or expansion is frequently used as a basis of comparison.

6.3 INTERMODULATION

In previous chapters the term "intermodulation component" was used, in a
general way, to refer to any nonlinear frequency component that resulted from
a frequency mix. In this broad sense, intermodulation encompasses all of the
nonlinear effects discussed in this chapter. However, having introduced the
categories of harmonic generation, gain compression/expansion, desensitization,
cross modulation, and spurious response, intermodulation is now given a more
restrictive interpretation. Specifically, intermodulation is used in this section
to refer to only those nonlinear frequency components, resulting from frequency mixes, that are not included in the above categories. For example, components produced by the frequency mixes $(f_1 + f_1 + f_1 + f_1 - f_1)$, $(f_1 + f_1 - f_1)$,
and $(f_2 + f_2 - f_1)$ are classified as belonging to the categories of harmonic generation, gain compression/expansion, and intermodulation, respectively. Intermodulation is known as a nonlinear distortion effect when only desired signal
components are involved in the frequency mix. Because intermodulation frequencies are not harmonically related to the signal frequencies, intermodulation
distortion is often more objectionable than is harmonic distortion. On the other
hand, when interfering signal components are involved in the frequency mix,
intermodulation is referred to as a nonlinear interference effect.

Over the years many different criteria have been introduced in order to characterize the intermodulation response of weakly nonlinear systems. In general,
the definitions involve ratios of load and input powers chosen in such a manner
as to yield analytical expressions that do not depend upon input amplitudes but
are functions only of system parameters. The criteria differ according to whether
average or available powers are employed and whether these powers are referred
to the output or the input. Unfortunately, the situation is confused in much of
the literature where either the same or very similar terminology is used for the
various criteria. To help clarify matters, we introduce a nomenclature in this
section that is suggestive of the underlying definitions.

6.3.1 Intermodulation Transducer Gain Ratio

Assume a weakly nonlinear system is excited by the sum of Q sinusoidal tones
as expressed in (4-5). Then the frequency mix represented by the vector

$$\mathbf{m} = (m_{-Q}, \ldots, m_{-1}, m_1, \ldots, m_Q) \tag{6-40}$$

generates a nonlinear frequency component at the intermodulation frequency

$$f_\mathbf{m} = (m_1 - m_{-1})f_1 + \cdots + (m_Q - m_{-Q})f_Q \tag{6-41}$$

where the positive integers in the vector $\mathbf{m}$ sum to n, the order of the frequency mix. The intermodulation transducer gain ratio, denoted by $\text{imtgr}_n(f_\text{m})$, is defined to be the ratio of the average power dissipated by this intermodulation component in the load to the product of the available powers from the source at the various input frequencies as shown below:

$$\text{imtgr}_n(f_\text{m}) = \frac{p_L(f_\text{m})}{[p_{AS}(f_1)]^{(m_1 + m_{-1})} \cdots [p_{AS}(f_Q)]^{(m_Q + m_{-Q})}}. \quad (6\text{-}42)$$

From (4-33) the average power dissipated in the load is expressed as

$$p_L(f_\text{m}) = \frac{(n; \mathbf{m})^2}{2^{2n-1}} \; [|E_1|^2]^{(m_1 + m_{-1})} \cdots [|E_Q|^2]^{(m_Q + m_{-Q})} |H_n(\mathbf{m})|^2 G_L(f_\text{m})$$

$$(6\text{-}43)$$

where $H_n(\mathbf{m})$ is the nth-order nonlinear transfer function defined in (4-23) and $G_L(f)$ is the real part of the load admittance. Let the real part of the source impedance be denoted by $R_S(f)$. Then, from (2-89), the available power from the source at frequency f_q is

$$p_{AS}(f_q) = \frac{|E_q|^2}{(2)^3 R_S(f_q)}; \quad q = 1, 2, \ldots, Q \quad (6\text{-}44)$$

from which it follows that

$$|E_q|^2 = (2)^3 p_{AS}(f_q) R_S(f_q). \quad (6\text{-}45)$$

In order to eliminate the squared magnitudes of the input complex voltages from (6-43), use is made of (6-45) to obtain

$$p_L(f_\text{m}) = 2^{n+1} (n; \mathbf{m})^2 [p_{AS}(f_1)]^{(m_1 + m_{-1})} \cdots [p_{AS}(f_Q)]^{(m_Q + m_{-Q})}$$
$$\cdot |H_n(\mathbf{m})|^2 [R_S(f_1)]^{(m_1 + m_{-1})} \cdots [R_S(f_Q)]^{(m_Q + m_{-Q})} G_L(f_\text{m}). $$

$$(6\text{-}46)$$

Finally, with reference to the definition in (6-42), it is seen that the intermodulation transducer gain ratio is given by

$$\text{imtgr}_n(f_\text{m}) = 2^{n+1} (n; \mathbf{m})^2 |H_n(\mathbf{m})|^2 [R_S(f_1)]^{(m_1 + m_{-1})}$$
$$\cdots [R_S(f_Q)]^{(m_Q + m_{-Q})} G_L(f_\text{m}). \quad (6\text{-}47)$$

Example 6.4

By way of illustration, (6-47) specializes to

$$\text{imtgr}_1 [f_{(0,0,1,0)}] = 4 |H_1(f_1)|^2 R_S(f_1) G_L(f_1)$$

$$\text{imtgr}_2 \left[f_{(0,0,1,1)} \right] = 32 \left| H_2(f_1, f_2) \right|^2 R_S(f_1) R_S(f_2) G_L(f_1 + f_2)$$

$$\text{imtgr}_3 \left[f_{(1,0,2,0)} \right] = 144 \left| H_3(f_1, f_1, -f_2) \right|^2 \left[R_S(f_1) \right]^2 R_S(f_2) G_L(2f_1 - f_2).$$

$$(6\text{-}48)$$

From the definition in (6-42), note that $\text{imtgr}_1 \left[f_{(0,0,1,0)} \right]$ is the ratio of the average power delivered to the load at frequency f_1 to the available power from the source at the same frequency. This is the transducer power gain defined for linear loaded 2-ports in Section 2.6. In fact, it can be seen that the expression for the transducer gain in (2-107) is identical to the first equation in (6-48) by observing that the linear transfer function is the ratio of the linear portion of the output voltage to the input voltage and the load conductance is given by (2-45). Thus transducer gain is seen to be a special case of the more general intermodulation transducer gain ratio.

As can been seen from (6-42), the dimensions of $\text{imtgr}_n(f_m)$ are watts$^{(1-n)}$. Consequently, as a first step in the conversion of the intermodulation transducer gain ratio to decibels, we rewrite (6-42) in terms of a reference power p_r as

$$[p_r]^{n-1} \text{imtgr}_n(f_m) = \frac{\dfrac{p_L(f_m)}{p_r}}{\left[\dfrac{p_{AS}(f_1)}{p_r} \right]^{(m_1 + m_{-1})} \cdots \left[\dfrac{p_{AS}(f_Q)}{p_r} \right]^{(m_Q + m_{-Q})}}.$$

$$(6\text{-}49)$$

This converts each term in (6-42) to a dimensionless quantity. The decibel form of the intermodulation transducer gain ratio is then given by

$$\text{IMTGR}_n(f_m) = P_L(f_m) - (m_1 + m_{-1}) P_{AS}(f_1) - \cdots - (m_Q + m_{-Q}) P_{AS}(f_Q)$$

$$(6\text{-}50)$$

where

$$\text{IMTGR}_n(f_m) = 10 \log_{10} \left[p_r^{n-1} \text{imtgr}_n(f_m) \right]$$

$$P_L(f_m) = 10 \log_{10} \left[p_L(f_m)/p_r \right]$$

$$P_{AS}(f_q) = 10 \log_{10} \left[p_{AS}(f_q)/p_r \right]; \quad q = 1, 2, \ldots, Q. \quad (6\text{-}51)$$

For a given system the intermodulation transducer gain ratio is a constant, as can be seen from (6-47). Consequently, as opposed to (6-50), it is often preferable to express $\text{IMTGR}_n(f_m)$ in terms of system parameters. For this purpose, it is convenient to write the reference power as

$$p_r = \tfrac{1}{2} \left| E_r \right|^2 G_r \qquad (6\text{-}52)$$

where E_r is a reference voltage and G_r is the real part of a reference admittance. From (6-52) and (6-47) it follows that

$$\{p_r^{n-1} \text{imtgr}_n(f_m)\} = \{|E_r|^{2n-2} |H_n(m)|^2\} \{4(n; m)^2 [R_S(f_1)]^{(m_1+m-1)}$$
$$\cdots [R_S(f_Q)]^{(m_Q+m-Q)} G_L(f_m) G_r^{n-1}\}. \qquad (6\text{-}53)$$

As is apparent from (4-32), the nth-order nonlinear transfer function has dimensions of volts$^{(1-n)}$. Therefore, all terms in (6-53) enclosed by brackets of the form $\{\cdot\}$ are dimensionless. As a result, the intermodulation transducer gain ratio may be expressed in decibels as

$$\text{IMTGR}_n(f_m) = 10 \log_{10} [p_r^{n-1} \text{imtgr}_n(f_m)] = 20 \log_{10} \{|E_r|^{n-1} |H_n(m)|\}$$
$$+ 10 \log_{10} \{4(n; m)^2 [R_S(f_1)]^{(m_1+m-1)}$$
$$\cdots [R_S(f_Q)]^{(m_Q+m-Q)} G_L(f_m) G_r^{n-1}\}. \qquad (6\text{-}54)$$

Example 6.5

With regard to the evaluation of (6-54), it is usually convenient to select the reference voltage level to be 1 V. Then $|E_r|^{n-1} = 1$. If the reference power is also chosen to be 1 mW, then the real part of the reference admittance is calculated from (6-52) to be $G_r = 2 \times 10^{-3}$ mho. Assuming both the load and source to be resistors of 50 Ω, $G_L(f_m) = \frac{1}{50}$ mho and $R_S(f_q) = 50 \Omega$ for each value of the index q. With these values the decibel expressions corresponding to the intermodulation transducer gain ratios given in (6-48) become

$$\text{IMTGR}_1 [f_{(0,0,1,0)}] = 20 \log_{10} [|H_1(f_1)|] + 6.0 \text{ dB}$$

$$\text{IMTGR}_2 [f_{(0,0,1,1)}] = 20 \log_{10} [|H_2(f_1, f_2)|] + 2.0 \text{ dB}$$

$$\text{IMTGR}_3 [f_{(1,0,2,0)}] = 20 \log_{10} [|H_3(f_1, f_1, -f_2)|] - 4.4 \text{ dB}. \qquad (6\text{-}55)$$

Because the nonlinear transfer functions are not dimensionless, it is important to remember that a reference voltage of 1 V has been assumed and that factors of $(1 \text{ V})^{n-1}$ have been suppressed in (6-55) for convenience. It is also important to remember that (6-55) is valid only when the reference power is 1 mW and the load and source impedances are 50 Ω. Unfortunately, expressions in the form of (6-55) sometimes appear without an explanation of the assumptions used in calculating the constants.

The intermodulation behavior of weakly nonlinear systems is usually determined in the laboratory from measurements of the average power dissipated in the load by various intermodulation components. From (6-50) the frequency mix represented by the vector m generates an intermodulation component having the average power

$$P_L(f_m) = (m_1 + m_{-1}) P_{AS}(f_1) + \cdots + (m_Q + m_{-Q}) P_{AS}(f_Q) + \text{IMTGR}_n(f_m).$$

(6-56)

Experimentally, it is convenient to adjust the available powers of the input tones to be equal. Since the components of the vector m sum to n, the expression in (6-56) then reduces to

$$P_L(f_m) = nP_{AS} + \text{IMTGR}_n(f_m)$$

(6-57)

where P_{AS} denotes the available power from the source at each input frequency.

Example 6.6

Assuming a two-tone input, (6-57) is plotted in Fig. 6.3 for intermodulation components at $f_1, f_1 + f_2$, and $2f_1 - f_2$ corresponding to frequency mixes represented by $m = (0, 0, 1, 0)$, $(0, 0, 1, 1)$, and $(1, 0, 2, 0)$, respectively. For the case of equal available input powers, the average power dissipated in the load by an intermodulation component of order n plots as a straight line of slope n. The

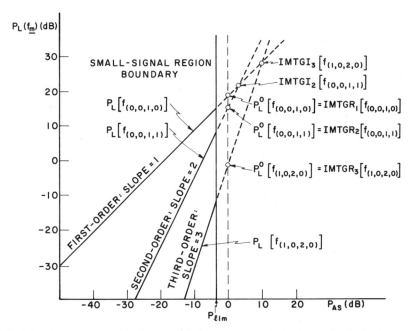

Fig. 6.3. Average power in intermodulation components for input signals having equal available input powers. (Variable with underbar in the figure has been set as a boldface italic term in text.)

equation of a straight line, of course, is uniquely determined from its slope and any one point on the line. For our purposes, a convenient point is that for which the available input power is 0 dB. Denoting the resulting average power in the intermodulation component by $P_L^0(f_m)$, (6-57) simplifies to

$$P_L^0(f_m) = \text{IMTGR}_n(f_m). \tag{6-58}$$

These values are indicated on the graph shown in Fig. 6.3. Since the intermodulation transducer gain ratio is a function of the input frequencies, as seen from (6-54), the effect of varying f_1 and f_2 is to translate the straight lines up and down. The slopes of the straight lines, however, are invariant to the choice of input frequencies.

It should be noted that the equation in (6-57) and the curves in Fig. 6.3 are theoretical in nature. Experimentally, the behavior predicted by the lower portion of the straight lines can be observed in the laboratory only if n is the lowest order frequency mix capable of generating the intermodulation frequency and the input amplitudes are small enough such that higher order effects are negligible. For example, the lowest order frequency mix capable of generating an intermodulation component at $2f_1 - f_2$ is given by $\mathbf{m} = (1, 0, 2, 0)$ for which $n = 3$. In addition, the fifth-order mixes represented by $\mathbf{m} = (1, 1, 3, 0)$ and $(2, 0, 2, 1)$ also produce responses at $2f_1 - f_2$. These contributions will be

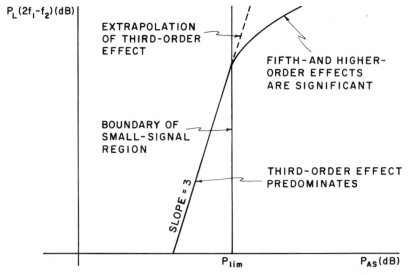

Fig. 6.4.   Average power in load dissipated at $2f_1 - f_2$ due to third-order and higher intermodulation components.

negligible with respect to the third-order response for small enough input powers. The average power in the load dissipated at $2f_1 - f_2$ will then follow the behavior predicted by the straight line in Fig. 6.3 with a slope equal to three. However, at larger input levels, higher order effects become significant and the line in Fig. 6.3 is no longer appropriate. Assuming equal available input powers at f_1 and f_2, a typical plot of $P_L(2f_1 - f_2)$ as a function of P_{AS} is shown in Fig. 6.4. The input power at which higher order effects become noticeable is denoted by $P_{\lim}$. The vertical line at $P_{\lim}$ serves to divide the curve into small- and large-signal regions. Recall that $P_L^0(f_m)$ equals the average power dissipated in the load when the available input power is 0 dB. When an input power of 0 dB falls outside of the small-signal region, as is the case in Fig. 6.3, $P_L^0(f_m)$ is obtained for each curve by extrapolating the straight lines into the large-signal region. Such extrapolations are shown in Fig. 6.3 as dashed lines.

6.3.2   Intermodulation Transducer Gain Intercept

The intermodulation transducer gain intercept is closely related to the intermodulation transducer gain ratio and is based upon extrapolation of the straight lines in Fig. 6.3 from the well-behaved small-signal region into the large-signal effects region. Recall that the curves of Fig. 6.3 assume equal available input powers to exist at each input frequency. The average output power contained in an intermodulation component generated by an nth-order frequency mix represented by the vector $\mathbf{m}$ is then given by (6-57). As demonstrated in Fig. 6.3, the higher order curves are guaranteed to intersect the first-order curve because the slopes of the straight lines increase with order. In decibels the intermodulation transducer gain intercept, denoted by $\text{IMTGI}_n(f_m)$ is defined to be the average output power at which the nth-order curve intersects the first-order curve. The intercepts for the frequency mixes represented by $\mathbf{m} = (0, 0, 1, 1)$ and $(1, 0, 2, 0)$ are indicated in Fig. 6.3 by $\text{IMTGI}_2\,[f_{(0,0,1,1)}]$ and $\text{IMGTI}_3\,[f_{(1,0,2,0)}]$, respectively.

The intercept concept is a relatively simple way to characterize intermodulation behavior in terms of the linear response of a system. In particular, given values for the linear transducer power gain and the intercepts, it is straightforward to reconstruct the straight lines in Fig. 6.3. The first-order curve, which is a plot of $P_L(f_1)$ versus P_{AS}, is drawn first. This is accomplished by recognizing that the linear transducer power gain equals $\text{IMTGR}_1\,[f_{(0,0,1,0)}]$ which, in turn, is identical to $P_L^0(f_1)$. Therefore, the first-order curve is a straight line of unity slope that passes through the point having coordinates $P_L(f_1) = \text{IMTGR}_1\,[f_{(0,0,1,0)}]$ and $P_{AS} = 0$. The higher order curves are drawn next. Specifically, the nth-order curve, corresponding to a plot of $P_L(f_m)$ versus P_{AS}, is a straight line of slope n that intersects the first-order curve at the ordinate where $P_L(f_m)$ equals the intercept $\text{IMGTI}_n(f_m)$.

From (6-54) and (6-58) the intermodulation transducer gain ratio $\text{IMTGR}_n(f_m)$ and, therefore, $P_L^o(f_m)$ are seen to become smaller as the magnitude of the nth-order nonlinear transfer function is decreased. Since a reduced value of $P_L^o(f_m)$ causes the nth-order curve in Fig. 6.3 to be lowered, the corresponding intercept increases in value. It follows that systems with relatively small non-linear transfer functions have relatively large intercepts and vice versa.

An analytical expression for $\text{IMTGI}_n(f_m)$ in terms of intermodulation trans-ducer gain ratios is readily obtained. In Fig. 6.3 the first-order curve intersects the nth-order curve when $P_L[f_{(0,0,1,0)}] = P_L(f_m)$. From (6-57) this is equiva-lent to

$$P_{AS} + \text{IMTGR}_1[f_{(0,0,1,0)}] = nP_{AS} + \text{IMTGR}_n(f_m). \tag{6-59}$$

Solution for the available input power yields

$$P_{AS} = \frac{1}{n-1}\{\text{IMTGR}_1[f_{(0,0,1,0)}] - \text{IMTGR}_n(f_m)\}. \tag{6-60}$$

However, by definition, $\text{IMTGI}_n(f_m)$ is the average output power at which the nth-order curve intersects the first-order curve. Therefore, making use of (6-60) in (6-57), we obtain

$$\text{IMTGI}_n(f_m) = \frac{n}{n-1}\text{IMTGR}_1[f_{(0,0,1,0)}] - \frac{1}{n-1}\text{IMTGR}_n(f_m). \tag{6-61}$$

Equation (6-61), of course, is valid only for n greater than or equal to 2.

Example 6.7

To illustrate application of (6-61), assume a reference power of 1 mW, a ref-erence voltage of 1 V, and resistive load and source impedances of 50 Ω, as was done in the calculation of (6-55). It follows that

$$\text{IMTGI}_2[f_{(0,0,1,1)}] = 40\log_{10}[|H_1(f_1)|] - 20\log_{10}[|H_2(f_1,f_2)|] + 10.0$$

$$= 20\log_{10}\frac{|H_1(f_1)|^2}{|H_2(f_1,f_2)|} + 10.0 \text{ dBm}$$

$$\text{IMTGI}_3[f_{(1,0,2,0)}] = 30\log_{10}[|H_1(f_1)|] - 10\log_{10}[|H_3(f_1,f_1,-f_2)|] + 11.2$$

$$= 10\log_{10}\left[\frac{|H_1(f_1)|^3}{|H_3(f_1,f_1,-f_2)|}\right] + 11.2 \text{ dBm}. \tag{6-62}$$

Equation (6-62) confirms the point that systems with smaller nonlinear transfer functions have larger intercepts.

6.3.3 Additional Intermodulation Criteria

A variety of different intermodulation criteria appear in the literature under an assortment of names such as intermodulation coefficient, intermodulation parameter, intermodulation distortion ratio, intermodulation factor, etc. All of these criteria can be derived from the expression for $p_L(f_m)$ given in (6-43). In order to illustrate the diverse multiplicity of criteria that exist, we develop some of the definitions in this section. In addition, we introduce names for the criteria that are suggestive of the underlying definitions. This is done to help clarify what is usually considered to be a confusing topic.

Intermodulation Average Gain Ratio: When deriving the expression for the intermodulation transducer gain ratio, available power was introduced to eliminate the squared magnitudes of the input complex voltages from (6-43). This was not necessary. The average input power at each input frequency could also have been used. These input powers are given by

$$p_S(f_q) = \tfrac{1}{2}|E_q|^2 \, G_{in}(f_q); \quad q = 1, 2, \ldots, Q \tag{6-63}$$

where $G_{in}(f)$ is the input conductance of the nonlinear system. Solving for $|E_q|^2$ from (6-63) and substituting into (6-43), the average power dissipated by the intermodulation component in the load becomes

$$p_L(f_m) = \frac{(n;\,m)^2}{2^{n-1}} \, [p_S(f_1)]^{(m_1+m_{-1})} \cdots [p_S(f_Q)]^{(m_Q+m_{-Q})}$$

$$\cdot \frac{|H_n(m)|^2 \, G_L(f_m)}{[G_{in}(f_1)]^{(m_1+m_{-1})} \cdots [G_{in}(f_Q)]^{(m_Q+m_{-Q})}}. \tag{6-64}$$

The intermodulation average gain ratio, denoted by $\mathrm{imagr}_n(f_m)$, is defined to be

$$\mathrm{imagr}_n(f_m) = \frac{p_L(f_m)}{[p_S(f_1)]^{(m_1+m_{-1})} \cdots [p_S(f_Q)]^{(m_Q+m_{-Q})}}. \tag{6-65}$$

We conclude that

$$\mathrm{imagr}_n(f_m) = \frac{(n;\,m)^2}{2^{n-1}} \frac{|H_n(m)|^2 \, G_L(f_m)}{[G_{in}(f_1)]^{(m_1+m_{-1})} \cdots [G_{in}(f_Q)]^{(m_Q+m_{-Q})}}. \tag{6-66}$$

Note that the average power gain defined for linear loaded 2-ports in Section 2.3 is identical to $\mathrm{imagr}_1 [f_{(0,0,1,0)}]$.

To convert to decibels, we again introduce the reference power defined in (6-52). Since the units of $\mathrm{imagr}_n(f_m)$ are watts$^{(1-n)}$, we consider

$$\{p_r^{n-1} \, \text{imagr}_n(f_{\text{m}})\} = \{|E_r|^{2n-2} \, |H_n(\text{m})|^2\}$$

$$\cdot \left\{ \frac{(n; \text{m})^2}{2^{2n-2}} \frac{G_L(f_{\text{m}}) \, G_r^{n-1}}{[G_{\text{in}}(f_1)]^{(m_1+m_{-1})} \cdots [G_{\text{in}}(f_Q)]^{(m_Q+m_{-Q})}} \right\}. \quad (6\text{-}67)$$

It follows that

$$\text{IMAGR}_n(f_{\text{m}}) = 10 \log_{10} \, [p_r^{n-1} \, \text{imagr}_n(f_{\text{m}})] = 20 \log_{10} \, [|E_r|^{n-1} \, |H_n(\text{m})|]$$

$$+ 10 \log_{10} \left\{ \frac{(n; \text{m})^2}{2^{2n-2}} \frac{G_L(f_{\text{m}}) \, G_r^{n-1}}{[G_{\text{in}}(f_1)]^{(m_1+m_{-1})} \cdots [G_{\text{in}}(f_Q)]^{(m_Q+m_{-Q})}} \right\}. \quad (6\text{-}68)$$

Example 6.8

For a 50-Ω system, in which $G_L(f) = G_{\text{in}}(f) = \frac{1}{50} = 2 \times 10^{-2}$ mho, and for a reference power of 1 mW with a reference voltage of 1 V, such that $G_r = 2 \times 10^{-3}$ mho, the intermodulation average gain ratios corresponding to $\text{m} = (0, 0, 1, 0), (0, 0, 1, 1)$, and $(1, 0, 2, 0)$ are given by

$$\text{IMAGR}_1 [f_{(0,0,1,0)}] = 20 \log_{10} \, [|H_1(f_1)|] \text{ dB}$$

$$\text{IMAGR}_2 [f_{(0,0,1,1)}] = 20 \log_{10} \, [|H_2(f_1, f_2)|] - 10.0 \text{ dB}$$

$$\text{IMAGR}_3 [f_{(1,0,2,0)}] = 20 \log_{10} \, [|H_3(f_1, f_1, -f_2)|] - 22.5 \text{ dB}. \quad (6\text{-}69)$$

As a final point, note that $\text{imagr}_n(f_{\text{m}})$ is identical to the intermodulation multiplier $c(\text{m})$ that was introduced in (3-60) and (4-36).

Intermodulation Input Gain Ratio: In some applications it is desirable to refer the average intermodulation power dissipated in the load back to the input. This is accomplished by focusing attention on the linear portion of the system and considering an equivalent on-tune excitation that generates the same amount of average output power as does the intermodulation component. Specifically, assume a sinusoidal input whose frequency coincides with the system-tuned frequency f_0. From (6-64) the average power in the load due to the linear portion of the response is

$$p_L(f_0) = p_S(f_0) |H_1(f_0)|^2 \frac{G_L(f_0)}{G_{\text{in}}(f_0)}. \quad (6\text{-}70)$$

If $P_S[f_0; f_{\text{m}}]$ denotes the average input power in the equivalent on-tune excitation, then equating $p_L(f_0)$ to $p_L(f_{\text{m}})$ results in

$$P_L(f_{\text{m}}) = P_S[f_0; f_{\text{m}}] |H_1(f_0)|^2 \frac{G_L(f_0)}{G_{\text{in}}(f_0)}. \quad (6\text{-}71)$$

Observe, in (6-71), that $p_L(f_m)$ has the interpretation of the average output power at frequency f_0 generated by the linear response of the system to an on-tune equivalent signal whose average input power is $P_S[f_0;f_m]$. Solving (6-71) for $P_S[f_0;f_m]$ and making use of (6-64), we obtain

$$P_S[f_0;f_m] = \frac{(n; m)^2}{2^{n-1}} [p_S(f_1)]^{(m_1+m_{-1})} \cdots [p_S(f_Q)]^{(m_Q+m_{-Q})}$$

$$\cdot \frac{|H_n(m)|^2}{|H_1(f_0)|^2} \frac{G_L(f_m)}{G_L(f_0)} \frac{G_{in}(f_0)}{[G_{in}(f_1)]^{(m_1+m_{-1})} \cdots [G_{in}(f_Q)]^{(m_Q+m_{-Q})}}. \quad (6\text{-}72)$$

The intermodulation input gain ratio, denoted by $\mathrm{imigr}_n(f_m)$, is defined to be

$$\mathrm{imigr}_n(f_m) = \frac{P_S[f_0;f_m]}{[p_S(f_1)]^{(m_1+m_{-1})} \cdots [p_S(f_Q)]^{(m_Q+m_{-Q})}}. \quad (6\text{-}73)$$

From (6-72), it follows that

$$\mathrm{imigr}_n(f_m) = \frac{(n; m)^2}{2^{n-1}} \frac{|H_n(m)|^2}{|H_1(f_0)|^2} \frac{G_L(f_m)}{G_L(f_0)}$$

$$\cdot \frac{G_{in}(f_0)}{[G_{in}(f_1)]^{(m_1+m_{-1})} \cdots [G_{in}(f_Q)]^{(m_Q+m_{-Q})}}. \quad (6\text{-}74)$$

Converting to decibels, we have

$$\mathrm{IMIGR}_n(f_m) = 20 \log_{10} \left[\frac{|E_r|^{n-1} |H_n(m)|}{|H_1(f_0)|} \right]$$

$$+ 10 \log_{10} \left\{ \frac{(n; m)^2}{2^{2n-2}} \frac{G_L(f_m)}{G_L(f_0)} \right.$$

$$\left. \cdot \frac{G_{in}(f_0) G_r^{n-1}}{[G_{in}(f_1)]^{(m_1+m_{-1})} \cdots [G_{in}(f_Q)]^{(m_Q+m_{-Q})}} \right\}. \quad (6\text{-}75)$$

Example 6.9

Using the same assumptions as in Example 6.8, there results

$$\mathrm{IMIGR}_1[f_{(0,0,1,0)}] = 20 \log_{10} \left[\frac{|H_1(f_1)|}{|H_1(f_0)|} \right] \mathrm{dB}$$

$$\mathrm{IMIGR}_2[f_{(0,0,1,1)}] = 20 \log_{10} \left[\frac{|H_2(f_1,f_2)|}{|H_1(f_0)|} \right] - 10.0 \, \mathrm{dB}$$

$$\mathrm{IMIGR}_3[f_{(1,0,2,0)}] = 20 \log_{10} \left[\frac{|H_3(f_1,f_1,-f_2)|}{|H_1(f_0)|} \right] - 22.5 \, \mathrm{dB}. \quad (6\text{-}76)$$

The constants in (6-76) are identical to those in (6-69) only because $G_L(f_0)$ was assumed equal to $G_{in}(f_0)$. When this is not the case, the constants will differ from each other.

———————————

The intermodulation input gain ratio is an attractive criterion because it enables the generation of intermodulation components by interfering signals to be interpreted in terms of an equivalent on-tune interference situation.

Intermodulation Output Gain Ratio: Another approach to the development of various intermodulation criteria stems from referring all powers to the output. With reference to the linear portion of the response, the average power dissipated in the load at the input frequency f_q is

$$p_L(f_q) = \tfrac{1}{2}|E_q|^2 \, |H_1(f_q)|^2 \, G_L(f_q); \qquad q = 1, 2, \ldots, Q. \qquad (6\text{-}77)$$

Therefore, the squared magnitude of the input complex voltage may be expressed as

$$|E_q|^2 = \frac{2p_L(f_q)}{|H_1(f_q)|^2 \, G_L(f_q)}. \qquad (6\text{-}78)$$

Substituting (6-78) into (6-43) yields for the average intermodulation load power

$$p_L(f_m) = \frac{(n; \mathbf{m})^2}{2^{n-1}} [p_L(f_1)]^{(m_1+m_{-1})} \cdots [p_L(f_Q)]^{(m_Q+m_{-Q})}$$

$$\cdot \frac{|H_n(\mathbf{m})|^2 \, G_L(f_m)}{[|H_1(f_1)|^2 \, G_L(f_1)]^{(m_1+m_{-1})} \cdots [|H_1(f_Q)|^2 \, G_L(f_Q)]^{(m_Q+m_{-Q})}}. \qquad (6\text{-}79)$$

Having obtained an expression for $p_L(f_m)$ in the form of (6-79), the intermodulation output gain ratio, denoted by $imogr_n(f_m)$, is defined to be

$$imogr_n(f_m) = \frac{p_L(f_m)}{[p_L(f_1)]^{(m_1+m_{-1})} \cdots [p_L(f_Q)]^{(m_Q+m_{-Q})}}. \qquad (6\text{-}80)$$

It follows from (6-79) that

$$imogr_n(f_m) = \frac{(n; \mathbf{m})^2}{2^{n-1}} \frac{|H_n(\mathbf{m})|^2}{[|H_1(f_1)|^2]^{(m_1+m_{-1})} \cdots [|H_1(f_Q)|^2]^{(m_Q+m_{-Q})}}$$

$$\cdot \frac{G_L(f_m)}{[G_L(f_1)]^{(m_1+m_{-1})} \cdots [G_L(f_Q)]^{(m_Q+m_{-Q})}}. \qquad (6\text{-}81)$$

Multiplying (6-81) by p_r^{n-1}, where p_r is the reference power defined in (6-52), and converting to decibels, we have

$$\text{IMOGR}_n(f_{\mathrm{m}}) = 20 \log_{10} \left\{ \frac{|E_r|^{n-1} |H_n(\mathbf{m})|}{[|H_1(f_1)|]^{(m_1+m_{-1})} \cdots [|H_1(f_Q)|]^{(m_Q+m_{-Q})}} \right\}$$

$$+ 10 \log_{10} \left\{ \frac{(n; \mathbf{m})^2}{2^{2n-2}} \frac{G_L(f_{\mathrm{m}}) G_r^{n-1}}{[G_L(f_1)]^{(m_1+m_{-1})} \cdots [G_L(f_Q)]^{(m_Q+m_{-Q})}} \right\}. \quad (6\text{-}82)$$

As usual, note that terms have been grouped into dimensionless quantities.

Example 6.10

Assuming the same reference power, reference voltage, and load conductance as in Example 6.8, the intermodulation output gain ratios corresponding to frequency mixes represented by $\mathbf{m} = (0, 0, 1, 0)$, $(0, 0, 1, 1)$, and $(1, 0, 2, 0)$ are given by

$$\text{IMOGR}_1 [f_{(0,0,1,0)}] = 0 \text{ dB}$$

$$\text{IMOGR}_2 [f_{(0,0,1,1)}] = 20 \log_{10} \left[\frac{|H_2(f_1, f_2)|}{|H_1(f_1)| \, |H_1(f_2)|} \right] - 10.0 \text{ dB}$$

$$\text{IMOGR}_3 [f_{(1,0,2,0)}] = 20 \log_{10} \left[\frac{|H_3(f_1, f_1, -f_2)|}{|H_1(f_1)|^2 \, |H_1(-f_2)|} \right] - 22.5 \text{ dB}.$$

$$(6\text{-}83)$$

From the definition in (6-80), the first-order intermodulation output gain ratio always equals unity. This accounts for the 0-dB result in (6-83).

———————————————

Many other intermodulation criteria are possible. For example, intercepts analogous to the intermodulation transducer gain intercept may be defined in conjunction with each criterion introduced in this section. A comparison between (6-55), (6-62), (6-69), (6-76), and (6-83) reveals strong similarities between various criteria. Yet, each criterion is different. Also, constants appearing in the expressions vary as a function of the reference power, reference voltage, input admittance, and load admittance. Because of these considerations, it is obvious that much care must be exercised if the different criteria are to be used properly.

6.4 DESENSITIZATION

In linear systems the sinusoidal response at a specific frequency is unaffected by the application of sinusoidal signals at other frequencies. However, in nonlinear systems, the sinusoidal response at frequency f_1 can be modified by the application of a second sinusoidal signal at frequency f_2. This phenomenon is referred to as desensitization.

To examine this effect analytically, let the input be given by

$$x(t) = |E_1| \cos (2\pi f_1 t + \theta_1) + |E_2| \cos (2\pi f_2 t + \theta_2). \tag{6-84}$$

For a weakly nonlinear system the nth-order portion of the response corresponding to a frequency mix represented by the vector $\mathbf{m}$ may be evaluated by means of (4-15). Because the excitation consists of a two-tone input, $\mathbf{m} = (m_{-2}, m_{-1}, m_1, m_2)$ where the components of $\mathbf{m}$ are nonnegative integers that sum to n, the order of the frequency mix. Hence for the input of (6-84), $y_n(t; \mathbf{m})$ becomes

$$y_n(t; \mathbf{m}) = \frac{(n; \mathbf{m})}{2^n} (E_2^*)^{m_{-2}} (E_1^*)^{m_{-1}} (E_1)^{m_1} (E_2)^{m_2}$$

$$\cdot H_n(\underbrace{-f_2, \ldots, -f_2}_{m_{-2}}, \underbrace{-f_1, \ldots, -f_1}_{m_{-1}}, \underbrace{f_1, \ldots, f_1}_{m_1}, \underbrace{f_2, \ldots, f_2}_{m_2})$$

$$\cdot \exp \{j2\pi[(m_1 - m_{-1})f_1 + (m_2 - m_{-2})f_2] t\}. \tag{6-85}$$

The total response at frequency f_1 includes contributions from all odd-order responses. The first- and third-order contributions result from the frequency mixes characterized by $\mathbf{m} = (0, 0, 1, 0)$, $(0, 1, 2, 0)$, and $(1, 0, 1, 1)$. Including only terms up to third order, it follows that the sinusoidal response at f_1 may be expressed as

$$y(t; f_1) = 2 \operatorname{Re} \{y_1(t; 0, 0, 1, 0) + y_3(t; 0, 1, 2, 0) + y_3(t; 1, 0, 1, 1)\}$$

$$= \operatorname{Re} \{ [E_1 H_1(f_1) + \tfrac{3}{4} E_1 |E_1|^2 H_3(-f_1, f_1, f_1)$$

$$+ \tfrac{3}{2} E_1 |E_2|^2 H_3(-f_2, f_1, f_2)] \exp (j2\pi f_1 t)\}. \tag{6-86}$$

The first term in (6-86) is the linear portion of the response, the second term is the third-order gain compression/expansion term, and the last term is the third-order desensitization term.

Because our objective in this section is to study desensitization, we assume $|E_1| \ll |E_2|$ so that the gain compression/expansion term is negligible with respect to the desensitization term. The complex output voltage at f_1 then reduces to

$$A_1 = E_1 H_1(f_1) + \tfrac{3}{2} E_1 |E_2|^2 H_3(-f_2, f_1, f_2). \tag{6-87}$$

Observe that (6-87) is valid only for small-signal inputs since effects above third order have been ignored. The voltage gain at frequency f_1 is given by

$$\text{vg} = \left| \frac{A_1}{E_1} \right| = |H_1(f_1)| \left| 1 + \frac{3}{2} |E_2|^2 \frac{H_3(-f_2, f_1, f_2)}{H_1(f_1)} \right|. \tag{6-88}$$

Because of the third-order nonlinearity, the gain is seen to depend nonlinearly upon the magnitude of the interfering signal at f_2. Departure of the voltage gain from that which would exist for a linear system is measured in terms of the desensitization ratio. This is defined to be

$$\text{dr} = \left| \frac{A_1}{E_1 H_1(f_1)} \right| = \left| 1 + \frac{3}{2} |E_2|^2 \frac{H_3(-f_2, f_1, f_2)}{H_1(f_1)} \right|. \tag{6-89}$$

A comparison of (6-89) with (6-22) reveals the desensitization ratio to be quite similar to the gain compression/expansion ratio. However, desensitization involves signals at two different frequencies whereas gain compression/expansion does not. Analogous to the vector diagrams illustrated in Fig. 6.2, the desensitization ratio may be interpreted as the magnitude of the resultant vector obtained by adding the vector $(\frac{3}{2}) |E_2|^2 H_3(-f_2, f_1, f_2)/H_1(f_1)$ to the unit vector. Depending upon the angle of $H_3(-f_2, f_1, f_2)$ relative to the angle of $H_1(f_1)$, the desensitization ratio may be greater or less than unity. Typically, the angles are approximately $180°$ out of phase and the ratio is less than unity.

Assuming

$$\frac{3}{2} |E_2|^2 \left| \frac{H_3(-f_2, f_1, f_2)}{H_1(f_1)} \right| \ll 1 \tag{6-90}$$

and applying the binomial theorem to (6-89), as was done in the derivation of (6-29), the desensitization ratio is approximated by

$$\text{dr} \approx 1 + \frac{3}{2} |E_2|^2 \text{ Re} \left\{ \frac{H_3(-f_2, f_1, f_2)}{H_1(f_1)} \right\}. \tag{6-91}$$

Obviously, when $\text{Re}\{H_3(-f_2, f_1, f_2)/H_1(f_1)\}$ is negative, $\text{dr} < 1$ and the ratio becomes smaller as $|E_2|$ is made larger. For $\text{dr} < 1$ the system is said to be desensitized by the interfering signal. The approximation in (6-91) should be used only to predict the onset of desensitization because of the assumptions leading to its development.

The available power in the signal at f_2 is given by

$$p_{AS}(f_2) = \frac{|E_2|^2}{8R_S(f_2)}. \tag{6-92}$$

Solving (6-92) for $|E_2|^2$ and substituting into (6-91), the desensitization ratio becomes

$$\text{dr} \approx 1 + 12R_S(f_2) p_{AS}(f_2) \text{ Re} \left\{ \frac{H_3(-f_2, f_1, f_2)}{H_1(f_1)} \right\}. \tag{6-93}$$

Converting (6-93) to decibels and successively making use of (6-33) and (6-35), there results

$$DR = 20 \log_{10} dr \approx 20 \log_{10} \left[1 + 12 R_S(f_2) p_{AS}(f_2) \, \text{Re} \left\{ \frac{H_3(-f_2, f_1, f_2)}{H_1(f_1)} \right\} \right]$$

$$= 8.6 \ln \left[1 + 12 R_S(f_2) p_{AS}(f_2) \, \text{Re} \left\{ \frac{H_3(-f_2, f_1, f_2)}{H_1(f_1)} \right\} \right]$$

$$\approx 103.2 R_S(f_2) p_{AS}(f_2) \, \text{Re} \left\{ \frac{H_3(-f_2, f_1, f_2)}{H_1(f_1)} \right\} dB. \qquad (6\text{-}94)$$

Finally, solving (6-94) for the available power in the interfering signal needed to produce a specified value of DR, we obtain

$$p_{AS}(f_2) \approx \frac{DR}{103.2 R_S(f_2) \, \text{Re} \left\{ \dfrac{H_3(-f_2, f_1, f_2)}{H_1(f_1)} \right\}}. \qquad (6\text{-}95)$$

In decibels referred to 1 mW, the available power is expressed as

$$P_{AS}(f_2) = 10 \log_{10} \frac{p_{AS}(f_2)}{10^{-3}}$$

$$\approx -10 \log_{10} \left[\frac{1}{DR} \, \text{Re} \left\{ \frac{H_3(-f_2, f_1, f_2)}{H_1(f_1)} \right\} \right] + 10 \log_{10} \left[\frac{9.7}{R_S(f_2)} \right] dBm.$$

$$(6\text{-}96)$$

Equation (6-96) should be used with care since only third-order effects have been included in our discussion. In addition, the inequality in (6-90) must be satisfied. In general, the approximations yield reasonable results whenever the desensitization ratio is less than 1 dB. For values larger than this, many higher order terms are usually needed in order to accurately predict system behavior. Under these conditions, the third-order approximation is not adequately corrected by simply adding fifth-order, and even seventh-order, effects. In such large-signal situations the nonlinear transfer function approach is not the appropriate analytical tool.

6.5   CROSS MODULATION

Cross modulation is the nonlinear effect whereby modulation from one signal is transferred to another. The ensuing crosstalk can cause serious degradation in system performance. When the modulation is speech, crosstalk interference is classified as either intelligible or unintelligible depending on whether the created interference is understandable. For the case of nonspeech modulations, the crosstalk is said to be intelligible when the transferred modulation is of the same type as the desired modulation.

In this section we investigate the cross-modulation phenomenon by considering the nonlinear interaction between a desired unmodulated carrier signal and an interfering signal that is amplitude modulated by a sinusoid. Specifically, let the input to a weakly nonlinear system be given by

$$x(t) = |E_S| \cos (2\pi f_S t + \theta_S) + |E_I| [1 + m_I \cos (2\pi f_m t + \theta_m)] \cos (2\pi f_I t + \theta_I)$$

(6-97)

where m_I is the modulation index of the interfering signal and f_m is the modulating frequency. To express the excitation as a sum of sinusoidal components, we make use of the trigonometric identity

$$\cos \alpha \cos \beta = \tfrac{1}{2} \cos (\alpha - \beta) + \tfrac{1}{2} \cos (\alpha + \beta).$$

(6-98)

It follows that

$$x(t) = |E_S| \cos (2\pi f_S t + \theta_S) + |E_I| \cos (2\pi f_I t + \theta_I)$$
$$+ \frac{m_I |E_I|}{2} \cos [2\pi (f_I - f_m) t + \theta_I - \theta_m]$$
$$+ \frac{m_I |E_I|}{2} \cos [2\pi (f_I + f_m) t + \theta_I + \theta_m].$$

(6-99)

To simplify matters further, we rewrite (6-99) as

$$x(t) = \sum_{q=1}^{4} |E_q| \cos (2\pi f_q t + \theta_q)$$

(6-100)

where the input frequencies have the interpretation

$f_1 = f_S =$ carrier frequency of desired signal

$f_2 = f_I =$ carrier frequency of interfering signal

$f_3 = f_I - f_m =$ lower sideband frequency of undesired signal

$f_4 = f_I + f_m =$ upper sideband frequency of undesired signal   (6-101)

and the complex voltages of the input tones are given by

$$E_1 = |E_S| e^{j\theta_S},$$
$$E_2 = |E_I| e^{j\theta_I},$$
$$E_3 = \frac{m_I |E_I|}{2} e^{j(\theta_I - \theta_m)},$$
$$E_4 = \frac{m_I |E_I|}{2} e^{j(\theta_I + \theta_m)}.$$

(6-102)

A pictorial representation of the input voltage spectrum, as would be seen with a spectrum analyzer, is shown in Fig. 6.5(a). Typically, the carrier frequencies f_I and f_S are much larger than the frequency of modulation f_m.

In response to the input of (6-100), there are many different terms contained in the output. For example, the third-order portion of the system generates sinusoidal components corresponding to 60 distinct frequency mixes. Fortunately, our analysis of cross modulation does not require that we evaluate each term in the output. We are interested only in the transfer of modulation from the interfering signal to the desired signal. As a result, we need evaluate only those frequency components near f_S. For simplicity, we assume nonlinear effects above third order to be negligible. The frequency mixes of interest are tabulated in Table 6.1. Because the excitation consists of four sinusoidal inputs, $\mathbf{m} = (m_{-4}, m_{-3}, m_{-2}, m_{-1}, m_1, m_2, m_3, m_4)$. Observe that only mixes that yield positive frequencies in the vicinity of f_S have been included in the table. To each vector $\mathbf{m}$ there is, of course, a vector $\mathbf{m}'$ that yields the negative frequency. The positive and negative frequency terms combine to produce real sinusoidal components. From Table 6.1 we see that sinusoidal outputs are generated at f_S, $f_S - f_m$, $f_S + f_m$, $f_s - 2f_m$, and $f_S + 2f_m$. The line spectrum of the output in the vicinity of f_S is shown in Fig. 6.5(b).

Let that portion of the response with frequency components in the vicinity of f_S be denoted by $y(t; f_S, f_S \pm f_m, f_S \pm 2f_m)$. With reference to (4-15) and

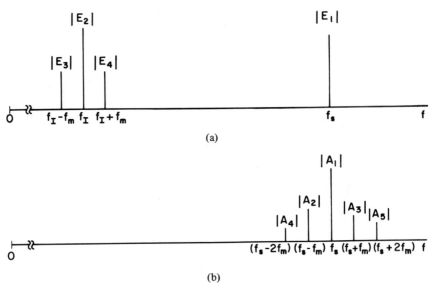

(a)

(b)

Fig. 6.5. Line spectra of (a) input voltage spectrum and (b) output voltage spectrum in the vicinity of f_S.

TABLE 6.1.   Frequency Mixes of Interest in the Cross-Modulation Analysis.

m_{-4}	m_{-3}	m_{-2}	m_{-1}	m_1	m_2	m_3	m_4	f_m
\multicolumn{8}{l}{$\mathbf{m} = (m_{-4}, m_{-3}, m_{-2}, m_{-1}, m_1, m_2, m_3, m_4)$}	
0	0	0	0	1	0	0	0	$f_1 = f_S$
1	0	0	0	1	0	0	1	$-f_4 + f_1 + f_4 = f_1 = f_S$
0	1	0	0	1	0	1	0	$-f_3 + f_1 + f_3 = f_1 = f_S$
0	0	1	0	1	1	0	0	$-f_2 + f_1 + f_2 = f_1 = f_S$
0	0	0	1	2	0	0	0	$-f_1 + 2f_1 = f_1 = f_S$
1	0	0	0	1	1	0	0	$-f_4 + f_1 + f_2 = f_S - f_m$
0	0	1	0	1	0	1	0	$-f_2 + f_1 + f_3 = f_S - f_m$
0	0	1	0	1	0	0	1	$-f_2 + f_1 + f_4 = f_S + f_m$
0	1	0	0	1	1	0	0	$-f_3 + f_1 + f_2 = f_S + f_m$
1	0	0	0	1	0	1	0	$-f_4 + f_1 + f_3 = f_S - 2f_m$
0	1	0	0	1	0	0	1	$-f_3 + f_1 + f_4 = f_S + 2f_m$

(4-22), $y(t; f_S, f_S \pm f_m, f_S \pm 2f_m)$ may be expressed as

$$y(t; f_S, f_S \pm f_m, f_S \pm 2f_m)$$
$$= \left|A_1\right| \cos\left[2\pi f_S t + \alpha_1\right] + \left|A_2\right| \cos\left[2\pi(f_S - f_m)t + \alpha_2\right]$$
$$+ \left|A_3\right| \cos\left[2\pi(f_S + f_m)t + \alpha_3\right] + \left|A_4\right| \cos\left[2\pi(f_S - 2f_m)t + \alpha_4\right]$$
$$+ \left|A_5\right| \cos\left[2\pi(f_S + 2f_m)t + \alpha_5\right] \tag{6-103}$$

where the complex amplitudes of the output are given by

$$A_1 = \left|A_1\right| e^{j\alpha_1} = E_1 H_1(f_1) + \tfrac{3}{2} E_1 \left|E_4\right|^2 H_3(-f_4, f_1, f_4)$$
$$+ \tfrac{3}{2} E_1 \left|E_3\right|^2 H_3(-f_3, f_1, f_3) + \tfrac{3}{2} E_1 \left|E_2\right|^2 H_3(-f_2, f_1, f_2)$$
$$+ \tfrac{3}{4} E_1 \left|E_1\right|^2 H_3(-f_1, f_1, f_1)$$
$$A_2 = \left|A_2\right| e^{j\alpha_2} = \tfrac{3}{2} E_4^* E_1 E_2 H_3(-f_4, f_1, f_2) + \tfrac{3}{2} E_2^* E_1 E_3 H_3(-f_2, f_1, f_3)$$
$$A_3 = \left|A_3\right| e^{j\alpha_3} = \tfrac{3}{2} E_2^* E_1 E_4 H_3(-f_2, f_1, f_4) + \tfrac{3}{2} E_3^* E_1 E_2 H_3(-f_3, f_1, f_2)$$
$$A_4 = \left|A_4\right| e^{j\alpha_4} = \tfrac{3}{2} E_4^* E_1 E_3 H_3(-f_4, f_1, f_3)$$
$$A_5 = \left|A_5\right| e^{j\alpha_5} = \tfrac{3}{2} E_3^* E_1 E_4 H_3(-f_3, f_1, f_4). \tag{6-104}$$

Note that the desired signal at f_S is surrounded by sidebands at $f_S \pm f_m$ and $f_S \pm 2f_m$. In effect, the modulation from the interferer has been transferred to the desired signal. However, it is a distorted version of the modulation that appears on the carrier at f_S. This is due to the fact that the original modulated signal consisted of one upper and lower sideband whereas the new modulated signal consists of two upper and lower sidebands. In addition, the frequency dependence of the magnitude and angle of the third-order transfer function may cause the magnitudes and angles of the sidebands to be unsymmetrical about f_S. This lack of symmetry can produce both angle modulation of the desired signal and increased amplitude distortion in the envelope. Of course, any unintentional modulation may be extremely troublesome in a system.

Although there are sideband components at $f_S \pm 2f_m$, they do not contribute to the modulation at the modulation frequency f_m and are usually ignored in cross-modulation criteria. For example, the modulation factor is defined to be

$$mf = \frac{|A_2| + |A_3|}{|A_1|} \tag{6-105}$$

while the cross-modulation ratio is defined as

$$cmr = \frac{\left(\begin{array}{c} \text{peak-to-peak variation in envelope of} \\ \text{desired signal due to sidebands at } f_S \pm f_m \end{array} \right)}{\left(\begin{array}{c} \text{amplitude of desired carrier with} \\ \text{interferer removed} \end{array} \right)}. \tag{6-106}$$

In general, simple analytical expressions are difficult to obtain for these criteria.

Example 6.11

The expressions for mf and cmr become considerably simplified when the third-order transfer function is approximately equal to a constant over the frequency range of interest. Denote this constant by k_3. Also denote $H_1(f_1)$ by k_1 and assume

$$\theta_S = \theta_m = \theta_I = 0. \tag{6-107}$$

Using these assumptions in (6-102) and (6-104), A_1, A_2, and A_3 become

$$A_1 = k_1 |E_S| + \frac{3}{2} k_3 |E_S| |E_I|^2 \left[1 + \frac{m_I^2}{2} + \frac{1}{2} \left| \frac{E_S}{E_I} \right|^2 \right]$$

$$A_2 = A_3 = \frac{3}{2} k_3 |E_S| |E_I|^2 \, m_I. \tag{6-108}$$

It follows that the modulation factor is given by

$$mf = \frac{3 |k_3| |E_I|^2 \, m_I}{\left| k_1 + \frac{3}{2} k_3 |E_I|^2 \left[1 + \frac{m_I^2}{2} + \frac{1}{2} \left| \frac{E_S}{E_I} \right|^2 \right] \right|}. \tag{6-109}$$

Typically, the second term in the denominator is negligible with respect to the first and the modulation factor is approximated by

$$\text{mf} \approx 3 \left| \frac{k_3}{k_1} \right| |E_I|^2 \, m_I. \tag{6-110}$$

The cross-modulation ratio is also readily determined assuming $\alpha_1 = \alpha_2 = 0$. Ignoring the frequency components at $f_S \pm 2f_m$ in (6-103), the envelope of the carrier at f_S is $|A_1| + 2|A_2| \cos 2\pi f_m t$. Consequently, the peak-to-peak variation in the envelope due to the sidebands at $f_S \pm f_m$ equals $4|A_2|$. Also, the amplitude of the desired carrier with the interferer removed is $\left| [k_1|E_S| + (\frac{3}{4}) k_3 |E_S|^3] \right|$. Hence the cross-modulation ratio is

$$\text{cmr} = \frac{6|k_3| \, |E_I|^2 \, m_I}{\left| k_1 + \frac{3}{4} k_3 |E_S|^2 \right|}. \tag{6-111}$$

The second term in the denominator of (6-111) is usually negligible with respect to the first. The cross-modulation ratio is then approximated by

$$\text{cmr} \approx 6 \left| \frac{k_3}{k_1} \right| |E_I|^2 \, m_I \approx 2\text{mf}. \tag{6-112}$$

With the assumptions of our discussion, we see that the cross-modulation ratio is typically twice the modulation factor. Although this result is widely quoted, observe that it represents a special case of the more general situation.

6.6 SPURIOUS RESPONSE

To many engineers the term "receiver" is synonymous with "superheterodyne receiver" because of the predominance of the superheterodyne design in existing equipments. The superheterodyne receiver employs one or more local oscillators to translate the carrier of a received signal to a fixed predetermined intermediate frequency, known as the IF frequency. In this text the term "spurious response" is used to refer to those undesired responses that result from an interfering signal, or one of its harmonics, mixing with the local oscillator signal, or one of its harmonics, to produce a signal that falls within the IF passband. The problem of spurious responses is one of the primary drawbacks to the superheterodyne design concept.

Spurious responses may be regarded as a special case of intermodulation where one of the two signals is associated with the local oscillator. Denote the frequency of the interfering signal by f_1, the frequency of the local oscillator by f_{Lo}, and let p and q be positive integers. A spurious response results whenever the intermodulation frequency given by

$$f_m = |\pm p f_{\text{Lo}} \pm q f_1| \tag{6-113}$$

falls within the IF passband. In the literature such a response is also referred to as the mixer p, q response.

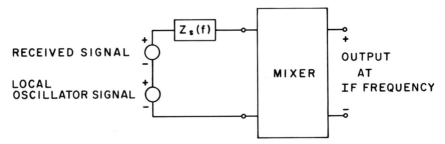

Fig. 6.6. Weakly nonlinear mixer with received signal applied in series with local oscillator signal.

The nonlinear transfer function approach, of course, is appropriate only for weakly nonlinear systems. Hence we consider mixer circuits that involve only mildly driven nonlinearities. Many diode, transistor, and vacuum-tube mixers, as well as varacter up-converters, can be modeled in this way. Switched mixers or mixers with large local oscillator drives are more efficiently treated by large-signal time-domain techniques. Also, because we have treated only single-port excitations in this volume, we assume the received signal is applied directly in series with the local oscillator signal, as shown in Fig. 6.6.

There are many different ways by which the intermodulation frequency indicated in (6-113) may be generated. To illustrate the variety of situations that may arise, we analyze four cases as follows.

Case 1: The received signal and local oscillator signal are harmonic free. Consequently, a $(p + q)$th-order mix is required to produce the spurious response.

Case 2: The local oscillator signal contains a pth harmonic and the received signal is harmonic free. Harmonics will be contained in the local oscillator signal whenever its waveform is a periodic distorted sine wave. In this case the spurious response is generated by a $(q + 1)$th-order frequency mix.

Case 3: The received signal contains a qth harmonic and the local oscillator signal is harmonic free. Nonlinearities in amplifier stages preceding the mixer are capable of creating a qth harmonic in the received signal. A $(p + 1)$th-order mix is now needed to cause the spurious response.

Case 4: The local oscillator signal contains a pth harmonic and the received signal contains a qth harmonic. For this situation only a second-order mix is needed to produce the spurious response.

Because all four cases can be analyzed in an identical manner, we discuss them simultaneously. The average power dissipated in the load by the spurious response may be expressed in terms of available source power by referring to (6-46). For the cases of interest, it follows that

Case 1 (signal at f_1 mixes with signal at f_{Lo}):

$$p_L(f_m) = 2^{p+q+1} \left[\frac{(p+q)!}{(p!)\,(q!)}\right]^2 [p_{AS}(f_{Lo})]^p \, [p_{AS}(f_1)]^q$$

$$\cdot \left| H_{p+q}(\underbrace{\pm f_{Lo}, \ldots, \pm f_{Lo}}_{p}, \underbrace{\pm f_1, \ldots, \pm f_1}_{q}) \right|^2$$

$$\cdot [R_S(f_{Lo})]^p \, [R_S(f_1)]^q \, G_L(f_m). \tag{6-114}$$

Case 2 (signal at f_1 mixes with signal at pf_{Lo}):

$$p_L(f_m) = 2^{q+2}(q+1)^2 \, [p_{AS}(pf_{Lo})] \, [p_{AS}(f_1)]^q$$

$$\cdot \left| H_{q+1}(\pm pf_{Lo}, \underbrace{\pm f_1, \ldots, \pm f_1}_{q}) \right|^2 [R_S(pf_{Lo})] \, [R_S(f_1)]^q \, G_L(f_m). \tag{6-115}$$

Case 3 (signal at qf_1 mixes with signal at f_{Lo}):

$$p_L(f_m) = 2^{p+2}(p+1)^2 \, [p_{AS}(f_{Lo})]^p \, [p_{AS}(qf_1)]$$

$$\cdot \left| H_{p+1}(\underbrace{\pm f_{Lo}, \ldots, \pm f_{Lo}}_{p}, \pm qf_1) \right|^2 [R_S(f_{Lo})]^p [R_S(qf_1)] \, G_L(f_m). \tag{6-116}$$

Case 4 (signal at qf_1 mixes with signal at pf_{Lo}):

$$p_L(f_m) = 2^3(2)^2 \, [p_{AS}(pf_{Lo})] \, [p_{AS}(qf_1)] \, |H_2(\pm pf_{Lo}, \pm qf_1)|^2$$

$$\cdot [R_S(pf_{Lo})] \, [R_S(qf_1)] \, G_L(f_m). \tag{6-117}$$

The spurious response transducer gain ratio, denoted by $\mathrm{srtgr}_n(f_m)$, is then defined as follows for each of the four cases.

Case 1 (signal at f_1 mixes with signal at f_{Lo}):

$$\mathrm{srtgr}_{p+q}(f_m) = \frac{p_L(f_m)}{[p_{AS}(f_{Lo})]^p \, [p_{AS}(f_1)]^q} = 2^{p+q+1} \left[\frac{(p+q)!}{(p!)\,(q!)}\right]^2$$

$$\cdot \left| H_{p+q}(\underbrace{\pm f_{Lo}, \ldots, \pm f_{Lo}}_{p}, \underbrace{\pm f_1, \ldots, \pm f_1}_{q}) \right|^2$$

$$\cdot [R_S(f_{Lo})]^p \, [R_S(f_1)]^q \, G_L(f_m). \tag{6-118}$$

Case 2 (signal at f_1 mixes with signal at pf_{Lo}):

$$\mathrm{srtgr}_{q+1}(f_m) = \frac{p_L(f_m)}{[p_{AS}(pf_{Lo})] \, [p_{AS}(f_1)]^q} = 2^{q+2}(q+1)^2$$

$$\cdot \left| H_{q+1}(\pm pf_{Lo}, \underbrace{\pm f_1, \ldots, \pm f_1}_{q}) \right|^2 [R_S(pf_{Lo})] \, [R_S(f_1)]^q \, G_L(f_m). \tag{6-119}$$

Case 3 (signal at qf_1 mixes with signal at f_{Lo}):

$$\text{srtgr}_{p+1}(f_m) = \frac{p_L(f_m)}{[p_{AS}(f_{Lo})]^p \, [p_{AS}(qf_1)]} = 2^{p+2}(p+1)^2$$

$$\cdot \, |H_{p+1}(\underbrace{\pm f_{Lo}, \ldots, \pm f_{Lo}}_{p}, \pm qf_1)|^2 \, [R_S(f_{Lo})]^p \, [R_S(qf_1)] \, G_L(f_m). \quad (6\text{-}120)$$

Case 4 (signal at qf_1 mixes with signal at pf_{Lo}):

$$\text{srtgr}_2(f_m) = \frac{p_L(f_m)}{[p_{AS}(pf_{Lo})] \, [p_{AS}(qf_1)]} = 2^5 |H_2(\pm pf_{Lo}, \pm qf_1)|^2$$

$$\cdot \, [R_S(pf_{Lo})] \, [R_S(qf_1)] \, G_L(f_m). \quad (6\text{-}121)$$

Following the same procedure used in obtaining (6-54), the spurious response transducer gain ratios are readily expressed in terms of decibels. As was done in Section 6.3 for intermodulation, many other spurious response criteria may be defined. Therefore, the user must be careful if each criterion is to be utilized properly. Also, the reader is reminded that the results presented in this section are appropriate only for mildly driven mixers in which the received and local oscillator signals appear in series with each other.

7

Nonlinear Circuit Modeling of Solid-State Devices

As discussed in Chapter 5, the analytical procedure for determining nonlinear transfer functions requires that all electronic devices contained in a network be modeled by equivalent nonlinear circuits. Of the many different circuit models available for each device, one must be selected for a given application. Models that are highly accurate over large ranges of amplitude and frequency tend to be rather complicated and computationally inefficient. Furthermore, their complete description frequently entails a considerable amount of effort in order to obtain values for the large number of model parameters typically required. Less sophisticated models tend to be more efficient and easier to specify. However, since they arise by either neglecting or approximating various physical effects, they are limited in their usefulness. Nevertheless, simpler models can provide acceptable results in many situations. For each electronic device there is no single model that can be claimed to be "best" in all possible applications. Clearly, the engineer should treat each case individually as he decides which of the available models is best suited for his purposes. In particular, he should evaluate the models according to their computational efficiency, the level of accuracy needed for the problem at hand, and the amount of time, money, and effort available for obtaining model parameters.

In this chapter equivalent nonlinear circuits are given for some solid-state devices commonly encountered in communications circuits. After a brief discussion of the approach taken in the modeling of electronic devices for weakly nonlinear effects, nonlinear incremental equivalent circuits are presented for the *p-n* junction diode, the bipolar junction transistor (BJT), the junction field-effect transistor (JFET), and the metal-oxide-semiconductor field-effect transistor (MOSFET). Although the models given are not the most sophisticated, they have been shown to yield good results in practice. In conjunction with each model, explanations are included on how to obtain values for the various model parameters. Because of space limitations, the physical theory underlying the operation of the electronic devices is not discussed. Still, if a model is to be properly utilized to its fullest extent, a clear understanding of its physical basis is essential. Consequently, an extensive list of references on the physical theory of electronic devices is provided in the Bibliography.

7.1 MODELING OF ELECTRONIC DEVICES FOR WEAKLY NONLINEAR EFFECTS

The operation of electronic devices is usually depicted in terms of their static characteristics. These are plots relating a device's terminal voltages and currents as obtained from measurements performed under dc or very slowly varying conditions. Therefore, the static characteristics of actual devices do not include the effects of energy-storage elements and frequency-dependent parameters. One approach to modeling an electronic device begins by first obtaining a dc resistive equivalent circuit that models the static characteristics. Capacitors and inductors are then added at one or more strategic locations to account for the device's frequency behavior.

A device is completely specified when its behavior, as a function of frequency, is known over all measurable voltages and currents. Corresponding models are referred to as global models. However, devices in communications circuits are typically operated in both a restricted frequency range and a localized region of their characteristics. Models for this more restrictive situation are known as localized models. In general, localized models are considerably simpler than their global counterparts.

The region of operation for an electronic device is established by employing a dc biasing circuit that determines the dc or quiescent operating point. The application of additional signals then results in the device voltages and currents varying in some neighborhood about the operating point. This is illustrated in Fig. 7.1 for a semiconductor diode. Recall that the static characteristic shown is valid only for dc and very low frequencies. In this example the quiescent operating point, denoted by Q, is located in the forward-biased region of the characteristic. Associated with the semiconductor diode is a junction capacitance whose effect is not accounted for by the static characteristic. In fact, at suf-

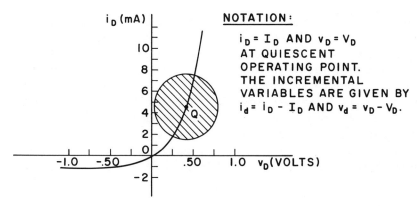

Fig. 7.1. Static characteristic of semiconductor diode illustrating quiescent operating point and region of operation.

ficiently high frequencies, the junction capacitance will cause the diode current and voltage to deviate from the static characteristic and fall somewhere in a neighborhood centered about Q, as indicated by the circular region in Fig. 7.1.

Localized models of electronic devices are adequate for our study of distortion and interference in weakly nonlinear systems. When developing the localized models, it is important to distinguish between total, dc, and incremental variables. In this text total variables are denoted by lowercase symbols with uppercase subscripts. For example, i_D and v_D represent the total current and voltage, respectively, of a diode. The dc value of a variable corresponds to that occurring at the quiescent operating point and is denoted by an uppercase symbol with an uppercase subscript. Therefore, $i_D = I_D$ and $v_D = V_D$ at the quiescent operating point of a diode. Incremental variables are denoted by lowercase symbols with lowercase subscripts and are defined to be the difference between the total and dc variables. For our diode example, the incremental current and voltage are given by

$$i_d = i_D - I_D$$

$$v_d = v_D - V_D. \tag{7-1}$$

Global models, which interrelate total instantaneous currents and voltages, are also the starting point for generating localized models used in the nonlinear transfer function approach. Specifically, localized models are developed by expanding device nonlinear relationships, involving total variables, into power series about the quiescent operating point. The series coefficients can be found either numerically or from Taylor series expansions. The numerical approach may be viewed as curve fitting of a polynomial to an experimental or theoretical curve. Since the coefficients vary from one operating point to the next, the procedure must be repeated for each different operating point. Nevertheless, for a polynomial of given degree, the numerical approach usually yields a closer approximation than does the corresponding truncated Taylor series. The Taylor series approach has the advantage that its coefficients are determined from analytical mathematical expressions. Therefore, wherever a device may be biased, the coefficients are readily determined analytically. In addition, fewer parameters are frequently needed to characterize the mathematical relationships of a model than are required in the numerical approach where a complete set of power series coefficients must be stored for each operating point. In our presentation the Taylor series approach is used to develop the localized models.

Example 7.1

As a simple example, consider, once again, the static characteristic of the semiconductor diode shown in Fig. 7.1. Let the corresponding current-voltage relationship be expressed as

$$i_D = g(v_D) \tag{7-2}$$

where $g(\cdot)$ is a single-valued zero-memory functional. As far as the static characteristic is concerned, a global model of the diode can be interpreted in terms of the nonlinear resistor indicated in Fig. 7.2(a). Observe that only total variables appear in the global model.

To obtain a localized model for the diode of Example 7.1, (7-2) is expanded in a Taylor series about the quiescent operating point. In general, the Taylor series expansion of a function $f(x)$ about the point x_0 is given by

$$f(x) = f(x_0) + \sum_{k=1}^{\infty} a_k (x - x_0)^k \tag{7-3}$$

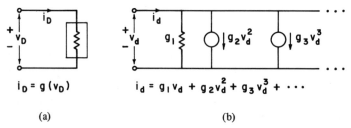

(a)   (b)

Fig. 7.2. Nonlinear circuit models for semiconductor diode based upon static characteristic. (a) Global model. (b) Localized model.

where the kth coefficient is determined from the relation

$$a_k = \frac{1}{k!} \frac{d^k f(x)}{dx^k}\bigg|_{x=x_0}.$$ (7-4)

The kth coefficient is seen to involve the kth derivative of $f(x)$, evaluated at the point about which the expansion occurs. It is clear from (7-3) and (7-4) that a function cannot have a Taylor series expansion about $x = x_0$ unless the function possesses finite derivatives of all orders at $x = x_0$. In practical applications the Taylor series is usually truncated to a finite number of terms. The remainder for a series truncated to N terms is defined to be

$$R_N(x) = f(x) - [f(x_0) + \sum_{k=1}^{N} a_k (x - x_0)^k].$$ (7-5)

An upper bound for this remainder can be shown to be given by the inequality

$$|R_N(x)| \leqslant \frac{M}{(N+1)!} |x - x_0|^{N+1}$$ (7-6)

where M is the maximum value of the $(N+1)$th derivative of $f(x)$ in the interval (x_0, x).

Example 7.2

A localized model is now developed for the diode of Example 7.1. The Taylor series expansion of (7-2) about the quiescent operating point becomes

$$i_D = g(V_D) + \sum_{k=1}^{\infty} g_k(v_D - V_D)^k$$ (7-7)

where the kth series coefficient is given by

$$g_k = \frac{1}{k!} \frac{d^k g(v_D)}{dv_D^k}\bigg|_{v_D = V_D}.$$ (7-8)

Providing a convenient analytic expression exists for the nonlinear functional $g(\cdot)$, the Taylor series coefficients are seen to be readily obtained analytically as a function of the quiescent operating point. At the quiescent operating point $i_D = I_D$ and $v_D = V_D$. It follows from (7-7) that $I_D = g(V_D)$. Making use of this identity, (7-7) may be rewritten as

$$i_D - I_D = \sum_{k=1}^{\infty} g_k(v_D - V_D)^k.$$ (7-9)

Finally, introducing the incremental variables of (7-1), (7-9) reduces to

$$i_d = \sum_{k=1}^{\infty} g_k v_d^k. \qquad (7\text{-}10)$$

The mathematical relationship expressed by (7-10) can be interpreted in terms of a nonlinear incremental equivalent circuit in which a linear conductance g_1 is in parallel with an infinite set of voltage-controlled current sources. This is illustrated in Fig. 7.2(b).

The circuit of Fig. 7.2(b) represents a localized model for the diode. As required by the analytical procedure in Chapter 5 for determining the nonlinear transfer functions, the nonlinear nature of the diode's static characteristic is modeled by controlled sources whose dependent variable is proportional to the independent variable raised to a power. Because we are interested in weakly nonlinear effects, the Taylor series can be truncated to a finite number of terms. For many applications a value of $N = 3$ is acceptable. A bound on the resulting error can be evaluated using (7-6). Observe that a linear incremental circuit model is obtained for the diode by retaining only the linear term in (7-10). This yields the linear conductance g_1 shown in Fig. 7.2(b).

7.2 SEMICONDUCTOR DIODE

A semiconductor diode consists of p- and n-type semiconductor materials that are joined together to form a junction. Since more sophisticated semiconductor devices are fabricated from p-n junctions, a knowledge of their behavior is useful in understanding the properties and characteristics of the more complicated devices. After presenting a global model for the diode in Section 7.2.1, nonlinear incremental equivalent circuits are developed under forward- and reverse-biased conditions in Sections 7.2.2 and 7.2.3, respectively. Sections 7.2.4 through 7.2.7 are devoted to a discussion of how the various model parameters are experimentally determined.

7.2.1 Global Model for Semiconductor Diode

A global model for the semiconductor diode is shown in Fig. 7.3 along with its schematic symbol. Note that the total diode current and voltage are denoted by i_D and v_D, respectively, while the total junction current and voltage are represented by i_J and v_J, respectively. The five elements in the global model are:

(1) R_S, the diode bulk resistance.
(2) R_L, the junction leakage resistance.
(3) R_J, the nonlinear resistor due to the diode junction. Its current-voltage

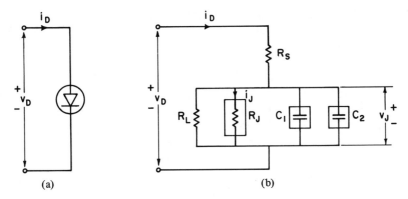

Fig. 7.3.   (a) Schematic symbol. (b) Global model of semiconductor diode.

relationship is given by

$$i_J = I_s \left[\exp\left(\frac{q v_J}{nkT} \right) - 1 \right]. \tag{7-11}$$

(4) C_1, the nonlinear transition capacitance of the depletion layer, also known as the barrier capacitance or the space-charge capacitance. it is given by

$$C_1 = \frac{C(0)}{\left(1 - \dfrac{v_J}{\phi} \right)^\mu}, \quad v_J < \phi. \tag{7-12}$$

(5) C_2, the nonlinear diffusion capacitance whose value depends linearly on the junction current in accordance with the relation

$$C_2 = C'_d (i_J + I_s). \tag{7-13}$$

The parameters that appear in (7-11), (7-12), and (7-13) are defined as follows:

I_s diode saturation current;
q electron charge = 1.6×10^{-19} C;
k Boltzmann's constant = 1.38×10^{-23} J/°K;
T junction absolute temperature in degrees Kelvin;
n diode ideality factor;
$C(0)$ transition capacitance for $v_J = 0$;
ϕ junction contact potential;
μ junction grading constant;
C'_d diffusion capacitance constant;

Near room temperature ($T = 290°$K), kT/q is approximately 25 mV. Also, the junction grading constant μ typically equals $\frac{1}{2}$ for an abrupt junction and $\frac{1}{3}$ for a graded junction. The junction leakage resistance R_L is usually larger than 1 MΩ. Finally, a typical value for the diode ideality factor is a number slightly larger than unity.

The global model may be simplified when the diode is either forward or reverse biased. We first consider forward-biased conditions where $v_J > 0$. Since the forward resistance of R_J is then much smaller than that of R_L, negligible current flows through R_L and the leakage resistance may be ignored. Also, when the diode is forward biased, the junction voltage v_J is essentially constant relative to the junction current i_J. Therefore, the transition capacitance C_1 may be treated as a constant. If V_J denotes the quiescent value of the junction voltage, C_1 is approximated by a linear capacitor, denoted by C_j, where

$$C_j = \frac{C(0)}{\left(1 - \dfrac{V_J}{\phi}\right)^\mu} , \quad V_J < \phi. \tag{7-14}$$

A global model for the diode, in forward bias, is shown in Fig. 7.4. Note that R_J and C_2 are the only nonlinear elements.

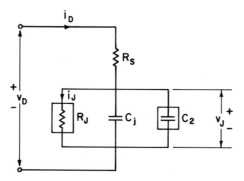

Fig. 7.4.   Forward-biased global model for the semiconductor diode ($v_J > 0$).

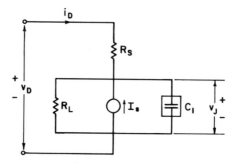

Fig. 7.5.   Reverse-biased global model for the semiconductor diode ($v_J < 0$).

When the diode is reverse biased, the junction voltage is negative and $\exp(qv_J/nkT)$ becomes approximately zero. The junction current, which reduces to $i_J \approx -I_s$, can then be modeled as an independent current source. Furthermore, as can be seen from (7-13), the diffusion capacitance becomes negligibly small. Therefore, under reverse-biased conditions, the diode global model is approximated by the circuit shown in Fig. 7.5. C_1 is now the only nonlinear element contained in the model.

7.2.2 Forward-Biased Nonlinear Incremental Equivalent Circuit for Semiconductor Diode

As explained in Section 7.1, nonlinear incremental equivalent circuits are obtained from global models by developing Taylor series expansions for the nonlinear elements. The forward-biased nonlinear incremental equivalent circuit begins with the global model of Fig. 7.4 which contains the nonlinear junction resistor R_J and the nonlinear diffusion capacitor C_2. Recall that I_J and V_J represent values of the junction current and voltage at the quiescent operating point while the corresponding total and incremental variables are denoted by i_J, v_J and i_j, v_j, respectively.

The current-voltage relation for R_J is given by (7-11). Making use of (7-3) and (7-4), the Taylor series expansion of (7-11) around the quiescent operating point may be written as

$$i_j = \sum_{l=1}^{\infty} g_l v_j^l \tag{7-15}$$

where

$$g_l = \frac{1}{l!} \left(\frac{q}{nkT}\right)^l I_s \exp\left(\frac{qV_J}{nkT}\right). \tag{7-16}$$

Observe from Fig. 7.4 that I_D, the dc value of the diode current, must equal I_J, the dc value of the junction current. Also, the saturation current I_s is typically negligible with respect to the junction current i_J when the diode is biased in its forward region. It follows that

$$I_D = I_J = I_s \left[\exp\left(\frac{qV_J}{nkT}\right) - 1\right] \approx I_s \exp\left(\frac{qV_J}{nkT}\right). \tag{7-17}$$

As a result, the lth Taylor series coefficient may be expressed as

$$g_l = \frac{1}{l!} \left(\frac{q}{nkT}\right)^l I_D = \frac{1}{l} \left(\frac{q}{nkT}\right) g_{l-1}. \tag{7-18}$$

The linear incremental resistance of the diode, denoted by r_d, is defined to be the reciprocal of g_1 and is given by

$$r_d = \frac{1}{g_1} = \frac{nkT}{qI_D}. \tag{7-19}$$

Therefore, in terms of r_d, an equivalent expression for g_l is

$$g_l = \frac{1}{l!}\left(\frac{1}{r_d I_D}\right)^l I_D = \frac{1}{l r_d I_D} g_{l-1}. \tag{7-20}$$

When a diode is to be modeled as a linear resistor, its resistance is commonly evaluated from (7-19). Note that r_d is inversely proportional to the diode quiescent operating current.

A relation for the nonlinear diffusion capacitance appears in (7-13). Eliminating the junction current with the aid of (7-11), the equation for C_2 becomes

$$C_2 = C_d' I_s \exp\left(\frac{qv_J}{nkT}\right). \tag{7-21}$$

The diffusion capacitance is actually an incremental capacitance, as defined in (5-21). Thus the total current through the capacitor is given by

$$i_{C_2} = C_2 \frac{dv_J}{dt}. \tag{7-22}$$

A Taylor series expansion of (7-21) around the operating point is

$$C_2 = c_{20} + \sum_{l=1}^{\infty} c_{2l} v_j^l \tag{7-23}$$

where the linear incremental capacitance is given by

$$c_{20} = C_d' I_s \exp\left(\frac{qV_J}{nkT}\right) = C_d' I_D \tag{7-24}$$

and

$$\begin{aligned}
c_{2l} &= \frac{1}{l!} C_d'\left(\frac{q}{nkT}\right)^l I_s \exp\left(\frac{qV_J}{nkT}\right) \\
&= \frac{1}{l!} C_d'\left(\frac{q}{nkT}\right)^l I_D = \frac{1}{l}\left(\frac{q}{nkT}\right) c_{2(l-1)} \\
&= \frac{1}{l!} C_d'\left(\frac{1}{r_d I_D}\right)^l I_D = \frac{1}{l r_d I_D} c_{2(l-1)} \\
&= C_d' g_l.
\end{aligned} \tag{7-25}$$

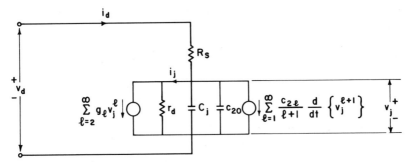

Fig. 7.6. Forward-biased nonlinear incremental equivalent circuit for semiconductor diode.

We see that the coefficients in the expansion for the diffusion capacitance are intimately related to those for R_J. With reference to (5-28), the incremental current through C_2 is

$$i_{c2} = \sum_{l=0}^{\infty} c_{2l} v_j^l \frac{dv_j}{dt}$$

$$= \sum_{l=0}^{\infty} \frac{c_{2l}}{l+1} \frac{d}{dt} \{v_j^{l+1}\}. \tag{7-26}$$

The forward-biased nonlinear incremental equivalent circuit for the semiconductor diode, which is shown in Fig. 7.6, follows in a straightforward manner from Fig. 7.4 and (7-15) and (7-26). Note that only incremental variables appear in the model. Nevertheless, r_d, C_j, c_{20}, and the series coefficients are functions of the quiescent operating point. The nonlinearities in the nonlinear incremental equivalent circuit appear as voltage-controlled current sources. When these sources are omitted, a linear incremental equivalent circuit results.

In practice, each series is truncated to a finite number of terms. We now illustrate application of the inequality in (7-6) for obtaining an upper bound on the truncation error.

Example 7.3

Let the series in (7-15) for the incremental junction current be terminated after N terms. The absolute value of the remainder then satisfies the inequality

$$|R_N(v_J)| \leqslant \frac{M}{(N+1)!} |v_J - V_J|^{N+1} = \frac{|v_j|^{N+1}}{(N+1)!} M \tag{7-27}$$

where M is the maximum value of the $(N+1)$th derivative of i_J over the interval (V_J, v_J). To simplify matters, assume $v_J > V_J$. Then the incremental junction voltage v_j is always positive and the absolute value signs may be removed from the right-hand side of (7-27).

For convenience, the dummy variable β is introduced into (7-11) so that we have

$$i_J = I_s \left[\exp \left(\frac{q\beta}{nkT} \right) - 1 \right].$$

(7-28)

The $(N + 1)$th derivative of i_J is then given by

$$\frac{d^{N+1} i_J}{d\beta^{N+1}} = \left(\frac{q}{nkT} \right)^{N+1} I_s \exp \left(\frac{q\beta}{nkT} \right).$$

(7-29)

Clearly, this derivative is maximized by maximizing β. By assumption $v_J > V_J$. Consequently, the maximum value of β in the interval (V_J, v_J) occurs for $\beta = v_J$. It follows that the inequality in (7-27) becomes

$$|R_N(v_J)| \leq \frac{(v_j)^{N+1}}{(N + 1)!} \left(\frac{q}{nkT} \right)^{N+1} I_s \exp \left(\frac{q v_J}{nkT} \right).$$

(7-30)

The normalized truncation error is defined as the remainder R_N of the series divided by the value of the function being expanded. In the forward-biased region the junction current is closely approximated by

$$i_J \approx I_s \exp \left(\frac{q v_J}{nkT} \right).$$

(7-31)

Hence the normalized truncation error is bounded by

$$\frac{|R_N(v_J)|}{i_J} \leq \frac{1}{(N + 1)!} \left(\frac{q v_j}{nkT} \right)^{N+1} = B_N.$$

(7-32)

Interestingly enough, the bound B_N depends only on the incremental voltage v_j and is independent of the operating point provided the approximation in (7-31) is valid. Observe that

$$\log_{10} B_N = \log_{10} \left[\frac{1}{(N + 1)!} \left(\frac{q}{nkT} \right)^{N+1} \right] + (N + 1) \log_{10} v_j.$$

(7-33)

Therefore, the bound plots as a straight line with slope $(N + 1)$ on log-log paper. For convenience, we assume the diode ideality factor n equals unity. Curves of B_N, the bound on the normalized truncation error, are shown in Fig. 7.7 for truncations to two, three, and four terms. Note that the bound with v_j equal to 25 mV is greater than 10 percent for $N = 2$, less than 10 percent for $N = 3$, and slightly less than 1 percent for $N = 4$. As long as the incremental junction voltage is less than 25 mV, we conclude that an error that is less than 10 percent is incurred by modeling the nonlinear junction resistor R_J only up to third-degree

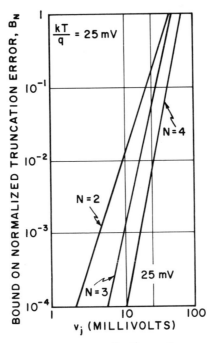

Fig. 7.7.   Bound on normalized truncation error, B_N.

terms. This is adequate for many communication circuits, as in a small-signal forward-biased mixer.

The normalized truncation error for other nonlinearities may also be determined in a similar manner.

7.2.3   Reverse-Biased Nonlinear Incremental Equivalent Circuit for Semiconductor Diode

The global model of Fig. 7.5 provides the basis for the reverse-biased nonlinear incremental equivalent circuit of the semiconductor diode. Since this model is an incremental model, the independent constant current source of value I_s may be ignored. In addition, the transition capacitance C_1 is the only nonlinear element for which a Taylor series must be developed. For this purpose, we follow the identical procedure used in determining the Taylor series expansion for the diffusion capacitance C_2.

By definition, the transition capacitance, whose dependence on junction voltage is expressed by (7-12), is an incremental capacitance. Therefore, the total current flowing through C_1 is given by

$$i_{C_1} = C_1 \frac{dv_J}{dt}.$$ (7-34)

A Taylor series expansion of (7-12) around the operating point is

$$C_1 = c_{10} + \sum_{l=1}^{\infty} c_{1l} v_j^l$$ (7-35)

where the linear incremental capacitance is given by

$$c_{10} = \frac{C(0)}{\left(1 - \dfrac{V_J}{\phi}\right)^{\mu}}$$ (7-36)

and

$$c_{1l} = \frac{c_{10}}{l!} \frac{\mu(\mu+1)\cdots(\mu+l-1)}{(\phi-V_J)^l} = \frac{(\mu+l-1)}{l(\phi-V_J)} c_{1(l-1)}.$$ (7-37)

Note that the quiescent value of the junction voltage appears in (7-36) and (7-37). The incremental current through C_1 is

$$i_{c1} = \sum_{l=0}^{\infty} c_{1l} v_j^l \frac{dv_j}{dt}$$

$$= \sum_{l=0}^{\infty} \frac{c_{1l}}{l+1} \frac{d}{dt} \{v_j^{l+1}\}.$$ (7-38)

The circuit shown in Fig. 7.8, which follows from Fig. 7.5 and (7-38), is the reverse-biased nonlinear incremental equivalent circuit for the semiconductor

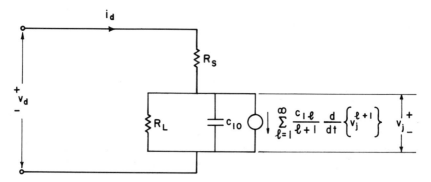

Fig. 7.8. Reverse-biased nonlinear incremental equivalent circuit for semiconductor diode.

diode. Once again, the nonlinearity is manifested as a controlled source. The infinite sum can be interpreted as a collection of voltage-controlled current sources, all connected in parallel. In practice, of course, only a finite number of sources are used in the model.

7.2.4 Determination of the Parameters I_s, n, and R_S

Numerical values for the saturation current I_s, the ideality factor n and the bulk resistance R_S are determined from the forward-region current-voltage characteristic of the diode. As pointed out earlier, at dc the junction current i_J is identical to the diode current i_D. Also, because the bulk resistance R_S is typically less than 100 Ω, the junction voltage v_J approximately equals the diode voltage v_D for suitably small currents. This can be deduced by examination of Fig. 7.4. It follows from (7-11) that the forward-current static charac-

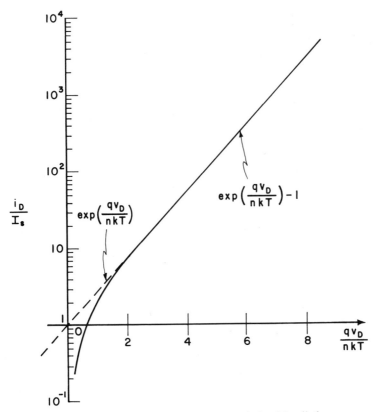

Fig. 7.9. Forward-current static characteristic of the diode.

teristic of the diode may be approximated by

$$\frac{i_D}{I_s} = \exp\left(\frac{qv_D}{nkT}\right) - 1. \tag{7-39}$$

Equation (7-39) is plotted on a semilog scale in Fig. 7.9. Note that the characteristic approaches $\exp(qv_D/nkT)$ for qv_D/nkT greater than 2. Therefore, for $v_D > 2nkT/q$,

$$\ln\left(\frac{i_D}{I_s}\right) = \frac{qv_D}{nkT}. \tag{7-40}$$

This plots as a straight line in Fig. 7.9. Observe that the straight line intersects the normalized current axis where the diode current equals the saturation current and $v_D = 0$. This fact can be useful in determining I_s from data measured in the forward-current region of the diode.

Example 7.4

A typical forward-current characteristic for a semiconductor diode is shown in Fig. 7.10. For the moment, ignore the curve at high current levels. Extrapolating the straight line to zero voltage, we see that the saturation current approximately equals 5 nA.

––––––––––––––––

The diode ideality factor is related to the slope of the straight line portion of the curve. If the coordinates of two points on the straight line are denoted by (i_{D1}, v_{D1}) and (i_{D2}, v_{D2}), it follows from (7-40) that

$$\ln\left(\frac{i_{D1}}{I_s}\right) = \frac{qv_{D1}}{nkT} \tag{7-41}$$

and

$$\ln\left(\frac{i_{D2}}{I_s}\right) = \frac{qv_{D2}}{nkT}. \tag{7-42}$$

The difference between (7-42) and (7-41) yields the equation

$$\frac{q}{nkT}(v_{D2} - v_{D1}) = \ln\left(\frac{i_{D2}}{I_s}\right) - \ln\left(\frac{i_{D1}}{I_s}\right)$$

$$= \ln\left(\frac{i_{D2}}{i_{D1}}\right). \tag{7-43}$$

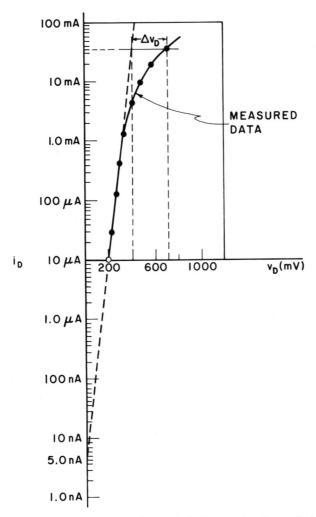

Fig. 7.10. Forward-current characteristic for a semiconductor diode.

Therefore, solving for the diode ideality factor, we have

$$n = \frac{q}{kT} \frac{(v_{D2} - v_{D1})}{\ln (i_{D2}/i_{D1})} . \tag{7-44}$$

Observe that it is not necessary to know the saturation current in order to evaluate n.

Example 7.5

Using the data from Fig. 7.10, we choose $i_{D2} = 468$ μA, $v_{D2} = 300$ mV and $i_{D1} = 10$ μA, $v_{D1} = 200$ mV. Near room temperature kT/q equals 25 mV. Substituting these values into (7-44), the ideality factor for the diode is given by

$$n = \frac{1}{25 \times 10^{-3}} \frac{(300 - 200) \times 10^{-3}}{\ln{(4.68 \times 10^{-4}/10^{-5})}} = \frac{4}{\ln{(46.8)}} = \frac{4}{3.85} = 1.04.$$

$$(7\text{-}45)$$

The ideality factor is a measure of the extent to which the physical diode differs from the ideal diode for which $n = 1$.

Example 7.6

To obtain a check on the validity of the value for the saturation current, the calculated value of n from Example 7.5 can be used in (7-39). For diode currents approximated by the straight line portion of the curve,

$$I_s \approx \frac{i_D}{\exp{\left(\dfrac{qv_D}{nkT}\right)}}.$$

$$(7\text{-}46)$$

Selecting $i_{D1} = 10$ μA, $v_{D1} = 200$ mV, and $n = 1.04$, (7-46) yields I_s equal to 4.65 nA which compares favorably to the graphically determined value of 5 nA.

Because of the voltage generated across the series bulk resistance of a physical diode, the diode characteristic deviates from a straight line at high forward currents. Let the voltage drop, at a fixed value of i_D, be denoted by Δv_D as indicated in Fig. 7.10. The bulk resistance is then calculated from the equation

$$R_S = \frac{\Delta v_D}{i_D}.$$

$$(7\text{-}47)$$

Example 7.7

For the diode of Fig. 7.10, Δv_D approximately equals 315 mV when i_D equals 35 mA. Therefore,

$$R_S \approx \frac{315 \times 10^{-3}}{35 \times 10^{-3}} = 9 \text{ }\Omega.$$

$$(7\text{-}48)$$

A word of caution is in order. The static characteristic of a junction diode is strongly dependent upon junction temperature. To avoid unwanted heating effects at the higher currents, it is recommended that a pulse measurement technique with a low-duty cycle be used. However, dc measurements are preferable for extremely small currents. To assure that the data taken by the two methods are consistent, a few points in the intermediate region should be measured by both techniques.

7.2.5 Determination of the Parameters C_j and C_d'

Figure 7.10 illustrates, under forward-biased conditions, that the diode voltage remains nearly constant over an extremely wide range of diode currents. Therefore, as indicated in (7-14), the transition capacitance may be approximated by the constant C_j. The total junction capacitance, denoted by C_T, equals the sum of the transition and diffusion capacitances. An expression for the diffusion capacitance is given by (7-13). In forward bias, i_J is normally several orders of magnitude larger than I_s. Also, at low frequencies, i_J approximately equals i_D. Consequently, the total junction capacitance may be expressed as

$$C_T = C_1 + C_2 \approx C_j + C_d' i_D. \qquad (7\text{-}49)$$

A typical plot of the total junction capacitance as a function of the diode current is shown in Fig. 7.11. Observe that the plot of capacitance closely approximates a straight line where C_j is the axis intercept for $i_D = 0$ and C_d' is the slope of the straight line.

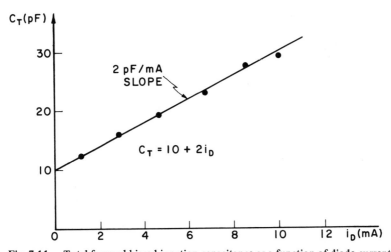

Fig. 7.11.   Total forward-biased junction capacitance as a function of diode current.

Example 7.8

For the curve plotted in Fig. 7.11, $C_j = 10$ pF and $C'_d = 2$ pF/mA.

7.2.6   Determination of the Parameter R_L

The junction leakage resistance R_L is determined from the reverse-current characteristic of the diode. The global model for the reverse-biased diode is shown in Fig. 7.5. Because the bulk resistance is so small relative to the leakage resistance, the voltage drop developed across R_S may be ignored. As a result, the diode voltage v_D approximately equals the junction voltage v_J. For dc inputs, the diode current is then approximately given by

$$i_D = \frac{v_D}{R_L} - I_s. \tag{7-50}$$

R_L is readily determined from (7-50).

Example 7.9

By measuring the reverse current for several values of reverse voltage, a plot such as that shown in Fig. 7.12 is obtained. After fitting a straight line to the measured data, the leakage resistance is evaluated from the reciprocal of the slope. For the plot of Fig. 7.12,

$$R_L = \frac{\Delta v_D}{\Delta i_D} = \frac{(-5) - (-15)}{(-30 \times 10^{-9}) - (-80 \times 10^{-9})} = \frac{10}{50 \times 10^{-9}} = 2 \times 10^8 = 200 \text{ M}\Omega.$$

$$\tag{7-51}$$

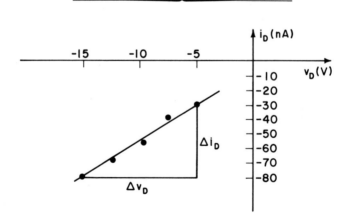

Fig. 7.12.   Reverse-current characteristic for a semiconductor diode.

7.2.7   Determination of the Parameters $C(0)$, μ, and ϕ

In general, the total junction capacitance equals the parallel combination of the transition and diffusion capacitances. However, under reverse-biased conditions, the diffusion capacitance is negligibly small. An expression for the transition capacitance is given by (7-12). Assuming reverse bias, v_D approximately equals v_J. The total junction capacitance may then be expressed as

$$C_T \approx C_1 \approx \frac{C(0)}{\left(1 - \dfrac{v_D}{\phi}\right)^{\mu}} \qquad (7\text{-}52)$$

where it is understood that $v_D \leqslant 0$. In this section we discuss methods for determining the parameters $C(0)$, μ, and ϕ.

From (7-52) it is apparent that $C(0)$ approximately equals C_T when v_D equals zero. Therefore, a numerical value for $C(0)$ is easily obtained by measuring the diode capacitance with the diode voltage constrained to zero.

The junction grading constant μ may be evaluated without knowledge of the junction contact potential ϕ by making use of capacitance measurements for which $-v_D \gg \phi$. Then

$$C_T \approx \frac{C(0)}{\left(\dfrac{-v_D}{\phi}\right)^{\mu}} = C(0)\left(\frac{-v_D}{\phi}\right)^{-\mu}. \qquad (7\text{-}53)$$

Taking the logarithm of C_T, if follows that

$$\log_{10} C_T = \log_{10} C(0) - \mu \log_{10} (-v_D) + \mu \log_{10} \phi. \qquad (7\text{-}54)$$

Consequently, as long as $-v_D \gg \phi$, the curve of the capacitance versus reverse diode voltage plots as a straight line on log-log paper. Only logarithms of real positive numbers are involved in (7-54) since C_T, $C(0)$, $-v_D$, and ϕ are all positive quantities.

Let C_{T1} and C_{T2} denote the capacitances measured at v_{D1} and v_{D2}, respectively. Utilizing (7-54), we have

$$\log_{10} C_{T1} = \log_{10} C(0) - \mu \log_{10} (-v_{D1}) + \mu \log_{10} \phi$$

$$\log_{10} C_{T2} = \log_{10} C(0) - \mu \log_{10} (-v_{D2}) + \mu \log_{10} \phi. \qquad (7\text{-}55)$$

Subtracting the two equations in (7-55), we obtain

$$\log_{10} C_{T2} - \log_{10} C_{T1} = -\mu[\log_{10} (-v_{D2}) - \log_{10} (-v_{D1})]. \qquad (7\text{-}56)$$

From (7-56), we conclude that

$$-\mu = \frac{\log_{10} C_{T2} - \log_{10} C_{T1}}{\log_{10} (-v_{D2}) - \log_{10} (-v_{D1})}. \qquad (7\text{-}57)$$

This is recognized as the slope of the straight line portion of the capacitance versus reverse-voltage curve when plotted on log-log paper.

The final parameter to be determined is the junction contact potential ϕ. Evaluating (7-52) for two different values of reverse voltage and dividing one of the resulting capacitances by the other yields

$$\frac{C_{T2}}{C_{T1}} = \left[\frac{1 - \dfrac{v_{D1}}{\phi}}{1 - \dfrac{v_{D2}}{\phi}}\right]^{\mu} = \left[\frac{\phi - v_{D1}}{\phi - v_{D2}}\right]^{\mu}. \tag{7-58}$$

For convenience, we introduce the parameter

$$A = \left[\frac{C_{T2}}{C_{T1}}\right]^{1/\mu} = \frac{\phi - v_{D1}}{\phi - v_{D2}}. \tag{7-59}$$

Solving (7-59) for the contact potential, it follows that

$$\phi = \frac{A v_{D2} - v_{D1}}{A - 1}. \tag{7-60}$$

Whereas (7-57) makes use of voltages for which $-v_D \gg \phi$, best results are obtained with (7-60) when v_{D2} and v_{D1} are both chosen to be small enough so that they correspond to a portion of the capacitance versus reverse-voltage curve which deviates from the straight line.

Example 7.10

The above procedure is now illustrated with a numerical example. A typical set of measured capacitances and reverse voltages is tabulated in Table 7.1 and

TABLE 7.1.   Measured Values of Transition Capacitance.

$-v_D(V)$	$C_T(pF)$
0	0.80
0.1	0.76
0.5	0.71
1.0	0.57
2.0	0.48
4.0	0.40
8.0	0.32
20.0	0.24
30.0	0.21

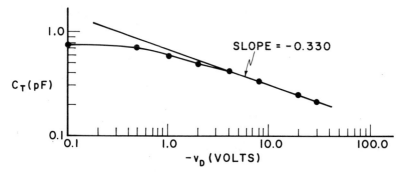

Fig. 7.13.   Measured diode capacitance as a function of reverse voltage.

a plot of the data is presented on log-log paper in Fig. 7.13. Focusing attention on the capacitance that was measured for $v_D = 0$, we conclude that

$$C(0) = 0.80 \text{ pF.} \tag{7-61}$$

To evaluate μ we choose the two largest values of $-v_D$ in order to minimize any errors resulting from the approximation in (7-53). With $-v_{D1} = 20.0$ V and $-v_{D2} = 30.0$ V, we obtain from (7-57)

$$-\mu = \frac{\log_{10}(0.21) - \log_{10}(0.24)}{\log_{10}(30) - \log_{10}(20)} = \frac{(9.322 - 10) - (9.380 - 10)}{1.477 - 1.301}$$

$$= \frac{-0.058}{0.176} = -0.330. \tag{7-62}$$

Note that this slope agrees with the slope of the straight line in Fig. 7.13. To evaluate ϕ we choose $-v_{D1} = 0$ V and $-v_{D2} = 1.0$ V. Making use of (7-59) and (7-60), it follows that

$$A = \left[\frac{0.57}{0.80}\right]^{1/0.330} = (0.713)^{3.03} = 0.359$$

$$\phi = \frac{(0.359)(-1.0) + (0)}{0.359 - 1} = \frac{-0.359}{-0.641} = 0.56 \text{ V.} \tag{7-63}$$

We conclude that the expression for the transition capacitance is

$$C_1 = \frac{0.8}{\left(1 - \dfrac{v_J}{0.56}\right)^{0.330}} \text{ pF.} \tag{7-64}$$

To check our result we calculate C_1 for $-v_J = 0.1$ V. This results in

$$C_1 = \frac{0.8}{\left(1 + \dfrac{0.1}{0.56}\right)^{0.330}} = \frac{0.8}{(1.179)^{0.330}} = \frac{0.8}{1.056} = 0.758 \text{ pF} \qquad (7\text{-}65)$$

which agrees closely with the tabulated value of 0.76 pF.

7.3 BIPOLAR JUNCTION TRANSISTOR (BJT)

Loosely speaking, a bipolar junction transistor (BJT) may be viewed as two p-n junctions coupled back to back. The p-n-p transistor consists of an n-type semiconductor material embedded between two p-type materials while the n-p-n transistor is comprised of a p-type material sandwiched between two n-type materials. The theory of operation is essentially the same for both. In fact, the theory of one may be obtained from that of the other by merely reversing all polarities and interchanging the roles played by holes and free electrons. A global model for the BJT is presented in Section 7.3.1 and is then specialized for the case in which the transistor is biased to operate in its amplification region. Considerable simplification of the global model is shown to result. A nonlinear incremental equivalent circuit for the BJT, when used as an amplifier, is developed in Section 7.3.2. Finally, methods for experimentally obtaining the model parameters are explained in Sections 7.3.3 through 7.3.8.

7.3.1 Global Model for Bipolar Junction Transistor

The schematic symbol for the n-p-n BJT is shown in Fig. 7.14(a). The terminals denoted by E, B, and C are referred to, respectively, as the emitter, base, and collector. Note that the arrow in the symbol points out of the emitter terminal for an n-p-n transistor. The Ebers–Moll model for the n-p-n BJT is shown in Fig. 7.14(b). This is a global model relating total variables. For a p-n-p BJT the Ebers–Moll model is identical except that i_{JE}, v_{JE}, i_{JC}, and v_{JC} are all reversed in polarity. This reversal is indicated in the p-n-p schematic symbol by having the emitter arrow pointed into the emitter terminal. Because the two models are essentially the same, we consider only the n-p-n BJT. The 13 elements in the global model are as follows.

(1) R_{SB}, the base bulk resistance.

(2) $\alpha_I i_{JC}$, the emitter current-controlled current source. Note that the control variable i_{JC} is the collector-base junction diode current.

(3) R_{LE}, the emitter-base junction leakage resistance.

(4) R_{JE}, the nonlinear resistor due to the emitter-base junction. Its current-

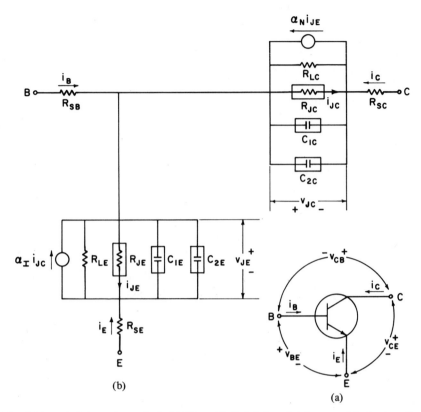

Fig. 7.14.   (a) Schematic symbol. (b) Ebers–Moll global model for *n-p-n* bipolar junction transistor.

voltage relationship is given by

$$i_{JE} = I_{se}\left[\exp\left(\frac{qv_{JE}}{n_E kT}\right) - 1\right]. \tag{7-66}$$

(5) C_{1E}, the nonlinear emitter-base junction transition capacitance, given by

$$C_{1E} = \frac{C_E(0)}{\left(1 - \dfrac{v_{JE}}{\phi_E}\right)^{\mu_E}}, \quad v_{JE} < \phi_E. \tag{7-67}$$

(6) C_{2E}, the nonlinear emitter-base junction diffusion capacitance whose value depends linearly on the emitter-base junction current in accordance with

$$C_{2E} = C'_{de}(i_{JE} + I_{se}). \tag{7-68}$$

(7) R_{SE}, the emitter bulk resistance.

(8) $\alpha_N i_{JE}$, the collector current-controlled current source. Note that the control variable i_{JE} is the emitter-base junction diode current.

(9) R_{LC}, the collector-base junction leakage resistance.

(10) R_{JC}, the nonlinear resistor due to the collector-base junction. Its current-voltage relationship is given by

$$i_{JC} = I_{sc}\left[\exp\left(\frac{qv_{JC}}{n_C kT}\right) - 1\right]. \qquad (7\text{-}69)$$

(11) C_{1C}, the nonlinear collector-base junction transition capacitance, given by

$$C_{1C} = \frac{C_C(0)}{\left(1 - \dfrac{v_{JC}}{\phi_C}\right)^{\mu_C}}, \qquad v_{JC} < \phi_C. \qquad (7\text{-}70)$$

(12) C_{2C}, the nonlinear collector-base junction diffusion capacitance whose value depends linearly on the collector-base junction current in accordance with

$$C_{2C} = C'_{dc}(i_{JC} + I_{sc}). \qquad (7\text{-}71)$$

(13) R_{SC}, the collector bulk resistance.

The parameters that appear in the controlled sources and in equations (7-66) through (7-71) are defined as follows:

α_I	common-base inverted-mode dc current gain = $-I_E/I_C$;
I_{se}	reverse-saturation current of the emitter-base diode;
q	electron charge = 1.6×10^{-19} C,
k	Boltzmann's constant = 1.38×10^{-23} J/°K;
T	junction absolute temperature in degree Kelvin;
n_E	emitter-base diode ideality factor;
$C_E(0)$	emitter-base junction transition capacitance for $v_{JE} = 0$;
ϕ_E	emitter-base junction contact potential;
μ_E	emitter-base junction grading constant;
C'_{de}	emitter-base junction diffusion capacitance constant;
α_N	common-base normal-mode dc current gain = $-I_C/I_E$;
I_{sc}	reverse-saturation current of the collector-base diode;
n_C	collector-base diode ideality factor;
$C_C(0)$	collector-base junction transition capacitance for $v_{JC} = 0$;
ϕ_C	collector-base junction contact potential;
μ_C	collector-base junction grading constant;
C'_{dc}	collector-base junction diffusion capacitance constant.

Depending on the polarities of the bias batteries, there are four possible regions in which a BJT may be operated. These are:

(1) the forward-active (or amplification) region in which the emitter-base junction is forward biased and the collector-base junction is reverse biased (i.e., $V_{EB} < 0$ and $V_{CB} > 0$ for an *n-p-n* transistor);

(2) the reverse-active region in which the emitter-base junction is reverse biased and the collector-base junction is forward biased (i.e., $V_{EB} > 0$ and $V_{CB} < 0$ for an *n-p-n* transistor);

(3) the cutoff region in which both junctions are reverse biased (i.e., $V_{EB} > 0$ and $V_{CB} > 0$ for an *n-p-n* transistor);

(4) the saturation region in which both junctions are forward biased (i.e., $V_{EB} < 0$ and $V_{CB} < 0$ for an *n-p-n* transistor).

In pulse and digital circuits, where the transistor is used as a switch, the transistor is ON when in the saturation region and is OFF when in the cutoff region. Since only weakly nonlinear effects are of concern in this text, we restrict our attention to applications where the transistor is confined to operate solely in the amplification region. Consequently, we are interested in modeling the BJT only when the emitter-base junction is forward biased and the collector-base junction is reverse biased.

With reference to Section 7.2.1, various simplifications are possible for the global model of a *p-n* junction that is either forward or reverse biased. Assuming the emitter-base junction of the BJT to be forward biased, it follows that the leakage resistance R_{LE} may be ignored. In addition, if V_{JE} is the quiescent value of the emitter-base junction voltage, the transition capacitance C_{1E} may be treated as a constant whose value is given by

$$C_{jE} = \frac{C_E(0)}{\left(1 - \frac{V_{JE}}{\phi_E}\right)^{\mu_E}}, \qquad V_{JE} < \phi_E. \tag{7-72}$$

Also, the emitter bulk resistance is excluded from the model because experience has shown that its effect is negligible in most applications.

Assuming the collector-base junction of the BJT is reverse biased, $i_{JC} \approx -I_{sc}$ and the nonlinear resistor R_{JC} may be replaced by an independent current source of value I_{sc}. Also, because I_{sc} is several orders of magnitude smaller than the emitter current, the current-controlled current source $\alpha_I i_{JC} \approx -\alpha_I I_{sc}$ may be removed from the emitter branch of the model. In addition, the diffusion capacitance C_{2C} and the collector bulk resistance R_{SC} may be omitted. The final modification deals with the current-controlled current source $\alpha_N i_{JE}$. In actuality α_N is not a constant but varies nonlinearly as a function of the collector current

i_C. Thus we rewrite α_N as $\alpha_N(i_C)$. Also, because of avalanche effects, the collector current is a nonlinear function of the collector-to-base voltage. This nonlinearity is incorporated into the Ebers–Moll model by including in the collector-controlled current source the avalanche multiplication factor given by

$$M = \frac{1}{1 - \left(\dfrac{v_{CB}}{V_{CBO}}\right)^{\eta}}. \tag{7-73}$$

V_{CBO} is the collector-to-base breakdown voltage, otherwise known as the avalanche voltage. It follows that the current source $\alpha_N i_{JE}$ is modified to be $\alpha_N(i_C) M i_{JE}$. The resulting global model in the amplification region for an n-p-n transistor is shown in Fig. 7.15.

It is possible to relate α_N to the dc h_{FE} of the transistor that is defined by

$$h_{FE} = \frac{I_C}{I_B}. \tag{7-74}$$

Since the dc currents into the transistor must sum to zero,

$$I_E + I_B + I_C = 0. \tag{7-75}$$

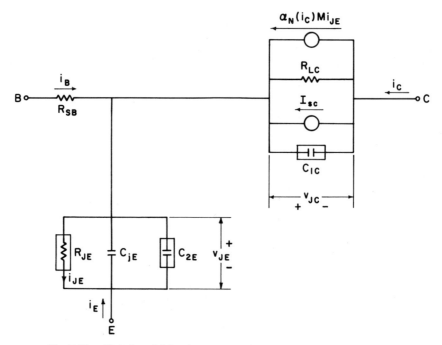

Fig. 7.15. Global model for the n-p-n transistor in the amplification region.

As mentioned earlier, α_N is the common-base normal-mode dc current gain. Therefore,

$$\alpha_N = -\frac{I_C}{I_E} = \frac{I_C}{I_B + I_C} = \frac{\dfrac{I_C}{I_B}}{1 + \dfrac{I_C}{I_B}} = \frac{h_{FE}}{1 + h_{FE}}. \tag{7-76}$$

Experimentally, it has been found that the variation of h_{FE} as a function of collector current can be expressed fairly accurately by the empirical relation

$$h_{FE} = \frac{h_{FE_{max}}}{1 + a \log_{10}^2\left(\dfrac{I_C}{I_{C_{max}}}\right)}. \tag{7-77}$$

Note that $I_{C_{max}}$ is the collector current at which h_{FE} equals its maximum value of $h_{FE_{max}}$ and a is a constant. By substitution of (7-77) into (7-76), we obtain

$$\alpha_N(i_C) = \frac{h_{FE_{max}}}{1 + h_{FE_{max}} + a \log_{10}^2\left(\dfrac{i_C}{I_{C_{max}}}\right)} \tag{7-78}$$

where the dc current I_C has been replaced by the total current i_C. This is valid provided (7-78) is interpreted in terms of the static transistor characteristics.

The current source $\alpha_N(i_C) M i_{JE}$ is in a form that is most convenient when the transistor is used in the common-base configuration. For a common-emitter connection, it is desirable to express the current source as a function of the base current. To accomplish this we consider dc signals so that $i_{JE} = -I_E$. Neglecting I_{sc} and the leakage resistance R_{LC}, the collector current is then given approximately by

$$I_C = -\alpha_N M I_E = \alpha_N M(I_B + I_C). \tag{7-79}$$

For convenience, we do not show in (7-79) the dependence of α_N on I_C. The solution of (7-79) for I_C results in

$$I_C = \frac{\alpha_N M}{1 - \alpha_N M} I_B = \frac{\dfrac{\alpha_N}{1 - (V_{CB}/V_{CBO})^\eta}}{1 - \dfrac{\alpha_N}{1 - (V_{CB}/V_{CBO})^\eta}} I_B \tag{7-80}$$

where use has been made of (7-73). Further simplification of (7-80) yields

$$I_C = \frac{\alpha_N}{1 - \alpha_N - (V_{CB}/V_{CBO})^\eta} I_B = \frac{\alpha_N}{1 - \alpha_N} \frac{1}{1 - \dfrac{1}{1 - \alpha_N}\left(\dfrac{V_{CB}}{V_{CBO}}\right)^\eta} I_B. \tag{7-81}$$

However, V_{CB} approximately equals V_{CE} for voltages where avalanche effects are noticeable. Also, from (7-76),

$$h_{FE} = \frac{\alpha_N}{1 - \alpha_N}. \tag{7-82}$$

Therefore, (7-81) may be written as

$$I_C = h_{FE} \frac{1}{1 - \left(\dfrac{V_{CE}}{V_{CEO}}\right)^\eta} I_B \tag{7-83}$$

where

$$V_{CEO} = (1 - \alpha_N)^{1/\eta} V_{CBO}. \tag{7-84}$$

V_{CEO} is known as the collector-to-emitter sustaining voltage. Note that V_{CEO} is not the avalanche voltage that was defined in (7-73). Once again, if we interpret (7-83) in terms of the static transistor characteristics, (7-83) may be written with total variables as

$$i_C = h_{FE} \frac{1}{1 - \left(\dfrac{v_{CE}}{V_{CEO}}\right)^\eta} i_B. \tag{7-85}$$

This completes our discussion of the global model. We are now in a position to develop a nonlinear incremental equivalent circuit suitable for the amplification region of an *n-p-n* transistor.

7.3.2 Nonlinear Incremental Equivalent Circuit for BJT in Amplification Region

The global model shown in Fig. 7.15 serves as our starting point. Because only incremental variables appear in the nonlinear incremental equivalent circuit, the constant current generator I_{sc} is not included. There are five nonlinearities for which power series expansions must be developed. These are as follows.

(1) R_{JE}, the nonlinear emitter-base junction resistor.
(2) C_{2E}, the nonlinear emitter-base junction diffusion capacitance.
(3) C_{1C}, the nonlinear collector-base junction transition capacitance.
(4) M, the nonlinear avalanche multiplication factor.
(5) $\alpha_N(i_C)$, the nonlinear functional relationship for α_N.

Some of the nonlinearities are voltage dependent while others are current dependent. Depending upon whether the equivalent circuit is to be used in a nodal, loop, or mixed-variable analysis, different nonlinear incremental equivalent circuits are needed. In particular, all of the nonlinearities are required to be expressed as functions of voltages in nodal analysis and as functions of currents in loop analysis. On the other hand, both voltages and currents may be used in mixed-variable analysis. For purposes of illustration, a model suitable for nodal analysis is developed in this section.

The current-voltage relation for R_{JE}, given by (7-66), is identical in form to (7-11). It follows from (7-15), (7-18), (7-19), and (7-20) that the Taylor series expansion of (7-66) around the quiescent operating point of the transistor may be written as

$$i_{je} = \sum_{l=1}^{\infty} g_{el} v_{je}^l \qquad (7\text{-}86)$$

where

$$g_{el} = \frac{1}{l!} \left(\frac{q}{n_E kT} \right)^l (-I_E) = \frac{1}{l} \left(\frac{q}{n_E kT} \right) g_{e(l-1)}$$

$$= \frac{1}{l!} \left[\frac{1}{r_e(-I_E)} \right]^l (-I_E) = \frac{1}{l r_e(-I_E)} g_{e(l-1)}, \qquad (7\text{-}87)$$

I_E is the dc value of the emitter current and r_e is the linear incremental emitter resistance of the transistor which is defined to be

$$r_e = \frac{1}{g_{e1}} = \frac{n_E kT}{q(-I_E)}. \qquad (7\text{-}88)$$

The minus sign associated with I_E arises because the currents i_{JE} and i_E point in opposite directions. In the amplification region I_E is negative for an n-p-n transistor and $(-I_E)$ represents a positive quantity.

Similarly, the expression for C_{2E} in (7-68) is identical in form to that for C_2 in (7-13). From equations (7-24) through (7-26) we conclude that the incremental current through C_{2E} may be expressed as

$$i_{c2e} = \sum_{l=0}^{\infty} \frac{c_{2el}}{l+1} \frac{d}{dt} \{v_{je}^{l+1}\} \qquad (7\text{-}89)$$

where the linear incremental diffusion capacitance is

$$c_{2eo} = C_{de}'(-I_E) = C_{de}'|I_E| \qquad (7\text{-}90)$$

and

$$c_{2el} = C_{de}' g_{el}. \qquad (7\text{-}91)$$

The incremental current through C_{1C} is determined in an analogous manner. The expression for C_{1C} in (7-70) is identical in form to that for C_1 in (7-12).

Therefore, with reference to equations (7-36) through (7-38), the incremental current through C_{1C} is given by

$$i_{c1c} = \sum_{l=0}^{\infty} \frac{c_{1cl}}{l+1} \frac{d}{dt} \{v_{jc}^{l+1}\} \tag{7-92}$$

where the linear incremental transition capacitance is

$$c_{1co} = \frac{C_C(0)}{\left(1 - \dfrac{V_{JC}}{\phi_C}\right)^{\mu_C}} \tag{7-93}$$

and

$$c_{1cl} = \frac{c_{1co}}{l!} \frac{\mu_C(\mu_C + 1) \cdots (\mu_C + l - 1)}{(\phi_C - V_{JC})^l} = \frac{(\mu_C + l - 1)}{l(\phi_C - V_{JC})} c_{1c(l-1)}. \tag{7-94}$$

Recall that V_{JC} denotes the dc value of the collector-base junction voltage.

The next nonlinearity to be handled is the avalanche multiplication factor M. An analytical expression for M is found in (7-73). Expanding (7-73) in a Taylor series about the dc value of the collector-to-base voltage, we obtain

$$M = m_0 + \sum_{k=1}^{\infty} m_k v_{cb}^k \tag{7-95}$$

where

$$m_0 = \left[1 - \left(\frac{V_{CB}}{V_{CBO}}\right)^{\eta}\right]^{-1}. \tag{7-96}$$

Unfortunately, a general expression for the kth coefficient m_k is difficult to derive. The first three coefficients are:

$$m_1 = \frac{1}{1!} \left.\frac{dM}{dv_{CB}}\right|_{v_{CB} = V_{CB}} = \frac{dm_0}{dV_{CB}} = \frac{\eta}{V_{CBO}^{\eta}} m_0^2 V_{CB}^{\eta-1},$$

$$m_2 = \frac{1}{2!} \left.\frac{d^2M}{dv_{CB}^2}\right|_{v_{CB} = V_{CB}} = \frac{1}{2} \frac{dm_1}{dV_{CB}}$$

$$= \frac{1}{2} \left[\frac{\eta}{V_{CBO}^{\eta}} \left(2m_0 \frac{dm_0}{dV_{CB}} V_{CB}^{\eta-1} + (\eta - 1) m_0^2 V_{CB}^{\eta-2}\right)\right]$$

$$= \frac{m_1^2}{m_0} + \frac{(\eta - 1) m_1}{2V_{CB}},$$

$$m_3 = \frac{1}{3!} \frac{d^3M}{dv_{CB}^3}\bigg|_{v_{CB}=V_{CB}} = \frac{1}{3} \frac{dm_2}{dV_{CB}}$$

$$= \frac{1}{3} \left[\frac{m_0 2m_1 \dfrac{dm_1}{dV_{CB}} - m_1^2 \dfrac{dm_0}{dV_{CB}}}{m_0^2} + \frac{(\eta-1)}{2} \frac{V_{CB} \dfrac{dm_1}{dV_{CB}} - m_1}{V_{CB}^2} \right]$$

$$= \frac{2}{3} m_2 \left[\frac{2m_1}{m_0} + \frac{(\eta-1)}{2V_{CB}} \right] - \frac{m_1}{3} \left[\frac{m_1^2}{m_0^2} + \frac{(\eta-1)}{2V_{CB}^2} \right]. \tag{7-97}$$

High order coefficients may be obtained in a similar manner.

The nonlinear functional relationship for α_N is the final nonlinearity to be discussed. Recall that α_N is dependent on the collector current. This complicates matters because the collector current is not easily expressed in terms of voltage. Since α_N is a dc current gain, the current-dependent nature of α_N reflects nonlinearities appearing in the static characteristics. Consequently, in this discussion we assume i_C varies slowly enough such that capacitive effects are negligible. We also assume that the current generator I_{sc} and the leakage resistance R_{LC} can be ignored. By inspection of Fig. 7.15, it follows that

$$i_C = \alpha_N(i_C) M i_{JE}. \tag{7-98}$$

Dividing through by $\alpha_N(i_C)$ and substituting from (7-78), (7-98) becomes

$$f(i_C) = \frac{i_C}{\alpha_N(i_C)} = i_C \left[\frac{1 + h_{FE_{max}} + a \log_{10}^2 \left(\dfrac{i_C}{I_{C_{max}}} \right)}{h_{FE_{max}}} \right] = M i_{JE}. \tag{7-99}$$

Observe that the left-hand side of the equation is a function only of the collector current.

The next step in the procedure is to expand $f(i_C)$ in a Taylor series about the quiescent operating point. This yields

$$f(i_C) = f_0 + \sum_{k=1}^{\infty} f_k i_c^k \tag{7-100}$$

where i_c is the incremental collector current and

$$f_0 = I_C \left[\frac{1 + h_{FE_{max}} + a \log_{10}^2 \left(\dfrac{I_C}{I_{C_{max}}} \right)}{h_{FE_{max}}} \right]. \tag{7-101}$$

Since a general expression for the kth coefficient f_k is not available, we simply generate the first three coefficients. For this purpose, we make use of the identity

$$\frac{d}{dx} \log_{10} u = \frac{\log_{10} e}{u} \frac{du}{dx}.$$
(7-102)

It follows that

$$f_1 = \frac{df_0}{dI_C} = \frac{1}{h_{FE_{max}}} \left[1 + h_{FE_{max}} + a \log_{10}^2 \left(\frac{I_C}{I_{C_{max}}} \right) + 2a \log_{10} \left(\frac{I_C}{I_{C_{max}}} \right) \log_{10} e \right]$$

$$f_2 = \frac{1}{2} \frac{df_1}{dI_C} = \frac{a \log_{10} e}{h_{FE_{max}} I_C} \left[\log_{10} \left(\frac{I_C}{I_{C_{max}}} \right) + \log_{10} e \right]$$

$$f_3 = \frac{1}{3} \frac{df_2}{dI_C} = -\frac{a \log_{10} e}{3 h_{FE_{max}} I_C^2} \log_{10} \left(\frac{I_C}{I_{C_{max}}} \right).$$
(7-103)

Higher order coefficients follow in a straightforward manner.

Inserting the Taylor series expansions for $f(i_C)$, M, and i_{JE} into (7-99), we obtain

$$f_0 + \sum_{k=1}^{\infty} f_k i_c^k = \left[m_0 + \sum_{k=1}^{\infty} m_k v_{cb}^k \right] \left[I_{JE} + \sum_{l=1}^{\infty} g_{el} v_{je}^l \right]$$

$$= m_0 I_{JE} + I_{JE} \sum_{k=1}^{\infty} m_k v_{cb}^k + m_0 \sum_{l=1}^{\infty} g_{el} v_{je}^l$$

$$+ \sum_{k=1}^{\infty} \sum_{l=1}^{\infty} m_k g_{el} v_{cb}^k v_{je}^l.$$
(7-104)

Note that

$$f_0 = m_0 I_{JE}$$
(7-105)

because the incremental variables are zero under dc conditions. For convenience, we define

$$y = I_{JE} \sum_{k=1}^{\infty} m_k v_{cb}^k + m_0 \sum_{l=1}^{\infty} g_{el} v_{je}^l + \sum_{k=1}^{\infty} \sum_{l=1}^{\infty} m_k g_{el} v_{cb}^k v_{je}^l.$$
(7-106)

Equating incremental quantities in (7-104), it is concluded that

$$y = \sum_{k=1}^{\infty} f_k i_c^k.$$
(7-107)

This gives a series expansion for y in terms of the incremental collector current. However, what is really wanted is a series expansion for i_c. This is accomplished by means of a series reversion of (7-107). It follows that

$$i_c = \sum_{k=1}^{\infty} \alpha_k y^k \tag{7-108}$$

where, from (5-13), (5-15), and (5-17), the first three coefficients are given by

$$\alpha_1 = \frac{1}{f_1} = \frac{h_{FE_{max}}}{1 + h_{FE_{max}} + a \log_{10}^2 \left(\dfrac{I_C}{I_{C_{max}}} \right) + 2a \log_{10} \left(\dfrac{I_C}{I_{C_{max}}} \right) \log_{10} e}$$

$$\alpha_2 = -\frac{f_2}{f_1^3} = -\alpha_1^3 \frac{a \log_{10} e}{3 h_{FE_{max}} I_C^2} \left[\log_{10} \left(\frac{I_C}{I_{C_{max}}} \right) + \log_{10} e \right]$$

$$\alpha_3 = \frac{2 f_2^2}{f_1^5} - \frac{f_3}{f_1^4} = \frac{2\alpha_2^2}{\alpha_1} + \alpha_1^4 \frac{a \log_{10} e}{3 h_{FE_{max}} I_C^2} \log_{10} \left(\frac{I_C}{I_{C_{max}}} \right)$$

$$= \frac{2\alpha_2^2}{\alpha_1} - \frac{\alpha_1 \alpha_2}{3 I_C} - \frac{\alpha_1^4 a \log_{10}^2 e}{3 h_{FE_{max}} I_C^2}. \tag{7-109}$$

In order to express the incremental collector current as a function of incremental voltages, (7-106) is substituted into (7-108). Evaluation of the resulting expression is cumbersome because of the large number of terms. To simplify notation, let $q_k(v_{cb}, v_{je})$ denote those terms in (7-108) of kth degree where the degree of a term is defined to be the sum of its exponents. Then

$$i_c = \sum_{k=1}^{\infty} q_k(v_{cb}, v_{je}). \tag{7-110}$$

Enumerating $q_k(v_{cb}, v_{je})$ for $k = 1, 2$, and 3, we obtain

$$q_1(v_{cb}, v_{je}) = \alpha_1 I_{JE} m_1 v_{cb} + \alpha_1 m_0 g_{e1} v_{je}$$

$$q_2(v_{cb}, v_{je}) = (\alpha_1 I_{JE} m_2 + \alpha_2 I_{JE}^2 m_1^2) v_{cb}^2$$

$$+ (\alpha_1 m_1 g_{e1} + 2\alpha_2 I_{JE} m_0 m_1 g_{e1}) v_{cb} v_{je}$$

$$+ (\alpha_1 m_0 g_{e2} + \alpha_2 m_0^2 g_{e1}^2) v_{je}^2$$

$$q_3(v_{cb}, v_{je}) = (\alpha_1 I_{JE} m_3 + 2\alpha_2 I_{JE}^2 m_1 m_2 + \alpha_3 I_{JE}^3 m_1^3) v_{cb}^3$$

$$+ (\alpha_1 m_2 g_{e1} + 2\alpha_2 I_{JE} m_0 m_2 g_{e1} + 2\alpha_2 I_{JE} m_1^2 g_{e1}) v_{cb}^2 v_{je}$$

$$+ (\alpha_1 m_1 g_{e2} + 2\alpha_2 I_{JE} m_0 m_1 g_{e2} + 2\alpha_2 m_0 m_1 g_{e1}^2) v_{cb} v_{je}^2$$

$$+ (\alpha_1 m_0 g_{e3} + 2\alpha_2 m_0^2 g_{e1} g_{e2} + \alpha_3 m_0^3 g_{e1}^3) v_{je}^3. \tag{7-111}$$

When the transistor is biased well outside of the avalanche region (i.e., $V_{CB} \ll V_{CBO}$), m_0 is approximately unity while m_1, m_2, and m_3 are extremely small. The terms in (7-111) involving v_{cb} can then be ignored.

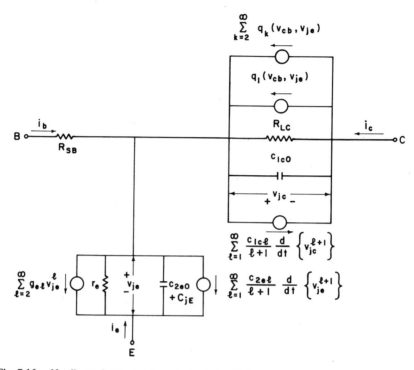

Fig. 7.16. Nonlinear incremental equivalent circuit for *n-p-n* transistor in amplification region.

The nonlinear incremental equivalent circuit for the *n-p-n* transistor follows directly from the global model of Fig. 7.15 and the Taylor series expansions developed in this section. The resulting model is shown in Fig. 7.16. This model is suitable for nodal analysis because the voltages v_{je}, v_{cb}, and v_{jc} are easily expressed in terms of node-to-datum voltages. Similar models can be developed for loop analysis and mixed-variable analysis.

7.3.3 Determination of the Parameters I_{se} and n_E

The measurements needed for determining numerical values of I_{se} and n_E are analogous to those discussed in Section 7.2.4. However, in this case, the negative of the emitter current is plotted versus the base-to-emitter voltage. For either dc or very slowly varying signals, it is apparent from the global model of Fig. 7.15 that $i_{JE} = -i_E$. Also, the voltage drop across the base bulk resistance R_{SB} is typically negligible with respect to v_{JE}. Therefore, $v_{JE} = v_{BE}$. It follows that a plot of $-i_E$ versus v_{BE} is equivalent to a plot of i_{JE} versus v_{JE}.

For $v_{BE} > 2n_E kT/q$, the curve plots as a straight line on a semilog scale. If the coordinates of two points on the straight line are denoted by $(-i_{E1}, v_{BE1})$ and $(-i_{E2}, v_{BE2})$, the emitter-base diode ideality factor is given by

$$n_E = \frac{q}{kT} \frac{(v_{BE2} - v_{BE1})}{\ln(-i_{E2}/-i_{E1})}. \tag{7-112}$$

Also, for emitter currents approximated by the straight line portion of the curve, the reverse-saturation current of the emitter diode may be evaluated from the relation

$$I_{se} = \frac{-i_E}{\exp\left(\dfrac{q v_{BE}}{n_E kT}\right)}. \tag{7-113}$$

Example 7.11

A typical plot of the forward-biased emitter-base junction static characteristic is shown in Fig. 7.17. Assuming the measurements to be made with the junction

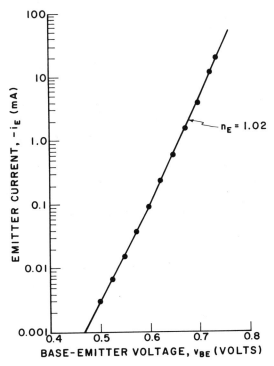

Fig. 7.17. Forward-biased emitter-base junction static characteristic of an *n-p-n* transistor.

near room temperature (i.e., $kT/q = 25$ mV) and selecting as coordinates $-i_{E2} = 10$ mA, $v_{BE2} = 0.722$ V and $-i_{E1} = 1.0$ mA, $v_{BE1} = 0.663$ V, we obtain

$$n_E = \frac{1}{25 \times 10^{-3}} \frac{(0.722 - 0.663)}{\ln(10)} = \frac{2.357}{2.302} = 1.02. \qquad (7\text{-}114)$$

Also, evaluating (7-113) with $-i_{E2} = 10$ mA and $v_{BE2} = 0.722$ V yields

$$I_{se} = \frac{10 \times 10^{-3}}{\exp\left(\dfrac{0.722}{1.02 \times 25 \times 10^{-3}}\right)} = \frac{10 \times 10^{-3}}{\exp(28.3)} = 5.1 \times 10^{-15} \text{ A.} \qquad (7\text{-}115)$$

This value can be checked by extrapolating the straight line in Fig. 7.17 to zero voltage. Note that the slope is 1 decade of emitter current per 0.059 V of base-to-emitter voltage. Beginning with $v_{BE2} = 0.722$ V, therefore, it is necessary to traverse $(0.722/0.059) = 12.2$ decades before the straight line intersects the current axis. This agrees approximately with the value calculated in (7-115). As in the junction diode measurement, a pulse measurement technique for the larger emitter currents may be desirable in order to avoid unwanted heating effects.

7.3.4   Determination of the Parameters C_{jE} and C'_{de}

The total emitter-base junction capacitance is denoted by C_{TE} and equals the sum of the transition and diffusion capacitances in the emitter branch. As indicated by (7-72), the transition capacitance may be treated as a constant C_{jE}. Also, from (7-115) I_{se} is seen to be an extremely small quantity that may be ignored in (7-68). Assuming either dc or slowly varying signals, $i_{JE} = -i_E$ and the total capacitance may be expressed as

$$C_{TE} = C_{jE} - C'_{de} i_E. \qquad (7\text{-}116)$$

Obviously, C_{TE} is a linear function of the emitter current.

Example 7.12

A typical plot of the emitter-base junction forward-bias capacitance for an n-p-n transistor is shown in Fig. 7.18. Since the straight line intersects the capacitance axis at 330 pF, we conclude that $C_{jE} = 330$ pF. Also, the slope of the straight line is 60 pF/mA which equals C'_{de}. For the curve shown we conclude that the total capacitance is given by

$$C_{TE} = 330 - 60 i_E \text{ pF.} \qquad (7\text{-}117)$$

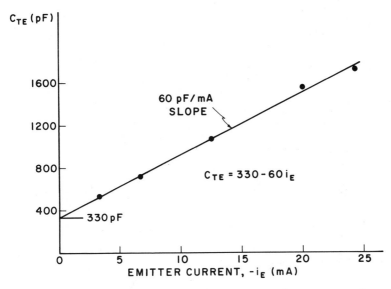

Fig. 7.18. Emitter-base junction forward-bias capacitance for an *n-p-n* transistor.

7.3.5   Determination of the Parameters R_{LC}, V_{CBO}, and η

The static collector characteristics of either the common-base or common-emitter configurations may be used to obtain numerical values for the collector-base junction leakage resistance R_{LC}, the collector-to-base breakdown voltage V_{CBO}, and the avalanche multiplication factor exponent η. After providing a general discussion of the two approaches, we illustrate some calculations in terms of the common-emitter characteristics.

In the common-base collector characteristics the collector current is plotted versus the collector-to-base voltage with the emitter current as a parameter, as shown in Fig. 7.19. The slope of the linear portion of the curves between the voltages zero and v_1 can be attributed to the leakage resistance R_{LC}. On the other hand, the curvature in the region between the voltages v_1 and V_{CBO} is due to avalanche breakdown in the collector-base junction. In the absence of these two effects the curves would plot as horizontal straight lines with values of I_1, I_2, I_3, etc.

Since we are dealing with static characteristics, the voltages and currents are assumed to be slowly varying such that capacitive effects can be ignored. Also, assuming the reverse saturation current I_{sc} to be negligible, it is obvious from Fig. 7.15 that the collector current equals the sum of the currents through R_{LC} and the current-dependent current source. If we focus attention on the particular curve in Fig. 7.19 for which the emitter current is constrained to be i_{E3}, it

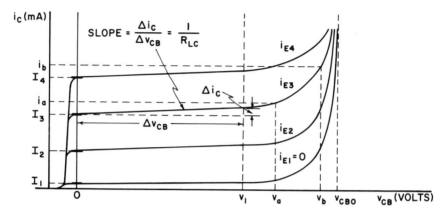

Fig. 7.19. Common-base collector characteristics for an *n-p-n* transistor.

follows that the collector current may be expressed as

$$i_C = \frac{v_{CB}}{R_{LC}} + MI_3 = \frac{v_{CB}}{R_{LC}} + \frac{I_3}{1 - \left(\dfrac{v_{CB}}{V_{CBO}}\right)^\eta} \tag{7-118}$$

where we have ignored the voltage drop across the base resistor, R_{SB}.

Two of the desired parameters are quickly determined from the characteristics. As indicated in Fig. 7.19, the leakage resistance is given by

$$R_{LC} = \frac{\Delta v_{CB}}{\Delta i_C} \tag{7-119}$$

which is the reciprocal slope of the straight line portion of the characteristic in the region between the voltages zero and v_1. Also, by inspection, V_{CBO} is the vertical asymptote corresponding to the open-circuited emitter (i.e., $i_{E1} = 0$) curve.

The exponent η is obtained by selecting two data points in the vicinity of avalanche, such as (v_a, i_a) and (v_b, i_b), which are shown in Fig. 7.19. From (7-118) we have

$$i_a = \frac{v_a}{R_{LC}} + \frac{I_3}{1 - \left(\dfrac{v_a}{V_{CBO}}\right)^\eta}$$

$$i_b = \frac{v_b}{R_{LC}} + \frac{I_3}{1 - \left(\dfrac{v_b}{V_{CBO}}\right)^\eta} \; . \tag{7-120}$$

At this point it is convenient to introduce the constants

$$k_a = i_a - \frac{v_a}{R_{LC}} = \frac{I_3}{1 - \left(\dfrac{v_a}{V_{CBO}}\right)^{\eta}}$$

$$k_b = i_b - \frac{v_b}{R_{LC}} = \frac{I_3}{1 - \left(\dfrac{v_b}{V_{CBO}}\right)^{\eta}} . \qquad (7\text{-}121)$$

Solving (7-121) for $(v_a/V_{CBO})^{\eta}$ and $(v_b/V_{CBO})^{\eta}$ and dividing one by the other, there results

$$\left(\frac{v_a}{v_b}\right)^{\eta} = \frac{k_b}{k_a}\left(\frac{k_a - I_3}{k_b - I_3}\right) . \qquad (7\text{-}122)$$

Finally, taking the logarithm of both sides, it follows that

$$\eta = \frac{\log_{10}\left[\dfrac{k_b}{k_a}\left(\dfrac{k_a - I_3}{k_b - I_3}\right)\right]}{\log_{10}\left(\dfrac{v_a}{v_b}\right)} . \qquad (7\text{-}123)$$

Observe that knowledge of V_{CBO} is not required for evaluation of either the constants k_a and k_b or the exponent η. However, from the first equation in (7-121), we have

$$V_{CBO} = v_a\left(\frac{k_a}{k_a - I_3}\right)^{1/\eta} . \qquad (7\text{-}124)$$

A check on the value for η is provided by comparing the previously determined value of V_{CBO} with that calculated from (7-124).

R_{LC}, V_{CBO}, and η can also be found from measurements of the common-emitter collector characteristics. These are plots of collector current versus collector-to-emitter voltage with the base current as a parameter. A typical set of characteristics is shown in Fig. 7.20. As with the common-base characteristics, we attribute the slope of the linear region between the voltages zero and v_1 to the leakage resistance R_{LC} and the curvature in the region between the voltages v_1 and V_{CEO} to avalanche effects.

Throughout this discussion it is assumed that the signals vary slowly enough such that capacitive effects can be ignored and $i_{JE} = -i_E$. We also assume the reverse-saturation current I_{sc} and the emitter-to-base junction voltage v_{JE} to be negligible. With these assumptions we conclude from Fig. 7.15 that the collector

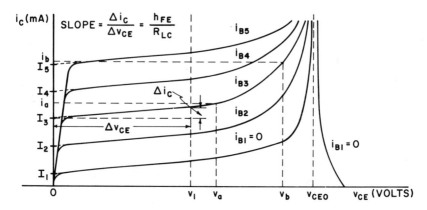

Fig. 7.20.   Common-emitter collector characteristics for an *n-p-n* transistor.

current can be written as

$$i_C = \frac{v_{CE}}{R_{LC}} - \alpha_N M i_E. \tag{7-125}$$

However, from Kirchhoff's current law, the emitter current is given by

$$i_E = -i_B - i_C. \tag{7-126}$$

Substituting (7-126) into (7-125) and solving for i_C, we obtain

$$
\begin{aligned}
i_C &= \frac{v_{CE}}{(1 - \alpha_N M) R_{LC}} + \frac{\alpha_N M i_B}{1 - \alpha_N M} \\[2mm]
&= \frac{v_{CE}}{(1 - \alpha_N M) R_{LC}} + \frac{h_{FE} i_B}{1 - \left(\dfrac{v_{CE}}{V_{CEO}}\right)^{\eta}}
\end{aligned}
\tag{7-127}
$$

where use has been made of the discussion leading to (7-85). Since the sustaining voltage V_{CEO} is generally much smaller than V_{CBO}, M is approximately unity for collector-to-emitter voltages less than V_{CEO}. This emphasizes, once again, that the transistor is not actually in avalanche breakdown when the common-emitter collector voltage is V_{CEO}. With $M = 1$, (7-127) becomes

$$i_C = \frac{v_{CE}}{(1 - \alpha_N) R_{LC}} + \frac{h_{FE} i_B}{1 - \left(\dfrac{v_{CE}}{V_{CEO}}\right)^{\eta}}. \tag{7-128}$$

Therefore, the slope $\Delta i_C / \Delta v_{CE}$ of the linear portion of the characteristic is equal to $1/(1 - \alpha_N) R_{LC}$. Noting that α_N is typically close to unity and making use of

(7-82), it follows that

$$R_{LC} = \frac{1}{(1 - \alpha_N)} \frac{\Delta v_{CE}}{\Delta i_C} \approx h_{FE} \frac{\Delta v_{CE}}{\Delta i_C} . \tag{7-129}$$

By inspection of Fig. 7.20, it is seen that the sustaining voltage V_{CEO} is the vertical asymptote of the open-circuited base (i.e., $i_{B1} = 0$) curve. From (7-84), the breakdown voltage V_{CBO} is then given by

$$V_{CBO} = V_{CEO}(1 - \alpha_N)^{-1/n}. \tag{7-130}$$

The exponent η is determined in a manner similar to that discussed for the common-base characteristics. For this purpose, consider the particular curve in Fig. 7.20 corresponding to the base current i_{B3}. The straight line portion of the curve, when extrapolated to $v_{CE} = 0$, intersects the vertical axis at $i_C = I_3$. Consequently, (7-128) may be rewritten as

$$i_C = \frac{h_{FE} v_{CE}}{R_{LC}} + \frac{I_3}{1 - \left(\dfrac{v_{CE}}{V_{CEO}}\right)^{\eta}} . \tag{7-131}$$

Choosing the two data points (v_a, i_a) and (v_b, i_b) shown in Fig. 7.20, we define the constants

$$l_a = i_a - \frac{h_{FE} v_a}{R_{LC}} = \frac{I_3}{1 - \left(\dfrac{v_a}{V_{CEO}}\right)^{\eta}}$$

$$l_b = i_b - \frac{h_{FE} v_b}{R_{LC}} = \frac{I_3}{1 - \left(\dfrac{v_b}{V_{CEO}}\right)^{\eta}} . \tag{7-132}$$

Because of the similarity between (7-132) and (7-121), it follows directly from (7-123) that η is given by

$$\eta = \frac{\log_{10}\left[\dfrac{l_b}{l_a}\left(\dfrac{l_a - I_3}{l_b - I_3}\right)\right]}{\log_{10}\left(\dfrac{v_a}{v_b}\right)} . \tag{7-133}$$

Also, in analogy with (7-124), the sustaining voltage may be calculated from

$$V_{CEO} = v_a \left(\frac{l_a}{l_a - I_3}\right)^{1/\eta} . \tag{7-134}$$

This second method for determining V_{CEO} can provide a check on the value obtained for η.

Example 7.13

The previous discussion is now illustrated with a numerical example based upon measurements, shown in Fig. 7.21, of the common-emitter collector characteristics for an *n-p-n* transistor. By inspection, the sustaining voltage V_{CEO} is approximately equal to 87 V. To determine the leakage resistance R_{LC} and the avalanche exponent η, we make use of the curve for which the base current equals 0.8 mA. The linear portion of this curve has a slope given by

$$\frac{\Delta i_C}{\Delta v_{CE}} = \frac{0.31 \text{ mA}}{110 \text{ V}} = 2.82 \times 10^{-6} \text{ mho.} \qquad (7\text{-}135)$$

Also, the h_{FE} of the transistor for $i_B = 0.8$ mA is

$$h_{FE} = \frac{i_C}{i_B} = \frac{5.44 \text{ mA}}{0.8 \text{ mA}} = 6.8 \qquad (7\text{-}136)$$

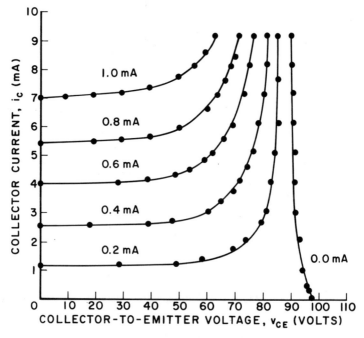

Fig. 7.21.   Measured common-emitter collector characteristics for an *n-p-n* transistor.

where the collector current is evaluated near $v_{CE} = 0$. It follows from (7-129) that

$$R_{LC} = \frac{6.8}{2.82 \times 10^{-6}} = 2.41 \times 10^6 \ \Omega. \tag{7-137}$$

Choosing the two data points $v_a = 60$ V, $i_a = 6.615$ mA and $v_b = 70$ V, $i_b = 8.538$ mA, the constants defined in (7-132) become

$$l_a = 6.615 \times 10^{-3} - \frac{(6.8)(60)}{2.41 \times 10^6} = 6.446 \text{ mA}$$

$$l_b = 8.538 \times 10^{-3} - \frac{(6.8)(70)}{2.41 \times 10^6} = 8.340 \text{ mA}. \tag{7-138}$$

Since $I_3 = 5.44$ mA, it follows from (7-133) that

$$\eta = \frac{\log_{10}\left[\frac{8.340}{6.446}\left(\frac{6.446 - 5.44}{8.340 - 5.44}\right)\right]}{\log_{10}\left(\frac{60}{70}\right)} = \frac{\log_{10}(0.45)}{\log_{10}(0.857)} = 5.18. \tag{7-139}$$

The breakdown voltage V_{CBO} is calculated from (7-130). Making use of (7-76)

$$\alpha_N = \frac{h_{FE}}{1 + h_{FE}} = \frac{6.8}{7.8} = 0.872. \tag{7-140}$$

Hence

$$V_{CBO} = 87(1 - 0.872)^{-1/5.18} = 129.5 \text{ V}. \tag{7-141}$$

Finally, we check the value of η by using (7-134) to evaluate V_{CEO}.

$$V_{CEO} = (60)\left(\frac{6.446}{1.006}\right)^{1/5.18} = 85.9 \text{ V}. \tag{7-142}$$

This compares favorably with the previously determined value of 87 V.

7.3.6   Determination of the Parameters $h_{FE_{max}}$, $I_{C_{max}}$, and a

The empirical expression for the common-emitter dc current gain h_{FE} is given by (7-77). Consequently, the parameters $h_{FE_{max}}$, $I_{C_{max}}$, and a are calculated from measurements of h_{FE} versus the dc collector current.

The most direct way to measure h_{FE} is simply to display the static common-emitter collector characteristics on a transistor curve tracer. h_{FE} is obtained as a function of I_C by plotting the ratio I_C/I_B for various values of I_C. Since the characteristics would plot as horizontal straight lines in the absence of the leakage resistance and avalanche, I_C should be measured by extrapolating the straight line portion of the curves to $V_{CE} = 0$. An alternate technique is to make use of pulsed measurements. This is necessary when the curve of h_{FE} versus I_C peaks at a relatively large value of collector current.

Example 7.14

Figure 7.22 shows a plot of h_{FE} that was obtained using the pulse measurement technique. Because of the large range of values covered by the collector current, I_C is plotted on a logarithmic scale. From (7-77) it is obvious that h_{FE} equals its maximum value $h_{FE_{max}}$ when I_C equals $I_{C_{max}}$. By inspection of Fig. 7.22, we conclude that $h_{FE_{max}} = 8.2$ and $I_{C_{max}} = 150$ mA. In addition, the solution of (7-77) for the parameter a results in

$$a = \frac{h_{FE_{max}} - h_{FE}}{h_{FE} \log_{10}^2 \left(\dfrac{I_C}{I_{C_{max}}}\right)}. \tag{7-143}$$

To evaluate a we select the data point corresponding to $I_C = 0.1$ mA and $h_{FE} = 3.677$. It follows that

$$a = \frac{8.2 - 3.677}{3.677 \log_{10}^2 \left(\dfrac{0.1}{150}\right)} = \frac{4.523}{(3.677)(-3.176)^2} = 0.122. \tag{7-144}$$

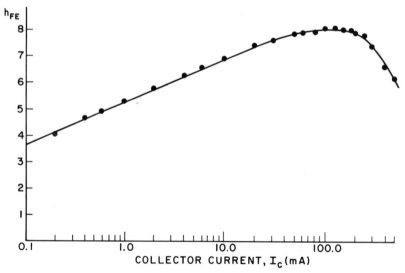

Fig. 7.22.   Common-emitter dc current gain h_{FE} as a function of collector current I_C.

Finally, substituting the calculated values of the parameters into (7-77), the empirical expression for h_{FE} becomes

$$h_{FE} = \frac{8.2}{1 + 0.122 \log_{10}^2 \left(\dfrac{I_C}{150}\right)} \qquad (7\text{-}145)$$

where I_C is expressed in milliamperes.

7.3.7 Determination of the Parameters $C_C(0)$, μ_C, and ϕ_C

An expression for the transition capacitance of the collector-base junction is given by (7-70). Assuming the voltage drop across the base bulk resistance R_{SB} to be negligible, $v_{JC} = -v_{CB}$. Then (7-70) may be expressed as

$$C_{1C} = \frac{C_C(0)}{\left(1 + \dfrac{v_{CB}}{\phi_C}\right)^{\mu_C}}, \qquad v_{CB} > -\phi_C. \qquad (7\text{-}146)$$

Because of the similarity between (7-70) and (7-52), the parameters $C_C(0)$, μ_C, and ϕ_C are obtained using the same techniques presented in Section 7.2.7.

$C_C(0)$, of course, equals the value of C_{1C} when v_{CB} equals zero. Therefore, a numerical value for $C_C(0)$ is determined by measuring the collector-to-base capacitance with the collector-to-base voltage constrained near zero.

The junction grading constant μ_C is evaluated from capacitance measurements for which $v_{CB} \gg \phi_C$. As in Section 7.2.7, let C_{T1} and C_{T2} denote the capacitances measured at v_{CB1} and v_{CB2}, respectively. In analogy with (7-57), μ_C is given by

$$-\mu_C = \frac{\log_{10} C_{T2} - \log_{10} C_{T1}}{\log_{10} v_{CB2} - \log_{10} v_{CB1}}. \qquad (7\text{-}147)$$

Hence when the capacitance versus collector-to-base voltage is plotted on log-log paper, $-\mu_C$ is the slope of the straight line portion of the curve.

The junction contact potential ϕ_C is evaluated from capacitance measurements for which v_{CB1} and v_{CB2} are both small enough so as to be in that region of the capacitance curve that deviates from a straight line. The parameter A is defined to be

$$A = \left[\frac{C_{T2}}{C_{T1}}\right]^{1/\mu_C}. \qquad (7\text{-}148)$$

In analogy with (7-60), the contact potential is determined from

$$\phi_C = \frac{-Av_{CB2} + v_{CB1}}{A - 1}.$$ (7-149)

Example 7.15

Experimental data for the collector-base transition capacitance are shown in Fig. 7.23. Also, some of the measurements are tabulated in Table 7.2 Since $C_{1C} = 34.3$ pF when $v_{CB} = 0$, it follows that

$$C_C(0) = 34.3 \text{ pF}.$$ (7-150)

Substituting the data points $v_{CB1} = 10$, $C_{T1} = 11.15$ pF and $v_{CB2} = 30$, $C_{T2} = 7.6$ pF into (7-147), the junction grading constant is given by

$$-\mu_C = \frac{\log_{10} 7.6 - \log_{10} 11.15}{\log_{10} 30 - \log_{10} 10} = \frac{0.881 - 1.047}{1.477 - 1.0} = -0.348.$$ (7-151)

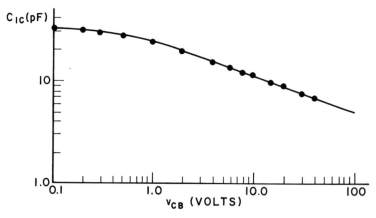

Fig. 7.23.   Measured collector-base transition capacitance for an *n-p-n* transistor.

TABLE 7.2.   Some Measured Values
of C_{1C}

$v_{CB}(V)$	$C_{1C}(pF)$
0	34.3
0.1	31.9
1.0	22.2
10.0	11.15
30.0	7.6

To evaluate ϕ_C, we choose $v_{CB1} = 0$, $C_{T1} = 34.3$ pF and $v_{CB2} = 1.0$, $C_{T2} = 22.2$ pF. Hence from (7-148) and (7-149) we have

$$A = \left[\frac{22.2}{34.3}\right]^{1/0.348} = (0.647)^{2.874} = 0.286$$

$$\phi_C = \frac{-(0.286)(1.0) + (0)}{0.286 - 1} = \frac{-0.286}{-0.714} = 0.40. \qquad (7\text{-}152)$$

Consequently, the expression for the collector-base transition capacitance is

$$C_{1C} = \frac{34.3}{\left(1 + \dfrac{v_{CB}}{0.40}\right)^{0.348}} \text{ pF.} \qquad (7\text{-}153)$$

As a check, we calculate the value of C_{1C} for $v_{CB} = 0.1$ V. Equation (7-153) yields

$$C_{1C} = \frac{34.3}{\left(1 + \dfrac{0.1}{0.4}\right)^{0.348}} = \frac{34.3}{(1.25)^{0.348}} = \frac{34.3}{1.08} = 31.76 \text{ pF.} \qquad (7\text{-}154)$$

This compares favorably with the tabulated value of 31.9 pF.

7.3.8   Determination of the Parameter R_{SB}

Accurate measurement of the bulk base resistance R_{SB} is not easily accomplished. For simplicity, two low-frequency techniques are discussed in this section. Both high- and low-frequency measurement schemes are possible. However, high-frequency techniques tend to involve both fairly sophisticated equipment and curve-fitting computer algorithms. In addition, distributed capacitance across the base resistance, which is not included in the Ebers–Moll model, introduces measurement uncertainties. On the other hand, low-frequency techniques suffer from the deficiency that effects due to the emitter and base bulk resistances cannot be distinguished. We treat this ambiguity by assuming all such effects to be attributed to R_{SB} (i.e., $R_{SE} = 0$). Although this assumption is adequate for many applications, it can result in significant error in some cases.

A dc technique for determining R_{SB} consists of measuring the dc base-to-emitter voltage as a function of the dc emitter current. The deviation from exponential behavior at high current levels is then ascribed to an $I_B R_{SB}$ voltage drop where I_B is the dc base current. This approach is analogous to that used in conjunction with (7-47) and Fig. 7.10 for obtaining the diode bulk resistance R_S.

The base resistance can also be obtained from an incremental measurement of the common-emitter short-circuited input admittance y_{ie}. Small-signal inputs

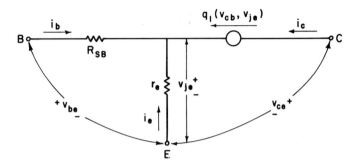

Fig. 7.24.   Transistor low-frequency linear incremental equivalent circuit.

at low frequencies are used such that the transistor behaves linearly and capacitive effects are negligible. Ignoring the current that flows through the leakage resistance R_{LC}, the nonlinear incremental equivalent circuit of Fig. 7.16 simplifies to the low-frequency linear incremental equivalent circuit shown in Fig. 7.24.

By definition, the common-emitter short-circuited input admittance is given by

$$y_{ie} = \frac{i_b}{v_{be}}\bigg|_{v_{ce}=0}.$$

(7-155)

With reference to (7-111), the collector current is

$$i_c = q_1(v_{cb}, v_{je}) = \alpha_1 I_{JE} m_1 v_{cb} + \alpha_1 m_0 g_{e1} v_{je}.$$

(7-156)

We assume the transistor to be biased such that $V_{CB} \ll V_{CBO}$. From (7-96) and (7-97) it follows that m_0 and m_1 are approximately equal to unity and zero, respectively. Also, as seen from (7-88), g_{e1} is the reciprocal of r_e. We conclude that

$$i_c \approx \alpha_1 g_{e1} v_{je} = \alpha_1 \frac{v_{je}}{r_e} = -\alpha_1 i_e.$$

(7-157)

Application of Kirchhoff's current law to the internal node of Fig. 7.24 results in

$$i_b + i_e + i_c = i_b + i_e - \alpha_1 i_e = 0.$$

(7-158)

Solving for the incremental emitter current, we obtain

$$i_e = -\frac{i_b}{(1 - \alpha_1)}.$$

(7-159)

The incremental base-to-emitter voltage v_{be}, with $v_{ce} = 0$, is given by

$$v_{be} = i_b R_{SB} - i_e r_e = i_b R_{SB} + i_b \frac{r_e}{(1 - \alpha_1)}. \qquad (7\text{-}160)$$

Therefore, the common-emitter short-circuited input admittance is

$$y_{ie} = \left. \frac{i_b}{v_{be}} \right|_{v_{ce}=0} = \left[R_{SB} + \frac{r_e}{(1 - \alpha_1)} \right]^{-1}. \qquad (7\text{-}161)$$

From (7-161) it is seen that

$$R_{SB} = \frac{1}{y_{ie}} - \frac{r_e}{(1 - \alpha_1)}. \qquad (7\text{-}162)$$

The values of r_e and α_1 are evaluated from (7-88) and (7-109), respectively.

Example 7.16

To illustrate the use of (7-162), assume $n_E = 1.02$, $kT/q = 25$ mV, $-I_E \approx I_C = 5$ mA, $h_{FE_{max}} = 8.2$, $I_{C_{max}} = 150$ mA, $a = 0.123$, and $y_{ie} = 16 \times 10^{-3}$ mho. From (7-88)

$$r_e = \frac{(1.02)(25 \times 10^{-3})}{(5 \times 10^{-3})} = 5.1 \ \Omega. \qquad (7\text{-}163)$$

Also, from (7-109)

$$\alpha_1 = \frac{8.2}{1 + 8.2 + (0.123) \log_{10}^2 \left(\frac{5}{150} \right) + 2(0.123) \log_{10} \left(\frac{5}{150} \right) \log_{10} (2.72)}$$

$$= 0.88 \qquad (7\text{-}164)$$

Substituting these values of r_e and α_1 into (7-162) along with the measured value of y_{ie}, we obtain

$$R_{SB} = \frac{1}{16 \times 10^{-3}} - \frac{5.1}{(1 - 0.88)} = 62.5 - 42.5 = 20 \ \Omega. \qquad (7\text{-}165)$$

___

As a final comment, the reader is cautioned that accurate determination of R_{SB} is an extremely difficult problem. The simplified approaches presented here are not always reliable.

7.4 JUNCTION FIELD-EFFECT TRANSISTOR (JFET)

Although various geometrical configurations are possible, it is convenient to think of an n-channel JFET as consisting of a bar of n-type semiconductor material sandwiched between two p-type materials. The n-type material is referred to as the channel. Terminals, called the source and drain, are located at each end of the channel while the two p-type regions are connected together to form an additional terminal known as the gate. In normal operation the p-n junctions formed by the p- and n-type materials are reverse biased so that negligible current flows through the gate lead. The device current is confined to the channel and flows between source and drain. This current is composed of the majority carriers in the channel and is regulated by the gate-to-source voltage. JFET's are also made with p-type channels and n-type gates. The two kinds of JFET's both operate in the same manner except that the channel majority carriers, which always flow from source to drain, are electrons for n-channel and holes for p-channel. As a result, the polarities of all voltages and currents are of opposite signs for the two transistors. To avoid unnecessary repetition, we consider only the n-channel JFET in this section.

The JFET global model is presented in Section 7.4.1. As was done with the bipolar junction transistor, the model is simplified to the special case for which the device is used as an amplifier. The corresponding nonlinear incremental equivalent circuit is developed in Section 7.4.2. Finally, Sections 7.4.3 through 7.4.6 contain discussions for determining the various model parameters.

7.4.1 Global Model for Junction Field-Effect Transistor

The schematic symbol for the n-channel JFET is shown in Fig. 7.25(a) where the source, gate, and drain terminals are denoted by S, G, and D, respectively. The p-channel JFET has the same symbol except that the arrow in the gate lead points in the opposite direction. Figure 7.25(b) illustrates a global model for the n-channel JFET. The four elements in the global model are as follows.

(1) R_S, the source bulk resistance.

(2) $i(v_{GJ})$, a nonlinear voltage-controlled current source. Note that the control variable v_{GJ} is the voltage between the gate terminal and the internal node denoted by J. A single analytical expression is not possible for $i(v_{GJ})$. Typically, it is specified piecewise for the various regions of operation.

(3) C_{GS}, the nonlinear gate-to-source capacitance, given by

$$C_{GS} = \frac{C_{GS}(0)}{\left(1 - \dfrac{v_{GJ}}{\phi}\right)^{\mu_S}}, \qquad v_{GJ} < \phi. \qquad (7\text{-}166)$$

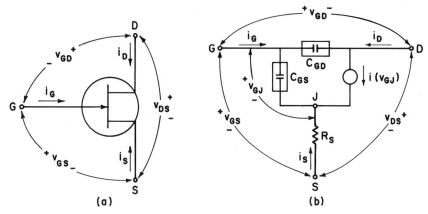

Fig. 7.25. (a) Schematic symbol. (b) Global model for n-channel junction field-effect transistor.

(4) C_{GD}, the nonlinear gate-to-drain capacitance, given by

$$C_{GD} = \frac{C_{GD}(0)}{\left(1 - \dfrac{v_{GD}}{\phi}\right)^{\mu_D}}, \qquad v_{GD} < \phi. \tag{7-167}$$

The parameters that appear in (7-166) and (7-167) are defined as follows:

$C_{GS}(0)$ gate-to-source capacitance for $v_{GJ} = 0$;

ϕ p-n junction contact potential;

μ_S exponent of the gate-to-source capacitance equation;

$C_{GD}(0)$ gate-to-drain capacitance for $v_{GD} = 0$;

μ_D exponent of the gate-to-drain capacitance equation.

Normal operation of an n-channel JFET requires v_{GS} to be negative and v_{DS} to be positive. For a fixed negative value of v_{GS}, the drain current i_D initially increases as a function of v_{DS}. However, the device eventually saturates. The constant current region, in which i_D is essentially a constant independent of v_{DS}, is known as the saturation region. The region below saturation, which occurs for small values of v_{DS}, is known as the ohmic region. When used as an amplifier, the JFET is usually operated in the saturation region.

In the saturation region the nonlinear voltage-controlled current source may be modeled as

$$i(v_{GJ}) = \frac{3 I_m \rho}{2} \left\{ -\rho + \sqrt{4\left[\frac{1}{3} - \left(\frac{v_{GJ} + \phi}{V_{po} + \phi}\right) + \frac{2}{3}\left(\frac{v_{GJ} + \phi}{V_{po} + \phi}\right)^{3/2} \right] + \rho^2} \right\}$$

$$\tag{7-168}$$

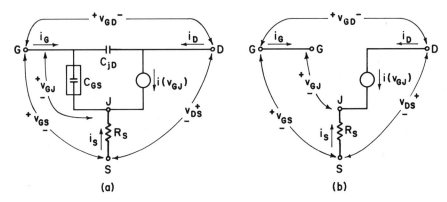

Fig. 7.26. (a) Global model. (b) Low-frequency global model for n-channel JFET in saturation region.

where ϕ is the p-n junction contact potential, V_{po} is the pinchoff voltage, and ρ and I_m are constants. This expression is based upon Fair's field-dependent mobility model.

In the saturation region a simplification is also possible relative to the nonlinear gate-to-drain capacitance C_{GD}. Note that v_{GD} is the difference between v_{GS} and v_{DS}. Because v_{GS} and v_{DS} have opposite polarities, $-v_{GD}$ is greater than $-v_{GJ}$. As a result, the capacitance C_{GD} is usually smaller and much less voltage dependent than the capacitance C_{GS}. If V_{GD} represents the quiescent value of v_{GD}, it follows that C_{GD} may be approximated by a linear capacitor with capacitance

$$C_{jD} = \frac{C_{GD}(0)}{\left(1 - \dfrac{V_{GD}}{\phi}\right)^{\mu_D}}. \tag{7-169}$$

In view of the previous discussion, a global model of the JFET which is valid in the saturation region is shown in Fig. 7.26(a). Note that only two nonlinear circuit elements are contained in this model. The corresponding low-frequency model is shown in Fig. 7.26(b). We see that the JFET is basically a high-input-impedance device and that the drain current i_D equals $i(v_{GJ})$ for low enough frequencies such that capacitive effects are negligible.

7.4.2   Nonlinear Incremental Equivalent Circuit for JFET in Saturation Region

Assuming the JFET to be operating as an amplifier in the saturation region, a nonlinear incremental equivalent circuit for the device is readily derived from the global model of Fig. 7.26(a). The only nonlinear elements are the dependent current source $i(v_{GJ})$ and the gate-to-source capacitance C_{GS}. Since both of these nonlinearities are voltage dependent, a distortion model that is suitable for nodal analysis results from a straightforward power series expansion of the nonlinearities.

The control equation for the voltage-controlled current source is given by (7-168). Expanding this expression in a Taylor series about the quiescent operating point, we obtain

$$i(v_{gj}) = \sum_{k=1}^{\infty} g_k v_{gj}^k \tag{7-170}$$

where v_{gj} is an incremental voltage. Because of the complexity of (7-168), it is not possible to derive a general analytical expression for the series coefficients. In practice, the coefficients are usually found using numerical differentiation of (7-168). A step size of 0.01 V works well in most cases.

Analytical relationships for the coefficients can be obtained through brute force. For this purpose, define

$$A = \rho^2 + 4\left[\frac{1}{3} - \left(\frac{V_{GJ} + \phi}{V_{po} + \phi} \right) + \frac{2}{3}\left(\frac{V_{GJ} + \phi}{V_{po} + \phi} \right)^{3/2} \right]$$

$$B = \frac{V_{GJ} + \phi}{V_{po} + \phi}. \tag{7-171}$$

The first three coefficients in (7-170) are then given by

$$g_1 = \frac{di(v_{GJ})}{dv_{GJ}}\bigg|_{v_{GJ} = V_{GJ}} = \frac{3I_m\rho}{V_{po} + \phi} \frac{-1 + B^{1/2}}{A^{1/2}}$$

$$g_2 = \frac{1}{2}\frac{dg_1}{dV_{GJ}} = \frac{3I_m\rho}{4(V_{po} + \phi)^2} \frac{AB^{-1/2} - 4[-1 + B^{1/2}]^2}{A^{3/2}}$$

$$g_3 = \frac{1}{3}\frac{dg_2}{dV_{GJ}} = \frac{I_m\rho}{8(V_{po} + \phi)^3} \frac{-A^2 B^{-3/2} - 12AB^{-1/2}[-1 + B^{1/2}] + 48[-1 + B^{1/2}]^3}{A^{5/2}}.$$

$$\tag{7-172}$$

The desirability of numerical differentiation is readily apparent from (7-172).

The expression for C_{GS} in (7-166) is identical in form to that for C_1 in (7-12). Therefore, with reference to equations (7-36) through (7-38), the incremental current through C_{GS} may be written as

$$i_{cgs} = \sum_{k=0}^{\infty} \frac{c_{gsk}}{k + 1} \frac{d}{dt} \{v_{gj}^{k+1}\} \tag{7-173}$$

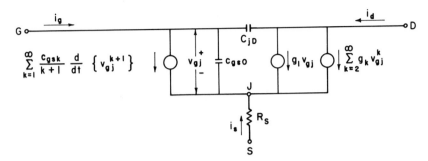

Fig. 7.27. Nonlinear incremental equivalent circuit for n-channel JFET in saturation region.

where the linear incremental gate-to-source capacitance is given by

$$c_{gso} = \frac{C_{GS}(0)}{\left(1 - \dfrac{V_{GJ}}{\phi}\right)^{\mu_S}} \tag{7-174}$$

and the kth coefficient is

$$c_{gsk} = \frac{c_{gso}}{k!} \frac{\mu_S(\mu_S + 1)\cdots(\mu_S + k - 1)}{(\phi - V_{GJ})^k} = \frac{(\mu_S + k - 1)}{k(\phi - V_{GJ})} c_{gs(k-1)}. \tag{7-175}$$

A nonlinear incremental equivalent circuit for the n-channel JFET when operating in the saturation region is shown in Fig. 7.27. This follows from the global model of Fig. 7.26 and the Taylor series given in (7-170) and (7-173). As with all of our nonlinear incremental equivalent circuits, the model involves only incremental variables. The quiescent operating point, however, does play a role in determining the series coefficients. Because the voltage v_{gj} is easily expressed in terms of node-to-datum voltages, this model is suited for nodal analysis.

7.4.3   Determination of the Parameters C_{jD}, $C_{GS}(0)$, μ_S, and ϕ

Because the procedures for obtaining the capacitive parameters C_{jD}, $C_{GS}(0)$, μ_S, and ϕ follow closely those previously discussed in Sections 7.2.7 and 7.3.7, only a brief presentation is given here.

A value for C_{jD} is determined by a direct measurement of the gate-to-drain capacitance. Typical values of C_{jD} fall in the vicinity of 1 pF.

An expression for the gate-to-source capacitance is given by (7-166). Assuming C_{GS} is measured under conditions for which the voltage drop across the source bulk resistance R_S is negligible, $v_{GJ} = v_{GS}$. The expression for C_{GS} may then be written as

$$C_{GS} = \frac{C_{GS}(0)}{\left(1 - \dfrac{v_{GS}}{\phi}\right)^{\mu_S}}, \qquad v_{GS} < \phi. \tag{7-176}$$

Thus $C_{GS}(0)$ is obtained by measuring C_{GS} with v_{GS} equal to zero. The parameters μ_S and ϕ are determined from a plot of C_{GS} versus $-v_{GS}$. When plotted on log-log paper, $-\mu_S$ is the slope of the straight line portion of the curve. For this straightline portion, let C_{T1} and C_{T2} denote the capacitances measured at $-v_{GS1}$ and $-v_{GS2}$, respectively. It follows that

$$-\mu_S = \frac{\log_{10} C_{T2} - \log_{10} C_{T1}}{\log_{10} (-v_{GS2}) - \log_{10} (-v_{GS1})}. \tag{7-177}$$

Finally, in analogy with (7-60), the contact potential ϕ is given by

$$\phi = \frac{A v_{GS2} - v_{GS1}}{A - 1} \tag{7-178}$$

where

$$A = \left[\frac{C_{T2}}{C_{T1}}\right]^{1/\mu_S} \tag{7-179}$$

and $-v_{GS2}$ and $-v_{GS1}$ are chosen to be small enough such that they fall in a region of the capacitance curve that deviates from a straight line.

We do not illustrate the parameter calculations involving equations (7-177) through (7-179) since they are identical to those carried out in Sections 7.2.7 and 7.3.7.

7.4.4 Determination of the Parameter R_S

The source bulk resistance R_S can be estimated from the incremental common-source input admittance y_{is} where

$$y_{is} = \left.\frac{i_g}{v_{gs}}\right|_{v_{ds}=0}. \tag{7-180}$$

With an ac short connected between the drain and source and assuming the effect of the controlled source $i(v_{GJ})$ to be negligible, the input admittance is approximated by C_{jD} in parallel with the series combination of C_{GS} and R_S. It follows that

$$y_{is} = \frac{j\omega C_{GS}}{1 + j\omega C_{GS} R_S} + j\omega C_{jD}$$

$$= \frac{\omega^2 C_{GS}^2 R_S}{1 + \omega^2 C_{GS}^2 R_S^2} + j\omega(C_{GS} + C_{jD}). \qquad (7\text{-}181)$$

Focusing attention on the real part of y_{is}, we have

$$\text{Re}\,\{y_{is}\} = \frac{\omega^2 C_{GS}^2 R_S^2}{1 + \omega^2 C_{GS}^2 R_S^2}. \qquad (7\text{-}182)$$

This results in the quadratic equation

$$R_S^2 - \frac{R_S}{\text{Re}\,\{y_{is}\}} + \frac{1}{\omega^2 C_{GS}^2} = 0. \qquad (7\text{-}183)$$

The solution for R_S is given by

$$R_S = \frac{1}{2\,\text{Re}\,\{y_{is}\}}\left[1 \pm \sqrt{1 - \left(\frac{2\,\text{Re}\,\{y_{is}\}}{\omega C_{GS}}\right)^2}\right] \qquad (7\text{-}184)$$

However, there is an ambiguity in (7-184) as to whether the plus or minus sign in front of the radical should be used. Typically, $\omega C_{GS} R_S < 1$. For extremely small values of $\omega C_{GS} R_S$, R_S may be approximated from (7-182) as

$$R_S \approx \frac{\text{Re}\,\{y_{is}\}}{\omega^2 C_{GS}^2}. \qquad (7\text{-}185)$$

Clearly the sign in (7-184) should be chosen so as to obtain a similar result. Assuming $\omega C_{GS} R_S \ll 1$ and applying the binomial expansion given in (6-26) to (7-184), we obtain

$$R_S \approx \frac{1}{2\,\text{Re}\,\{y_{is}\}}\left[1 \pm \left(1 - \frac{1}{2}\frac{4\,\text{Re}^2\,\{y_{is}\}}{\omega^2 C_{GS}^2}\right)\right]. \qquad (7\text{-}186)$$

This reduces to (7-185) provided the minus sign in front of the radical is selected. We conclude that

$$R_S = \frac{1}{2\,\text{Re}\,\{y_{is}\}}\left[1 - \sqrt{1 - \left(\frac{2\,\text{Re}\,\{y_{is}\}}{\omega C_{GS}}\right)^2}\right]. \qquad (7\text{-}187)$$

Example 7.17

By way of illustration of (7-187), assume $f = 100$ MHz, $C_{GS} = 2.6$ pF, and Re $\{y_{is}\} = 1.5 \times 10^{-4}$ mho. Substituting these values into (7-187), R_S is approximately equal to 57 Ω.

7.4.5   Determination of the Parameter V_{po}

According to Cobbold, the JFET current-voltage relation in the saturation region
can be approximated by

$$i_D = I_{DO} \left(1 - \frac{v_{GS} + \phi}{V_{po} + \phi}\right)^n. \tag{7-188}$$

Although this relation is not adequate for accurately predicting nonlinear effects,
it does provide a means for determining the pinchoff voltage V_{po}.

Making use of (7-188), the linear incremental transconductance of the JFET
is given by

$$g_m = \frac{di_D}{dv_{GS}} = nI_{DO} \left(1 - \frac{v_{GS} + \phi}{V_{po} + \phi}\right)^{n-1} \left(-\frac{1}{V_{po} + \phi}\right). \tag{7-189}$$

Dividing (7-188) by (7-189), we obtain

$$\frac{i_D}{g_m} = -\frac{V_{po} + \phi}{n} \left(1 - \frac{v_{GS} + \phi}{V_{po} + \phi}\right)$$

$$= -\frac{1}{n}(V_{po} + \phi - v_{GS} - \phi) = -\frac{1}{n}(V_{po} - v_{GS}). \tag{7-190}$$

Thus if i_D/g_m is plotted as a function of v_{GS}, the curve should plot as a straight
line with a zero-current intercept equal to the pinchoff voltage V_{po}.

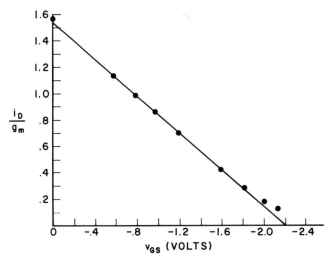

Fig. 7.28.   Plot of i_D/g_m versus v_{GS} for determining the pinchoff voltage V_{po}.

Example 7.18

A typical experimental plot of (7-190) is shown in Fig. 7.28. Note that most of the data falls on a straight line. Extrapolating the straight line to a zero value for i_D/g_m, the pinchoff voltage is seen to equal -2.2 V. Observe that this technique does not require knowledge of the contact potential ϕ.

7.4.6 Determination of the Parameters ρ and I_m

The control equation for $i(v_{GJ})$ is given by (7-168). Assuming that R_S, V_{po}, and ϕ have been previously determined, the only unknowns are ρ and I_m. To obtain these parameters, measurements are made of the drain current at two different values of gate-to-source voltage.

Provided the measurements are made at low enough frequencies that capacitive effects are negligible, the drain current is identical to $i(v_{GJ})$. If the two gate-to-source voltages are denoted by v_{GS1} and v_{GS2}, the corresponding values of v_{GJ} are given by

$$v_{GJ1} = v_{GS1} - i(v_{GJ1})R_S$$

$$v_{GJ2} = v_{GS2} - i(v_{GJ2})R_S. \tag{7-191}$$

Using these values in (7-168) to obtain expressions for $i(v_{GJ1})$ and $i(v_{GJ2})$ and dividing one by the other results in

$$\frac{i(v_{GJ1})}{i(v_{GJ2})} = \frac{-\rho + \sqrt{\rho^2 + 4\left[\dfrac{1}{3} - \left(\dfrac{v_{GJ1} + \phi}{V_{po} + \phi}\right) + \dfrac{2}{3}\left(\dfrac{v_{GJ1} + \phi}{V_{po} + \phi}\right)^{3/2}\right]}}{-\rho + \sqrt{\rho^2 + 4\left[\dfrac{1}{3} - \left(\dfrac{v_{GJ2} + \phi}{V_{po} + \phi}\right) + \dfrac{2}{3}\left(\dfrac{v_{GJ2} + \phi}{V_{po} + \phi}\right)^{3/2}\right]}}. \tag{7-192}$$

Because of the nonlinear nature of (7-192), an analytical expression for ρ is not possible. Nevertheless, an iterative scheme, such as Newton–Raphson iteration, can be used to determine ρ numerically. Once ρ has been found, I_m can be evaluated from

$$I_m = \frac{\dfrac{2i(v_{GJ1})}{3\rho}}{-\rho + \sqrt{\rho^2 + 4\left[\dfrac{1}{3} - \left(\dfrac{v_{GJ1} + \phi}{V_{po} + \phi}\right) + \dfrac{2}{3}\left(\dfrac{v_{GJ1} + \phi}{V_{po} + \phi}\right)^{3/2}\right]}}. \tag{7.193}$$

Possible values for the parameters are $\rho = 0.20$ and $I_m = 45$ mA.

7.5   METAL-OXIDE-SEMICONDUCTOR FIELD-EFFECT TRANSISTOR (MOSFET)

In addition to the JFET, there is a second type of field-effect transistor known as the MOSFET. The letters MOS are derived from the transistor's metal-oxide-semiconductor fabrication. MOSFET operation is basically similar to that of the JFET. Current flows between the source and drain terminals and is regulated by the gate-to-source voltage. MOSFET's operate either in depletion or enhancement modes depending upon whether majority-type carriers are depleted from an existing channel or a channel is induced by enhancing minority-type carriers into the semiconductor material between the source and drain. A depletion-type MOSFET can be operated in both the depletion and enhancement modes while the operation of the enhancement-type MOSFET is restricted to the enhancement mode. Because of the strong similarity between the JFET and MOSFET, only a brief discussion is presented in this section.

The schematic symbol for a p-channel enhancement MOSFET is shown in Fig. 7.29(a). The direction of the arrow is reversed for an n-channel MOSFET. When a MOSFET is operated in the depletion mode, the graphic symbol for the JFET is frequently used. A global model that is applicable to both depletion- and enhancement-type MOSFET's is presented in Fig. 7.29(b). This model is very similar to the JFET model given in Fig. 7.25(b). However, because measurements on several discrete MOSFET's reveal the gate-to-drain and gate-to-source capacitances to be essentially constant, they are modeled as linear capacitors for the MOSFET. Also, the functional dependence of the nonlinear voltage-controlled current source, $i(v_{GJ})$, is different for the MOSFET.

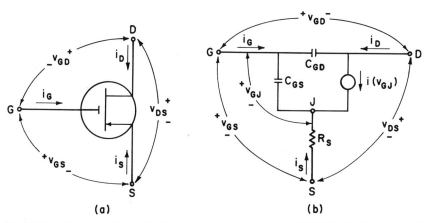

Fig. 7.29. (a) Schematic symbol for p-channel enhancement MOSFET. (b) Global model for both depletion- and enhancement-type MOSFET's.

Observe that the only nonlinear element in the MOSFET global model is $i(v_{GJ})$. As with the JFET, the MOSFET is operated in the saturation region when used as an amplifier. In the saturation region the drain-current dependence on the gate-to-source voltage is usually modeled in terms of a square-law relationship. However, it has been found experimentally that this is inadequate for predicting such interference effects as cross modulation and intermodulation. At present there are no well-developed theoretical nonlinear MOSFET models suitable for this purpose. Consequently, the incremental current for the nonlinear controlled source is modeled by the Taylor series

$$i(v_{gj}) = \sum_{k=1}^{\infty} g_k v_{gj}^k \qquad (7\text{-}194)$$

where the coefficients are determined empirically. A nonlinear incremental equivalent circuit for the MOSFET is shown in Fig. 7.30. At low enough frequencies, the two capacitors may be replaced by open circuits.

The parameters of the nonlinear incremental equivalent circuit are readily obtained. C_{GS} and C_{GD} are determined by direct measurements of the gate-to-source and gate-to-drain capacitances. As explained in Section 7.4.4, the source bulk resistance R_S is estimated from the incremental common-source input admittance y_{is}. Finally, the power series coefficients g_k are evaluated from measurements of the transconductance, g_m, as a function of the gate-to-source voltage. This latter procedure is described next.

By definition, the transconductance is given by

$$g_m = \left.\frac{\partial i_D}{\partial v_{GS}}\right|_{v_{DS} = \text{constant}} \approx \left.\frac{i_d}{v_{gs}}\right|_{v_{DS} = \text{constant}} . \qquad (7\text{-}195)$$

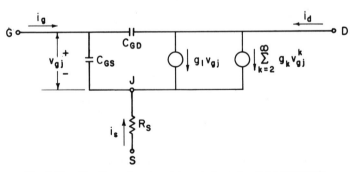

Fig. 7.30.   Nonlinear incremental equivalent circuit for MOSFET.

Therefore, the drain current at a specific value of v_{GS} is the area under the transconductance curve according to the relation

$$i_D = f(v_{GS}) = \int g_m(v_{GS})\, dv_{GS}. \tag{7-196}$$

Assuming the measurements are made at low enough frequencies such that i_d equals $-i_s$, it follows that

$$v_{gs} = v_{gj} + i_d R_S. \tag{7-197}$$

Substituting (7-194) into (7-197), we obtain

$$v_{gs} = v_{gj} + \left(\sum_{k=1}^{\infty} g_k v_{gj}^k \right) R_S$$

$$= (1 + g_1 R_S)\, v_{gj} + g_2 R_S v_{gj}^2 + g_3 R_S v_{gj}^3 + \cdots. \tag{7-198}$$

Using the series reversion technique described in Section 5.1 results in

$$v_{gj} = \sum_{k=1}^{\infty} a_k v_{gs}^k \tag{7-199}$$

where the first three coefficients are given by

$$a_1 = \frac{1}{1 + g_1 R_S}$$

$$a_2 = \frac{-g_2 R_S}{(1 + g_1 R_S)^3}$$

$$a_3 = \frac{2(g_2 R_S)^2 - (1 + g_1 R_S) g_3 R_S}{(1 + g_1 R_S)^5}. \tag{7-200}$$

On the other hand, expansion of (7-196) in a Taylor series about the quiescent operating point yields

$$i_d = \sum_{k=1}^{\infty} b_k v_{gs}^k. \tag{7-201}$$

Substituting (7-201) into (7-197) and solving for v_{gj}, we have

$$v_{gj} = (1 - b_1 R_S)\, v_{gs} - b_2 R_S v_{gs}^2 - b_3 R_S v_{gs}^3 - \cdots. \tag{7-202}$$

Equating the coefficients in (7-199) and (7-202) and making use of (7-200), it is concluded that

$$g_1 = \frac{b_1}{1 - b_1 R_S}$$

$$g_2 = \frac{b_2}{(1 - b_1 R_S)^3}$$

$$g_3 = \frac{b_3}{(1 - b_1 R_S)^4} + \frac{2 b_2^2 R_S}{(1 - b_1 R_S)^5}. \tag{7-203}$$

Higher order coefficients are determined in like manner.

The procedure, therefore, consists first of measuring the transconductance, g_m, as a function of the gate-to-source voltage. A least-squares polynomial is then fitted to the curve. Since fifth-order terms may be significant, it is recommended that the polynomial be at least of fifth degree. This expression is then integrated mathematically to obtain the i_D versus v_{GS} curve. The coefficients b_k are determined next. These are then used in (7-203) in order to obtain the coefficients g_k that are required for the nonlinear incremental equivalent circuit. The transconductance curve is measured in preference to a direct measurement of the i_D versus v_{GS} curve since the higher order nonlinearities are more accurately determined by this technique.

APPENDIX A

Response of Weakly Nonlinear Circuits to Arbitrary Inputs

For the purpose of mathematical simplicity, discussion in the main body of the text is restricted to the sinusoidal steady-state analysis of weakly nonlinear circuits. However, the Volterra approach is applicable to more general excitations. The response of weakly nonlinear circuits to arbitrary inputs is considered in this Appendix. To simplify the presentation, the general input–output relation is developed with the aid of several intuitive plausibility arguments. We begin our discussion by reviewing the impulse, an "ideal" function which plays an important role in the development.

A.1 THE IMPULSE FUNCTION

A unit impulse, occurring at $t = 0$, can be interpreted as the limiting form of the rectangular pulse, $p(t)$, plotted in Fig. A.1. Note that the area under $p(t)$ equals unity irrespective of the value of $\Delta\tau$. In the limit, as $\Delta\tau$ approaches zero, the height of the rectangular pulse approaches infinity while its width approaches zero. Nevertheless, throughout the limiting process the area remains equal to unity. The unit impulse is "defined" to be the limiting form of $p(t)$ and is denoted by $\delta(t)$. An impulse of area A, occurring at $t = t_1$, is denoted by $A\delta(t - t_1)$ and is plotted as shown in Fig. A.2. Observe that $A\delta(t - t_1)$ is the limiting form of

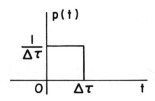

Fig. A.1.   Rectangular pulse whose limit approaches a unit impulse as $\Delta\tau$ approaches zero.

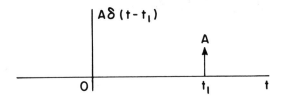

Fig. A.2.   Impulse of area A which occurs at $t = t_1$.

$Ap(t - t_1)$. Since integration can be interpreted in terms of the area under the integrand, it follows that

$$\int_{-\infty}^{\infty} A\delta(t - t_1) \, dt = A. \tag{A-1}$$

Another useful property of the impulse is that

$$\int_{-\infty}^{\infty} g(t) \, A\delta(t - t_1) \, dt = Ag(t_1) \tag{A-2}$$

for each function $g(t)$ which is continuous at $t = t_1$. This result is readily demonstrated. Consider $g(t)$ and $Ap(t - t_1)$ as plotted in Fig. A.3. Assume $\Delta\tau$ is sufficiently small so that $g(t)$ approximately equals $g(t_1)$ over the interval $t_1 \leqslant t \leqslant t_1 + \Delta\tau$. An approximation to the integrand in (A-2), which is given by the product of $g(t)$ and $Ap(t - t_1)$, is shown in Fig. A.4. In terms of the area

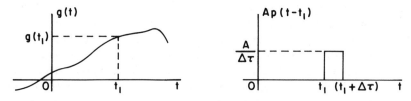

Fig. A.3.   Plots of $g(t)$ and $Ap(t - t_1)$.

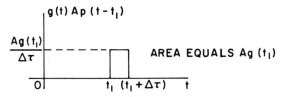

Fig. A.4. Approximation to the integrand in (A-2).

under the integrand, it is concluded that

$$\int_{-\infty}^{\infty} g(t)\, A\delta(t - t_1)\, dt \approx \int_{-\infty}^{\infty} g(t)\, Ap(t - t_1)\, dt = Ag(t_1). \qquad \text{(A-3)}$$

We see that an integral which contains an impulse in the integrand is especially easy to evaluate.

A.2 THE CONVOLUTION INTEGRAL FOR LINEAR CIRCUITS

The input–output relation for a weakly nonlinear circuit can be viewed as a generalization of the convolution integral which arises in the analysis of linear circuits. Consequently, the convolution integral for linear circuits is developed in this section.

Consider a linear circuit with input $x(t)$ and output $y(t)$ as shown in Fig. A.5. Assuming all initial conditions of the circuit to be zero, let the circuit response be denoted by $h(t)$ for the special case in which the excitation is the unit impulse $\delta(t)$. $h(t)$ is referred to as the impulse response. A linear time-invariant circuit has the property that the response to the input $A\delta(t - t_0)$ is given by $Ah(t - t_0)$. In other words, the response to an impulse which is delayed in time by an amount t_0 and has area A is the impulse response delayed by t_0 and multiplied by the constant A.

Now consider a typical input $x(t)$, as shown in Fig. A.6. As indicated in the figure, let $x(t)$ be approximated by a succession of elementary rectangular pulses, each adjacent to the next, and applied at intervals $\Delta\tau$ along the time axis. In

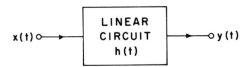

Fig. A.5. Linear circuit with impulse response $h(t)$.

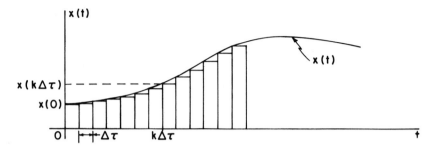

Fig. A.6.   Approximation to $x(t)$ by a succession of rectangular pulses.

terms of the rectangular pulse of Fig. A.1, the input can be approximated as

$$x(t) \approx \sum_{k=0}^{\infty} x(k\Delta\tau)\, \Delta\tau p(t - k\Delta\tau). \tag{A-4}$$

Observe that the kth term in the summation corresponds to a rectangular pulse of width $\Delta\tau$ whose leading edge occurs at $k\Delta\tau$ and whose height is $x(k\Delta\tau)$.

Since the circuit is linear, superposition applies and the total response is the sum of the responses to the individual pulses. For sufficiently small values of $\Delta\tau$, $p(t)$ may be interpreted as an adequate approximation to the unit impulse $\delta(t)$. Therefore, the response to $p(t)$ may be approximated by the impulse response $h(t)$ and the response to $x(k\Delta\tau)\,\Delta\tau p(t - k\Delta\tau)$ may be approximated by $x(k\Delta\tau)\,\Delta\tau h(t - k\Delta\tau)$. Let K denote the number of rectangular pulses occurring between 0 and some positive time t. It follows that the circuit output at time t may be approximated as

$$y(t) \approx \sum_{k=0}^{K} x(k\Delta\tau)\, h(t - k\Delta\tau)\, \Delta\tau. \tag{A-5}$$

The approximation becomes more accurate as the pulsewidth approaches zero.

In particular, let $\Delta\tau$ decrease and k increase in such a manner that their product is equal to ν. In the limit, as $\Delta\tau$ approaches the differential element $d\nu$, the summation in (A-5) becomes an integration and the response at time t is given by

$$y(t) = \int_{0}^{t} x(\nu)\, h(t - \nu)\, d\nu. \tag{A-6}$$

Equation (A-6) is a convolution integral. We say that the response of a linear time-invariant circuit can be obtained by convolving the input with the impulse response. The impulse response is seen to be a complete characterization of the linear circuit in the sense that knowledge of the impulse response is sufficient to enable determination of the circuit response to any input.

It is convenient to extend the limits of integration in (A-6). Since the input is assumed to be zero for negative time, $x(\nu) = 0$ for $\nu < 0$ and the lower limit may be extended to $-\infty$. In addition, a realizable circuit does not respond to an impulse before it is applied. Therefore, the impulse response is zero for negative arguments. This implies that $h(t - \nu) = 0$ for $\nu > t$ and the upper limit in (A-6) may be extended to $+\infty$. Consequently, the response can be rewritten as

$$y(t) = \int_{-\infty}^{\infty} x(\nu)\, h(t - \nu)\, d\nu. \tag{A-7}$$

Now consider the change of variables, $\tau = t - \nu$. Note that $\nu = t - \tau$, $d\nu = -d\tau$, $\tau = -\infty$, when $\nu = +\infty$, and $\tau = +\infty$ when $\nu = -\infty$. It follows that the response can also be expressed as

$$y(t) = \int_{-\infty}^{\infty} h(\tau)\, x(t - \tau)\, d\tau. \tag{A-8}$$

Equation (A-8) is also in the form of a convolution integral.

The approach used in this section is readily extended to weakly nonlinear circuits. For this purpose, it is desirable to examine (A-5) from a different point of view. Let

$$x_k = x(k\Delta\tau)$$

$$a_k = h(t - k\Delta\tau)\, \Delta\tau. \tag{A-9}$$

In terms of the new notation, (A-5) becomes

$$y(t) \approx \sum_{k=0}^{K} a_k x_k. \tag{A-10}$$

Observe that x_k is a parameter of the input signal whereas, for a specified value of t, a_k is a parameter of the circuit. Therefore, the response, as given by (A-10), can be interpreted as a function of the $(K + 1)$ input variables $x_0, x_1, \ldots, x_K$. This is emphasized by expressing the response as

$$y(t) \approx f(x_0, x_1, \ldots, x_K). \tag{A-11}$$

For a linear system, $f(\cdot)$ is a linear function of the $(K + 1)$ input variables, as given by (A-10). However, for a nonlinear system $f(\cdot)$ is a nonlinear function of the $(K + 1)$ input variables. This observation is the starting point for development of the Volterra series representation given in the next section.

A.3 VOLTERRA SERIES REPRESENTATION OF WEAKLY NONLINEAR CIRCUITS

Consider a weakly nonlinear circuit with input $x(t)$ and output $y(t)$, as shown in Fig. A.7. Let the input be approximated by a succession of elementary rectangular pulses, as illustrated in Fig. A.6. As before, denote the height of the kth pulse by

$$x_k = x(k\Delta\tau) \tag{A-12}$$

and let K denote the number of rectangular pulses occurring between 0 and some positive time t. As with the linear system, it is reasonable to expect the response to be a function of the $(K + 1)$ input variables $x_0, x_1, \ldots, x_K$. Consequently, we write

$$y(t) \approx f(x_0, x_1, \ldots, x_K) \tag{A-13}$$

where $f(\cdot)$ is now a nonlinear function of the $(K + 1)$ input variables.

Assuming $f(\cdot)$ can be expanded into a $(K + 1)$-dimensional Taylor series, it is convenient to group terms having identical order. In particular, $y(t)$ can be expressed in the form

$$y(t) \approx \sum_{k_1=0}^{K} a_{k_1} x_{k_1} + \sum_{k_1=0}^{K} \sum_{k_2=0}^{K} a_{k_1 k_2} x_{k_1} x_{k_2}$$

$$+ \sum_{k_1=0}^{K} \sum_{k_2=0}^{K} \sum_{k_3=0}^{K} a_{k_1 k_2 k_3} x_{k_1} x_{k_2} x_{k_3} + \cdots. \tag{A-14}$$

Alternatively, the output may be written as

$$y(t) = \sum_{n=1}^{\infty} y_n(t) \tag{A-15}$$

where the nth-order term is approximated by

$$y_n(t) \approx \sum_{k_1=0}^{K} \sum_{k_2=0}^{K} \cdots \sum_{k_n=0}^{K} a_{k_1 k_2 \cdots k_n} x_{k_1} x_{k_2} \cdots x_{k_n}. \tag{A-16}$$

$$x(t) \circ\!\!\longrightarrow \boxed{\begin{array}{c} \textbf{WEAKLY} \\ \textbf{NONLINEAR} \\ \textbf{CIRCUIT} \end{array}} \longrightarrow\!\circ y(t)$$

Fig. A.7. Weakly nonlinear circuit with input $x(t)$ and output $y(t)$.

For a linear circuit, $y_n(t) = 0$ for $n \geqslant 2$ and the total response is given by

$$y(t) = y_1(t) \approx \sum_{k_1=0}^{K} a_{k_1} x_{k_1} . \tag{A-17}$$

This agrees with the result obtained in (A-10).

The general term in the approximation to $y_n(t)$, given by

$$a_{k_1 k_2 \cdots k_n} x_{k_1} x_{k_2} \cdots x_{k_n},$$

can be interpreted as the nth-order response to n elementary rectangular pulses applied at $k_1 \Delta\tau, k_2 \Delta\tau, \ldots, k_n \Delta\tau$. Consequently, in analogy with development of the convolution integral for linear circuits, it is reasonable to let

$$a_{k_1 k_2 \cdots k_n} = h_n(t - k_1 \Delta\tau, t - k_2 \Delta\tau, \ldots, t - k_n \Delta\tau)(\Delta\tau)^n . \tag{A-18}$$

Substitution of (A-12) and (A-18) into (A-16) results in

$$y_n(t) \approx \sum_{k_1=0}^{K} \sum_{k_2=0}^{K} \cdots \sum_{k_n=0}^{K} h_n(t - k_1 \Delta\tau, t - k_2 \Delta\tau, \ldots, t - k_n \Delta\tau)$$
$$\cdot x(k_1 \Delta\tau) x(k_2 \Delta\tau) \cdots x(k_n \Delta\tau)(\Delta\tau)^n . \tag{A-19}$$

We now consider the limit as $\Delta\tau$ approaches zero. Specifically, for $j = 1, 2, \ldots, n$, let $\Delta\tau$ decrease and k_j increase in such a manner that their product is equal to v_j. In the limit, the k-fold summation in (A-19) reduces to a k-fold integration and the nth-order term becomes

$$y_n(t) = \int_0^t \int_0^t \cdots \int_0^t h_n(t - v_1, t - v_2, \ldots, t - v_n) x(v_1) x(v_2)$$
$$\cdots x(v_n) \, dv_1 \, dv_2 \cdots dv_n . \tag{A-20}$$

With this integral representation for $y_n(t)$, the infinite series in (A-15) is known as a Volterra series. Observe that

$$y_1(t) = \int_0^t h_1(t - v_1) x(v_1) \, dv_1 \tag{A-21}$$

which is the form of the response obtained in (A-6) for a linear circuit.

Since the input is assumed to be zero for negative time, the lower limits of integration in (A-20) may be extended to $-\infty$. In addition, for a realizable circuit $h_n(t - v_1, t - v_2, \ldots, t - v_n) = 0$ whenever any of its n arguments are negative. Therefore, the upper limits in (A-20) may be extended to $+\infty$. Hence, the nth-order term can be expressed as

$$y_n(t) = \int_{-\infty}^{\infty} \int_{-\infty}^{\infty} \cdots \int_{-\infty}^{\infty} h_n(t - \nu_1, t - \nu_2, \ldots, t - \nu_n) x(\nu_1) x(\nu_2)$$

$$\cdots x(\nu_n) \, d\nu_1 \, d\nu_2 \cdots d\nu_n. \qquad (A\text{-}22)$$

An alternate form for $y_n(t)$ is obtained through the changes of variables $\tau_j = t - \nu_j, j = 1, 2, \ldots, n$. Then

$$y_n(t) = \int_{-\infty}^{\infty} \int_{-\infty}^{\infty} \cdots \int_{-\infty}^{\infty} h_n(\tau_1, \tau_2, \ldots, \tau_n) x(t - \tau_1) x(t - \tau_2)$$

$$\cdots x(t - \tau_n) \, d\tau_1 \, d\tau_2 \cdots d\tau_n. \qquad (A\text{-}23)$$

This is the conventional form for the nth-order term in a Volterra functional series. $y_n(t)$ is of nth-order in the sense that multiplication of the input $x(t)$ by a constant A results in multiplication of $y_n(t)$ by the constant A^n.

In practice, the Volterra series representation is of most use when the total response can be adequately approximated by including only the first few terms of the series. The output can then be expressed as the finite sum

$$y(t) = \sum_{n=1}^{N} y_n(t) \qquad (A\text{-}24)$$

where the nth-order term is given by (A-23) and terms above Nth-order have been omitted from the infinite series because they are assumed to contribute negligibly to the output. It follows that a weakly nonlinear circuit can be modeled as shown in Fig. A.8. This model consists of the parallel combination of N blocks with each block having, as a common input, the circuit excitation $x(t)$. The total response is obtained by summing the outputs of the individual blocks. The Volterra kernels, $h_n(\tau_1, \tau_2, \ldots, \tau_n)$, are determined solely by the weakly nonlinear circuit and are independent of the circuit excitation. The invariance of the Volterra kernels to circuit input is a highly desirable feature of the approach. The first N Volterra kernels completely characterize the weakly nonlinear circuit in the sense that knowledge of the Volterra kernels is sufficient to enable determination of the circuit response to any input. Of particular interest is the response of weakly nonlinear circuits to sinusoidal inputs. This is discussed next.

A.4   SINUSOIDAL INPUTS AND NONLINEAR TRANSFER FUNCTIONS

To study the sinusoidal steady-state response of weakly nonlinear circuits, let the input to the model shown in Fig. A.8 be the sum of Q sinusoidal signals. With reference to equations (3-3) through (3-6) the excitation can be expressed

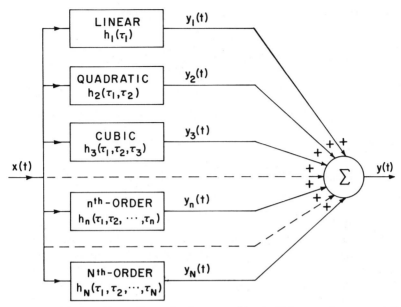

Fig. A.8. Model of weakly nonlinear circuit suggested by the Volterra series representation.

as

$$x(t) = \sum_{q=1}^{Q} |E_q| \cos (2\pi f_q t + \theta_q) = \tfrac{1}{2} \sum_{q=-Q}^{Q} E_q \exp (j2\pi f_q t). \quad \text{(A-25)}$$

Substitution of (A-25) into (A-23) results in

$$y_n(t) = \frac{1}{2^n} \int_{-\infty}^{\infty} \int_{-\infty}^{\infty} \cdots \int_{-\infty}^{\infty} h_n(\tau_1, \tau_2, \ldots, \tau_n) \sum_{q_1=-Q}^{Q} E_{q_1} \exp [j2\pi f_{q_1}(t - \tau_1)]$$

$$\cdot \sum_{q_2=-Q}^{Q} E_{q_2} \exp [j2\pi f_{q_2}(t - \tau_2)] \cdots \sum_{q_n=-Q}^{Q} E_{q_n}$$

$$\cdot \exp [j2\pi f_{q_n}(t - \tau_n)] \, d\tau_1 \, d\tau_2 \cdots d\tau_n. \quad \text{(A-26)}$$

Interchanging the order of summation and integration and rearranging terms, $y_n(t)$ becomes

$$y_n(t) = \frac{1}{2^n} \sum_{q_1=-Q}^{Q} \sum_{q_2=-Q}^{Q} \cdots \sum_{q_n=-Q}^{Q} E_{q_1} E_{q_2} \cdots E_{q_n}$$

$$\cdot \exp [j2\pi (f_{q_1} + f_{q_2} + \cdots + f_{q_n}) t] \int_{-\infty}^{\infty} \int_{-\infty}^{\infty} \cdots \int_{-\infty}^{\infty} h_n(\tau_1, \tau_2, \ldots, \tau_n)$$

$$\cdot \exp [-j2\pi (f_{q_1}\tau_1 + f_{q_2}\tau_2 + \cdots + f_{q_n}\tau_n)] \, d\tau_1 \, d\tau_2 \cdots d\tau_n. \quad \text{(A-27)}$$

The nth-order nonlinear transfer function is defined to be

$$H_n(f_{q_1}, f_{q_2}, \ldots, f_{q_n}) = \int_{-\infty}^{\infty} \int_{-\infty}^{\infty} \cdots \int_{-\infty}^{\infty} h_n(\tau_1, \tau_2, \ldots, \tau_n)$$

$$\cdot \exp\left[-j2\pi(f_{q_1}\tau_1 + f_{q_2}\tau_2 + \cdots + f_{q_n}\tau_n)\right]$$

$$\cdot d\tau_1 \, d\tau_2 \cdots d\tau_n. \tag{A-28}$$

Consequently, the nth-order response may be rewritten as

$$y_n(t) = \frac{1}{2^n} \sum_{q_1=-Q}^{Q} \cdots \sum_{q_n=-Q}^{Q} E_{q_1} \cdots E_{q_n} H_n(f_{q_1}, \ldots, f_{q_n})$$

$$\cdot \exp\left[j2\pi(f_{q_1} + \cdots + f_{q_n})\, t\right]. \tag{A-29}$$

This is identical to (4-7) which served as a starting point in our discussion of the nonlinear transfer function approach to the sinusoidal steady-state analysis of weakly nonlinear circuits.

Observe that the first-order transfer function is given by

$$H_1(f_1) = \int_{-\infty}^{\infty} h_1(\tau_1) \exp\left[-j2\pi f_1 \tau_1\right] d\tau_1. \tag{A-30}$$

This is recognized as the Fourier transform of the impulse response of the linear portion of the circuit. Therefore, $H_1(f_1)$ is a conventional linear transfer function. With reference to (A-2), $H_n(f_1, f_2, \ldots, f_n)$ is the n-dimensional Fourier transform of $h_n(\tau_1, \tau_2, \ldots, \tau_n)$. As a matter of fact, $h_n(\tau_1, \tau_2, \ldots, \tau_n)$ can be recovered from $H_n(f_1, f_2, \ldots, f_n)$ according to the inverse Fourier transform relation

$$h_n(\tau_1, \tau_2, \ldots, \tau_n) = \int_{-\infty}^{\infty} \int_{-\infty}^{\infty} \cdots \int_{-\infty}^{\infty} H_n(f_1, f_2, \ldots, f_n)$$

$$\cdot \exp\left[j2\pi(f_1\tau_1 + f_2\tau_2 + \cdots + f_n\tau_n)\right] df_1 \, df_2 \cdots df_n. \tag{A-31}$$

Therefore, specification of the Volterra kernels, $h_n(\tau_1, \tau_2, \ldots, \tau_n)$, is equivalent to specification of the nonlinear transfer functions, $H_n(f_1, f_2, \ldots, f_n)$, and vice versa. As a result, the model of Fig. A.8 is, in every way, equivalent to the model of Fig. 4.1.

Bibliography

1. VOLTERRA SERIES—THEORY

Alper, P., "A Consideration of the Discrete Volterra Series," *IEEE Trans. Automat. Contr.*, Vol. AC-10, pp. 322–327, July 1965.

Ambati, S. and H. A. Barker, "Nonlinear Sampled-Data System Analysis by Multidimensional Z-Transforms," *Proc. Inst. Elec. Eng.*, Vol. 119, No. 9, pp. 1407–1413, 1972.

Barrett, J. F., "The Use of Functionals in the Analysis of Nonlinear Physical Systems," *J. Electron. Contr.*, Vol. 15, No. 6, pp. 567–615, Dec. 1963.

Barrett, J. F., "The Use of Volterra Series to Find the Region of Stability of a Nonlinear Differential Equation," *Int. J. Contr.*, Vol. 1, No. 3, pp. 209–216, March 1965.

Bedrosian, E. and S. O. Rice, "The Output Properties of Volterra Systems Driven by Harmonic and Gaussian Inputs," *Proc. IEEE*, Vol. 59, pp. 1688–1707, Dec. 1971.

Brilliant, M. B., "Theory of the Analysis of Nonlinear Systems," M.I.T. Res. Lab. of Electron., Tech. Rep. 345, March 3, 1958.

Bush, A. M., "Some Techniques for the Synthesis of Nonlinear Systems," M.I.T. Res. Lab. of Electron., Tech. Rep. 441, March 25, 1966.

Chesler, D. A., "Nonlinear Systems with Gaussian Inputs," M.I.T. Res. Lab. of Electron., Tech. Rep. 366, Feb. 15, 1960.

Crum, L. A. and J. A. Heinen, "Simultaneous Reduction and Expansion of Multidimensional Laplace Transformation Kernels," *SIAM J. Appl. Math.*, Vol. 26, No. 4, pp. 753–771, 1974.

Deutsch, R., *Non-Linear Transformations of Random Processes.* Englewood Cliffs, N.J.: Prentice-Hall, Inc., 1962.

Flake, R. F., "Volterra Series Representation of Nonlinear Systems," *AIEE Trans. Appl. Ind.*, Vol. 81, pp. 330–335, Jan. 1963.

Frank, P. and R. McFee, "Determining Input–Output Relationships of Nonlinear Systems by Inversions," *IEEE Trans. Circuit Theory*, Vol. CT-10, pp. 168–180, June 1963.

Fu, F. C. and J. B. Farison, "Analysis of a Class of Nonlinear Discrete-Time Systems by the Volterra Series," *Int. J. Contr.*, Vol. 18, No. 3, pp. 545–551, March 1973.

Fu, F. C. and J. B. Farison, "On the Volterra Series Functional Evaluation of the Response of Nonlinear Discrete-Time Systems," *Int. J. Contr.*, Vol. 18, No. 3, pp. 553–558, March 1973.

Fu, F. C. and J. B. Farison, "On the Volterra Series Functional Identification of Nonlinear Discrete-Time Systems," *Int. J. Contr.*, Vol. 18, No. 6, pp. 1281–1289, Dec. 1973.

George, D. A., "Continuous Nonlinear Systems," M.I.T. Res. Lab. of Electron., Tech. Rep. 355, July 24, 1959.

Halme, A. and J. Orava, "Generalized Polynomial Operators for Nonlinear Systems Analysis," *IEEE Trans. Automat. Contr.*, Vol. AC-17, pp. 226–228, April 1972.

Ku, Y. H. and A. A. Wolf, "Volterra-Wiener Functionals for the Analysis of Nonlinear Systems," *J. Franklin Inst.*, Vol. 281, No. 1, pp. 9–26, Jan. 1966.

Ku, Y. H. and C. C. Su, "Volterra Functional Analysis of Nonlinear Time-Varying Systems," *J. Franklin Inst.*, Vol. 284, No. 6, pp. 344–365, Dec. 1967.

McFee, R., "Determining the Response of Nonlinear Systems to Arbitrary Inputs," *AIEE Trans.*, Vol. 80, pp. 189–193, Sept. 1961.

Palm, G. and T. Poggio, "The Volterra Representation and the Wiener Expansion: Validity and Pitfalls," *SIAM J. Appl. Math.*, Vol. 33, No. 2, pp. 195–216, Sept. 1977.

Parente, R. B., "Functional Analysis of Systems Characterized by Nonlinear Differential Equations," M.I.T. Res. Lab. of Electron., Tech. Rep. 444, July 15, 1966.

Parente, R. B., "Nonlinear Differential Equations and Analytic System Theory," *SIAM J. Appl. Math.*, Vol. 18, No. 1, pp. 41–66, Jan. 1970.

Rice, S. O., "Volterra Systems with More Than One Input Port-Distortion in a Frequency Converter," *Bell Syst. Tech. J.*, Vol. 52, No. 8, pp. 1255–1270, Oct. 1973.

Schetzen, M., "A Theory of Nonlinear Systems Identification," *Int. J. Contr.*, Vol. 20, No. 4, pp. 577–592, Oct. 1974.

Schetzen, M., "Theory of *P*th Order Inverses of Nonlinear Systems," *IEEE Trans. Circuits and Syst.*, Vol. CAS-23, No. 5, pp. 285–294, May 1976.

Smets, H. B., "Analysis and Synthesis of Nonlinear Systems," *IRE Trans. Circuit Theory*, Vol. CT-7, pp. 459–469, Dec. 1960.

Volterra, V., *Theory of Functionals and of Integral and Integro-Differential Equations*. New York: Dover, 1959.

Wiener, N., *Nonlinear Problems in Random Theory*. Cambridge, Mass.: M.I.T. Press, 1958.

Zadeh, L. A., "Optimum Nonlinear Filters," *J. Appl. Phys.*, Vol. 24, No. 4, pp. 396–404, April 1953.

Zadeh, L. A., "A Contribution to the Theory of Nonlinear Systems," *J. Franklin Inst.*, Vol. 254, pp. 387–408, May 1953.

Zames, G., "Nonlinear Operators for System Analysis," M.I.T. Res. Lab. of Electron., Tech. Rep. 370, Aug. 25, 1960.

Zames, G., "Functional Analysis Applied to Nonlinear Feedback Systems," *IEEE Trans. Circuit Theory*, Vol. CT-19, pp. 392–404, Sept. 1963.

2. VOLTERRA SERIES—APPLICATIONS

Bedrosian, E. and S. O. Rice, "Distortion and Crosstalk of Linearly Filtered Angle-Modulated Signals," *Proc. IEEE*, Vol. 56, No. 1, pp. 2–13, Jan. 1968.

Bussgang, J. J., L. Ehrman, and J. W. Graham, "Analysis of Nonlinear Systems with Multiple Inputs," *Proc. IEEE*, Vol. 62, No. 8, pp. 1088–1119, Aug. 1974.

Chang, K. Y., "Intermodulation Noise and Products Due to Frequency Dependent Nonlinearities in CATV Systems," *IEEE Trans. Commun.*, Vol. COM-23, No. 1, pp. 142–155, Jan. 1975.

Crippa, G., "Evaluation of Distortion and Intermodulation in Nonlinear Transmission Systems by Means of Volterra Series Expansion," *Alta Freq.*, Vol. 38, pp. 332–336, May 1969.

Goldman, J., "A Volterra Series Description of Crosstalk Interference in Communications Systems," *Bell Syst. Tech. J.*, Vol. 52, No. 5, pp. 649–668, May-June 1973.

Graham, J. W. and L. Ehrman, "Nonlinear System Modeling and Analysis with Applications to Communications Receivers," Rome Air Development Center, Rome, N.Y., Tech. Rep. RADC-TR-73-178, DDC Document AD 766 278, June 1973.

Kuo, Y. L., "Frequency-Domain Analysis of Weakly Nonlinear Networks, Part 1," *IEEE Circuits and Syst. Soc. Mag.*, pp. 2–8, Aug. 1977.

Kuo, Y. L., "Frequency-Domain Analysis of Weakly Nonlinear Networks, Part 2," *IEEE Circuits and Syst. Soc. Mag.*, pp. 2–6, Oct. 1977.

Landau, M. and C. T. Leondes, "Application of the Volterra Series to the Analysis and Design of an Angle Track Loop," *IEEE Trans. Aerosp. Electron. Syst.*, Vol. AES-8, No. 3, pp. 306–318, May 1972.

Lawless, W. J. and M. Schwartz, "Binary Signaling over Channels Containing Quadratic Nonlinearities," *IEEE Trans Commun.*, Vol. COM-22, No. 3, pp. 288–297, March 1974.

Lee, Y. W. and M. Schetzen, "Measurement of the Wiener Kernels of a Nonlinear System by Cross-Correlation," *Int. J. Contr.*, Vol. 2, No. 3, pp. 237–254, Sept. 1965.

Marchesini, G. and G. Picci, "On the Functional Identification of Nonlinear Systems from Input-Output Data Records," *IEEE Trans. Automat. Contr.*, Vol. AC-14, pp. 757–759, Dec. 1969.

Maurer, R. E. and S. Narayanan, "Noise Loading Analysis of a Third-Order Nonlinear System With Memory," *IEEE Trans. Commun. Technol.*, Vol. COM-16, No. 5, pp. 701–712, Oct. 1968.

Meyer, R. G., M. J. Shensa, and R. Eschenbach, "Cross Modulation and Intermodulation in Amplifiers at High Frequencies," *IEEE J. Solid-State Circuits*, Vol. SC-7, No. 1, pp. 16–23, Feb. 1972.

Mircea, A. and H. Sinnreich, "Distortion Noise in Frequency-Dependent Nonlinear Networks," *Proc. Inst. Elec. Eng.*, Vol. 116, pp. 1644–1648, Oct. 1969.

Narayanan, S., "Transistor Distortion Analysis Using Volterra Series Representation," *Bell Syst. Tech. J.*, Vol. 46, pp. 991–1024, May-June 1967.

Narayanan, S., "Intermodulation Distortion of Cascaded Transistors," *IEEE J. Solid-State Circuits*, Vol. SC-4, No. 3, pp. 97–106, June 1969.

Narayanan, S., "Application of Volterra Series to Intermodulation Distortion Analysis of Transistor Feedback Amplifiers," *IEEE Trans. Circuit Theory*, Vol. CT-17, No. 4, pp. 518–527, Nov. 1970.

Narayanan, S. and H. C. Poon, "An Analysis of Distortion in Bipolar Transistors Using Integral Charge-Control Model and Volterra Series," *IEEE Trans. Circuit Theory*, Vol. CT-20, No. 4, pp. 341–351, July 1973.

Rashed, A. M. H. and G. S. Christenson, "Response of Nonlinear Sampled-Data Systems with Nonzero Initial Conditions and Response for In-Between Sampling Via Volterra Series," *IEEE Trans. Automat. Contr.*, Vol. AC-16, No. 3, pp. 269–270, June 1971.

Rudko, M. and D. D. Weiner, "Volterra Systems with Random Inputs: A Formalized Approach," *IEEE Trans. Commun.*, Vol. COM-26, No. 2, pp. 217–227, Feb. 1978.

Sarkar, T. and D. D. Weiner, "Scattering Analysis of Nonlinearly Loaded Antennas," *IEEE Trans. Antennas Propagat.*, Vol. AP-24, No. 2, pp. 125–131, March 1976.

Sarkar, T., D. D. Weiner, and R. Harrington, "Analysis of Nonlinearly Loaded Multiport Antenna Structures Over an Imperfect Ground Plane Using the Volterra Series Method," *IEEE Trans. Electromagn. Compat.*, Vol. EMC-20, No. 2, pp. 278–287, May 1978.

Schetzen, M., "Measurement of the Kernels of a Nonlinear System of Finite Order," *Int. J. Contr.*, Vol. 1, No. 3, pp. 251–263, March 1965.

Schetzen, M., "Synthesis of a Class of Nonlinear Systems," *Int. J. Contr.*, Vol. 1, pp. 401–414, May 1965.

Schetzen, M., "Power-Series Equivalence of Some Functional Series with Applications," *IEEE Trans. Circuit Theory*, Vol. CT-17, pp. 305–313, 1970.

Shanmugam, K. S., "Analysis and Synthesis of a Class of Nonlinear Systems," *IEEE Trans. Circuits and Systems*, Vol. CAS-23, No. 1, pp. 17–25, Jan. 1976.

Thomas, E. J., "Some Considerations on the Application of the Volterra Representation of Nonlinear Networks in Adaptive Echo Cancellers," *Bell Syst. Tech. J.*, Vol. 50, No. 8, pp. 2797–2805, Oct. 1971.

Van Trees, H. L., "Functional Techniques for the Analysis of the Nonlinear Behavior of Phase-Locked Loops," *Proc. IEEE*, Vol. 52, No. 8, pp. 894–911, Aug. 1964.

Weiner, D. D. and G. Naditch, "A Scattering Variable Approach to the Volterra Analysis of Nonlinear Systems," *IEEE Trans. Microwave Theory Tech.*, Vol. MTT-24, No. 7, pp. 422–433, July 1976.

3. SEMICONDUCTOR AND VACUUM-TUBE DEVICE PHYSICS AND MODELING

Abraham, H. E. and R. G. Meyer, "Transistor Design for Low Distortion at High Frequencies," *IEEE Trans. Electron Devices*, Vol. ED-23, No. 12, pp. 1290–1297, Dec. 1976.

Cobbold, R. S. C., *Theory and Applications of Field-Effect Transistors*. New York: John Wiley & Sons, Inc., 1970.

Fair, R. B., "Harmonic Distortion in the Junction Field-Effect Transistor with Field-Dependent Mobility," *IEEE Trans. Electron Devices*, Vol. ED-19, No. 1, pp. 9–13, Jan. 1972.

Gray, P. E., D. DeWitt, A. R. Boothroyd, and J. F. Gibbons, *Physical Electronics and Circuit Models of Transistors*. New York: John Wiley & Sons, Inc., 1964.

Gray, T. S., *Applied Electronics*. New York: John Wiley & Sons, Inc., 1954.

Gummel, H. K., "A Charge Control Relation for Bipolar Transistors," *Bell Syst. Tech. J.*, Vol. 49, pp. 115–120, Jan. 1970.

Gummel, H. K. and H. C. Poon, "An Integral Charge-Control Model of Bipolar Transistors," *Bell Syst. Tech. J.*, Vol. 49, pp. 827–852, May-June 1970.

Hamilton, D. J., F. A. Lindholm, and A. H. Marshak, *Principles and Applications of Semiconductor Device Modeling*. New York: Holt, Rinehart, and Winston, Inc., 1971.

Lindholm, F. A., "Unified Modeling of Field-Effect Devices," *IEEE J. Solid-State Circuits*, Vol. SC-6, No. 4, pp. 250–259, Aug. 1971.

Linvill, J. G., *Models of Transistors and Diodes*. New York: McGraw-Hill Book Company, 1963.

Lotsch, H., "Theory of Nonlinear Distortion Produced in Semiconductor Diodes," *IEEE Trans. Electron Devices*, Vol. ED-15, No. 5, pp. 294–307, May 1968.

Mar, J., "A Time-Domain Method of Measuring Transistor Parameters," *IEEE J. Solid-State Circuits*, Vol. SC-6, No. 4, pp. 223–226, Aug. 1971.

Müller, O., "Ultralinear UHF Power Transistors for CATV Applications," *Proc. IEEE*, Vol. 58, No. 7, pp. 1112–1121, July 1970.

Poon, H. C., "Modeling of Bipolar Transistor Using Integral Charge-Control Model with Application to Third-Order Distortion Studies," *IEEE Trans. Electron Devices*, Vol. ED-19, No. 6, pp. 719–731, June 1972.

Poon, H. C. and H. K. Gummel, "Modeling of Emitter Capacitance," *Proc. IEEE*, Vol. 57, No. 12, pp. 2181–2182, Dec. 1969.

Poon, H. C. and J. C. Meckwood, "Modeling of Avalanche Effect in Integral Charge Control Model," *IEEE Trans. Electron Devices*, Vol. ED-19, No. 1, pp. 90–97, Jan. 1972.

Riva, M., P. J. Bénétéau, and E. D. Volta, "Amplitude Distortion in Transistor Amplifiers," *Proc. Inst. Elec. Eng.*, Vol. 111, No. 3, pp. 481–490, March 1964.

Sansen, W. M. C. and R. G. Meyer, "Characterization and Measurement of the Base and Emitter Resistances of Bipolar Transistors," *IEEE J. Solid-State Circuits*, Vol. SC-7, No. 6, pp. 492–498, Dec. 1972.

Searle, C. L., et al., *Elementary Circuit Properties of Transistors*, SEEC Vol. 3. New York: John Wiley & Sons, Inc., 1964.

Sevin, L. J., Jr., *Field-Effect Transistors*. New York: McGraw-Hill Book Company, 1965.

Spangenberg, K. R., *Vacuum Tubes*. New York: McGraw-Hill Book Company, 1948.

Sparkes, J. J., "Device Modeling," *IEEE Trans. Electron Devices*, Vol. ED-14, No. 5, pp. 229–232, May 1967.

4. NONLINEAR INTERFERENCE AND DISTORTION EFFECTS—CLASSICAL POWER SERIES

Akgün, M. and M. J. O. Strutt, "Cross Modulation and Nonlinear Distortion in RF Transistor Amplifiers," *IRE Trans. Electron Devices*, Vol. ED-6, pp. 457–467, Oct. 1959.

Duff, W. G. and D. R. J. White, *Prediction and Analysis Techniques*, Vol. 5 of A Handbook Series in Electromagnetic Interference and Compatibility. Germantown, Maryland. Don White Consultants, 1972.

Gretsch, W. R., "The Spectrum of Intermodulation Generated in a Semiconductor Diode Junction," *Proc. IEEE*, Vol. 54, No. 11, pp. 1528–1535, Nov. 1966.

Hilling, A. E. and S. K. Salmon, "Intermodulation in Common Emitter Transistor Amplifiers," *Electron Eng.*, Vol. 40, pp. 360–364, July 1968.

Jones, B. L., "Cross Modulation in Transistor Amplifiers," *Solid State Design*, Vol. 3, pp. 31–34, Nov. 1962.

Lambert, W. H., "Second-Order Distortion in CATV Push-Pull Amplifiers," *Proc. IEEE*, Vol. 58, No. 7, pp. 1057–1062, July 1970.

Lieberman, D., "Cross-Modulation Figure of Merit for Transistor Amplifier Stages," *Proc. IEEE*, Vol. 58, No. 7, pp. 1063–1071, July 1970.

Lotsch, H., "Third-Order Distortion and Cross-Modulation in a Grounded Emitter Transistor Amplifier," *IRE Trans. Audio*, pp. 49–58, March-April 1961.

Miller, D. M. and R. G. Meyer, "Nonlinearity and Cross Modulation in Field-Effect Transistors," *J. Solid-State Circuits*, Vol. SC-6, No. 4, pp. 244–250, Aug. 1971.

Reynolds, J., "Nonlinear Distortions and Their Cancellation in Transistors," *IEEE Trans. Electron Devices*, Vol. ED-12, No. 11, pp. 595–599, Nov. 1965.

Simons, K. A., "The Decibel Relationship Between Amplifier Distortion Products," *Proc. IEEE*, Vol. 58, No. 7, pp. 1071–1086, July 1970.

Steiner, J. W., "An Analysis of Radio Frequency Interference Due to Mixer Intermodulation Products," *IEEE Trans. Electromagn. Compat.*, Vol. EMC-6, No. 1, pp. 62–68, Jan. 1964.

Thomas, L. C., "Eliminating Broadband Distortion in Transistor Amplifiers," *Bell Syst. Tech. J.*, pp. 315–342, March 1968.

Tuil, J., "Intermodulation in Aerial Amplifiers," *Electronic Applications*, Vol. 28, pp. 6–21, Jan. 1968.

Wass, C. A. A., "A Table of Intermodulation Products," *IEE J.*, Vol. 95, Part 3, pp. 31–39, Jan. 1948.

5. MISCELLANEOUS PAPERS DEALING WITH NONLINEAR SYSTEMS

Aprille, T. J., Jr. and T. N. Trick, "Steady-State Analysis of Nonlinear Circuits with Periodic Inputs," *Proc. IEEE*, Vol. 60, pp. 108–114, Jan. 1972.

Aprille, T. J., Jr. and T. N. Trick, "A Computer Algorithm to Determine the Steady-State Response of Nonlinear Oscillators," *IEEE Trans. Circuit Theory*, Vol. CT-19, pp. 354–360, July 1972.

Chisholm, S. H. and L. W. Nagel, "Efficient Computer Simulation of Distortion in Electronic Circuits," *IEEE Trans. Circuit Theory*, Vol. CT-20, pp. 742–745, Nov. 1973.

Chua, L. O., *Introduction to Nonlinear Network Theory*. New York: McGraw-Hill Book Company, 1969.

Collon, F. R. and T. N. Trick, "Fast Periodic Steady-State Analysis for Large-Signal Electronic Circuits," *IEEE J. Solid-State Circuits*, Vol. SC-8, No. 4, pp. 260–269, Aug. 1973.

Gallman, P. G., "An Iterative Method for the Identification of Nonlinear Systems Using a Uryson Model," *IEEE Trans. Automat. Contr.*, Vol. AC-20, No. 6, pp. 771–775, Dec. 1975.

Idleman, T. E., F. S. Jenkins, W. J. McCalla, and D. O. Pederson, "SLIC–A Simulator for Linear Integrated Circuits," *IEEE J. Solid-State Circuits*, Vol. SC-6, No. 4, pp. 188–203, Aug. 1971.

Jenkins, F. S. and S. P. Fan, "TIME–A Nonlinear DC and Time-Domain Circuit-Simulation Program," *IEEE J. Solid-State Circuits*, Vol. SC-6, No. 4, pp. 182–188, Aug. 1971.

Kuo, Y. L., "Distortion Analysis of Bipolar Transistor Circuits," *IEEE Trans. Circuit Theory*, Vol. CT-20, No. 11, pp. 709–716, Nov. 1973.

Nagel, L. and R. Rohrer, "Computer Analysis of Nonlinear Circuits, Excluding Radiation (CANCER)," *IEEE J. Solid-State Circuits*, Vol. SC-6, No. 4, pp. 166–182, Aug. 1971.

Neill, T. B. M., "Improved Method of Analyzing Nonlinear Electrical Networks," *Electron Lett.*, Vol. 5, No. 1, pp. 13–15, Jan. 9, 1969.

Porter, W. A., "An Overview of Polynomic System Theory," *Proc. IEEE*, Vol. 64, No. 1, pp. 18–23, Jan. 1976.

Rugh, W. J. and W. W. Smith, "Structure of a Class of Nonlinear Systems," *IEEE Trans. Automat. Contr.*, Vol. AC-19, No. 6, pp. 701–706, Dec. 1974.

Index